NORTON / LIBRARY OF CONGRESS VISUAL SOURCEBOOKS IN ARCHITECTURE, DESIGN AND ENGINEERING

CHRISTINE MACY

W. W. Norton & Company, New York and London | Library of Congress, Washington D.C.

DAMS

DEDICATED TO THE MEMORY OF J. B. JACKSON (1909–1996)

Teacher and advocate of the human-made landscape

Printed in The United States of America
FIRST EDITION

For information about permission to reproduce selections from this book, write to Permissions, 500 Fifth Avenue, New York, NY 10110

Manufacturing by Edwards Brothers
Series Design by Kristina Kachele Design LLC
Composition and Paging by Ken Gross
Production manager: Leeann Graham
Indexing by Bob Elwood

Library of Congress Cataloging-in-Publication Data
Macy, Christine.
Dams / Christine Macy. — 1st ed.
p. cm.
(Norton/Library of Congress visual sourcebooks in architecture, design and engineering)
Includes bibliographical references and index.
ISBN 978-0-393-73139-2 (hardcover)
1. Dams. I. Title.
TC547.M23 2009
627'.8—dc22 2007039485

W. W. Norton & Company, Inc.,
500 Fifth Avenue
New York, NY 10110
www.wwnorton.com

W. W. Norton & Company Ltd.,
Castle House, 75/76 Wells St.
London W1T 3QT

0 9 8 7 6 5 4 3 2 1

Center for Architecture, Design and Engineering

The Norton/Library of Congress Visual Sourcebooks in Architecture, Design and Engineering series is a project of the Center for Architecture, Design and Engineering in the Library of Congress, established through a bequest from the distinguished American architect Paul Rudolph. The Center's mission is not only to support the preservation of the Library's enormously rich collections in these subject areas, but also to increase public knowledge of and access to them. Paul Rudolph hoped that others would join him in supporting these efforts. To further this progress, and to support additional projects such as this one, the Library of Congress is therefore pleased to accept contributions to the Center for Architecture, Design and Engineering Fund or memorials in Mr. Rudolph's name as additions to the Paul Rudolph Trust.

For further information on the Center for American Architecture, Design and Engineering, you may visit its Web site: www.loc.gov/rr/print/adecenter/adecent.html

C. FORD PEATROSS
CURATOR OF ARCHITECTURE, DESIGN AND ENGINEERING

The Center for Architecture, Design and Engineering and the Publishing Office of the Library of Congress are pleased to join with W. W. Norton & Company to publish the pioneering series of the Norton / Library of Congress Visual Sourcebooks in Architecture, Design and Engineering

Based on the unparalleled collections of the Library of Congress, this series of handsomely illustrated books draws from the collections of the nation's oldest federal cultural institution and the largest library in the world, with more than 134 million items on approximately 615 miles of bookshelves. The collections include more than 20.5 million books, 2.8 million recordings, 12 million photographs, 5.2 million maps, and 59 million manuscripts.

The subjects of architecture, design, and engineering are threaded throughout the rich fabric of this vast archive, and the books in this new series will serve not only to introduce researchers to the illustrations selected by their authors, but also to build pathways to adjacent and related materials, and even entire archives—to millions of photographs, drawings, prints, views, maps, rare publications, and written information in the general and special collections of the Library of Congress, much of it unavailable elsewhere.

Each volume serves as a portal to the collections, providing a treasury of select visual material (much of it in the public domain) for students, scholars, teachers, researchers, historians of art, architecture, design, and technology, and practicing architects, engineers, and designers of all kinds.

A CD-ROM accompanying each volume contains high-quality, downloadable versions of all the illustrations. It offers a direct link to the Library's online, searchable catalogs and image files, including the hundreds of thousands of high-resolution photographs, measured drawings, and data files in the Historic American Buildings Survey, Historic American Engineering Record, and the Historic American Landscape Survey. The Library's Web site has rapidly become one of the most popular and valuable locations on the Internet, visited 111 million times in 2006 and comprising more than 184 terrabytes of knowledge, and serving audiences ranging from school children to the most advanced scholars throughout the world, with a potential usefulness that has only begun to be explored.

Among the subjects to be covered in this series are building types, materials, and details; historical periods and movements; landscape architecture and garden design; interior and ornamental design and furnishings; and industrial design. *Dams* is another exemplar of the goals and possibilities on which this series is based.

JAMES H. BILLINGTON
THE LIBRARIAN OF CONGRESS

HOW TO USE THIS BOOK

The introduction to this book provides an overview of the history and development of dams in the United States. It is a view that is broad and inspired by the depth and quality of the resources of the Library of Congress. The balance of the book, containing approximately 800 images which can also be found on the CD at the back of the book, is organized into eight sections. Figure-number prefixes designate the section.

Captions give the essential identifying information, where known: subject, location, date, creator(s) of the image, the Library of Congress call number and digital ID number, which can be used to find the image online. Note that a link to the Library of Congress Web site may be found on the CD.

Abbreviations used in captions

AEP	American Environmental Photographs / University of Chicago Library Collection
AIPNW	American Indians of the Pacific Northwest
ALAD	American Landscape and Architectural Design / Harvard Graduate School of Design Collection
AP	Associated Press
DETR (formerly DPCC)	Detroit Publishing Company Collection
DN	Chicago Daily News Collection
DPCC	Detroit Publishing Company Collection
FSA/OWI	Farm Security Administration/Office of War Information Collection
G&M	Geography and Map Division
HABS	Historic American Building Survey
HAER	Historic American Engineering Record
HAW/DPL	History of the American West / Denver Public Library Collection
LC	Library of Congress
LOOK	Look Magazine Collection
NYWTS	New York World Telegram & Sun Newspaper Photograph Collection
P&P	Prints and Photographs Division
S	Stereograph File
SOUV	Souvenir File
SSF	Subject Specific File
UPI	United Press International
WC	Wittemann Collection

CONTENTS

DAMS ACROSS AMERICA

DAMS ARE A MONUMENTAL PRESENCE on the American landscape. They divert and restrain mighty rivers that have run for millennia. They impound vast artificial lakes. Water from dams has turned deserts into orchards, slaked the thirst of millions of metropolitan citizens, and powered wartime production from the Southeast to the Northwest; but dams have also prevented salmon from spawning, flooded forests and fields, displaced populations, and required graves to be exhumed. It is not surprising, then, that dam building inspires powerful emotions.

Most people, when asked about American dams, think of one of the massive federal projects built between the 1930s and the 1970s, such as Hoover Dam or the Grand Coulee. Yet according to the National Research Council, there are over 2.5 million dams in the United States, most of which are small, privately owned structures.[1] Only a very small number—six thousand, to be precise—are large dams over 50 feet high, and only a small portion of these have been built by the federal government.[2] It is a bit daunting, then, to present a picture of dams across America. Which dams should one talk about? The most typical or the most exceptional ones? The structurally innovative, politically contentious, or newsworthy ones? The exemplary, influential, or precedent-setting ones?

Histories of engineering talk about dams in superlatives. They are the longest, highest, or most massive; they have the biggest reservoir or the highest head drop (distance from the reservoir surface to the powerhouse), or they were made with the least or the most amount of material. They were the first or the latest to set records in any of these ways. Dams imply a kind of engineering Olympics, a measure of people against nature. Like that other great technological achievement, the skyscraper, one can always build a

IN-001. Steel worker, Grand Coulee Dam, Columbia Basin Reclamation Project, Columbia River, between Douglas and Grant Counties, Washington, 1933–1942. F. B. Pomeroy, photographer, 1942. P&P,FSA/OWI,No. 6649-CB.

IN-002

IN-002. Norris Dam, Clinch River, Knoxville Vic., Anderson County, Tennessee, 1933–1936. Barbara Wright, photographer, 1945. P&P,LC-dig-ppmsca-17371.

This was the first dam built by the Tennessee Valley Authority.

bigger dam. Certainly, the biggest dams in the United States were colossal undertakings that took years to complete. The architectural critic Lewis Mumford called them "democratic pyramids" and compared them to the greatest constructions of antiquity.[3] The cultural historian David Nye suggests that because dams inspire feelings of awe, they engender national pride.[4] If newspaper coverage is any measure, this would certainly seem to be the case with the dams that span the great rivers of the country, like Hoover Dam across the Colorado (see 6-092–6-105), the Grand Coulee Dam across the Columbia (see 7-076–7-085), and Fort Peck Dam across the Missouri (see 4-031).

But the history of dam design is a history of invention and development that proceeded by trial and error in some cases, and by calculation and engineering in others. It is an iterative process in which each new design incorporates lessons from an earlier, similar dam. Dam design has to take into account precedents, engineering science, and the specific circumstances of site and construction process—such as

IN-003. Workers on spillway, Hoover Dam, Colorado River, Boulder City, between Clark County, Nevada, and Mohave County, Arizona, 1930–1936. Bureau of Reclamation, 1934. P&P,LC-dig-ppmsca-17404.

IN-003

whether the labor force is large or small, whether the ground conditions are solid or porous, and whether it is easy or difficult to get construction materials to the site.

In social terms, dams have always held the promise of using technology to harness nature for the benefit of people. This dream is perhaps best exemplified by the dams of the Tennessee Valley Authority (see 2-081–2-124). Yet today, dams are a much-maligned, even vilified, presence in the country's western landscapes. They are criticized as destroyers of animal habitats, usurpers of native lands, boondoggles for land speculators, or subsidies for wealthy farmers. Yet ranchers, farmers, industrialists, unions, municipalities, states, and federal agencies have all, over the years, vied energetically for federal support in dam building, because dams require a huge mobilization of capital, manpower, and resources. And they have similarly huge effects, both desirable and collateral.

IN-004

IN-004. Worker on face of dam, Shasta Dam, Central Valley Reclamation Project, Sacramento River Redding Vic., Shasta County, California, 1938–1945. Bureau of Reclamation, ca. 1940. P&P,LC-dig-ppmsca-17305.

IMAGES OF DAMS

We can learn a great deal about dams by looking at the images people have made of them. In the process, we also learn about the people who made the images—how they saw dams, and what they felt about them. The large collection of images in this book tells us not only how people have viewed dams over the years in the United States, but also how the country has developed.

Stereoscopic photographs, lithographs from travel brochures, and picture postcards depict dams as if they were scenes from a trip—which indeed they were, after the advent of railroad and, later, automobile tourism. In these images, the technological accomplishment involved in creating a dam is often juxtaposed against a wild and seemingly remote nature, a sleight-of-hand that transforms a built artifact into a natural wonder (IN-005–IN-006). Newspapers or illustrated journals, in contrast, represent dam building

IN-005. Arrowrock Dam, Boise Irrigation Project, Boise River, Ada County, Idaho, 1911–1915. Walter J. Lubken, photographer, 1916. P&P,US GEOG FILE-Idaho-Arrowrock Dam.

IN-005

(or dam disasters) as newsworthy events. Such images focus on groundbreakings, dedication ceremonies, or other momentous events in the life of a dam, and they almost always include people in the frame to give scale to the enormous project and humanize it (IN-007).

Government agencies involved in dam building have carried out their own documentation, usually very thoroughly and with an emphasis on the technological innovations involved. While the National Archives holds all the collections of the U.S. Army Corps of Engineers and the Bureau of Reclamation (the two largest dam-building agencies in the country), the Library of Congress has some government photographs of dams, most notably the New Deal photographs of Roy Stryker's team for the Farm Services Administration and, later, the Office of War Information. These Depression-era photographs record the epic dam-building efforts of the Works Progress Administration, the Tennessee Valley Authority, and many Reclamation and Army Corps projects as well, focusing particularly on the labor and heroic effort involved (IN-009).

HARNESSING THE WATERS

The colonization of North America took place at the same time as the Industrial Revolution in Europe. But while emerging British industries could take advantage of a ready pool of labor in farmers driven from the countryside by the Enclosure Acts, North

IN-006

IN-006. Roosevelt Dam, Salt River, Gila County, Arizona, 1903–1911, modified 1936, 1989–1996. WC. P&P,LC-dig-ppmsca-17309.

Americans comprised much more dispersed populations and confronted natural resources on a much vaster scale. Labor was scarce if not expensive. In this context, early settlers quickly saw the advantages of putting water to work for them. They developed mills of all kinds—for grinding grains, sawing wood, pumping water, and running machinery. The settlers also had to transport their goods to markets and reach sources for raw materials that were often at great distances.

The dams that fed the mills and supplied the canals were made of wood, and wood continues to be a material of choice even today for small or temporary dams. These structures were rarely engineered but relied on the experience of their builders—who might have been millwrights with excellent carpentry skills, or maybe just lumbermen who could hastily assemble a dam when they had to. Such dams did not necessarily completely span a river. Many early structures were simply wing dams that projected into a stream, diverting water to feed mill races or serve navigation canals. Others did span rivers to raise water levels high enough so that shallow barges could navigate downstream or be towed upstream. Such overflow dams, or weirs, allowed water to flow over the dam crest. They were rarely the kind of dams that impounded water, nor did they need to be, since water was a readily available resource (IN-008).

By the first half of the nineteenth century, construction of larger and more durable dams began. Building dams of this size required significant concentrations of capital. Sometimes a consortium of entrepreneurs would pool their resources to finance such large projects, as was the case with the Great Stone Dam (Lawrence, 1831) and the

IN-007. Outlet valves, O'Shaughnessy Dam, Tuolumne River, Yosemite National Park, Tuolumne County, California, 1913–1923. Acme, 1934. P&P,LC-dig-ppmsca-17255.

IN-007

massive wood-crib Holyoke Dam (1848), both in Massachusetts (see 1-013, 1-015). Both of these large-scale endeavors supplied power canals that served textile factories. Other large masonry dams in this period provided reliable water supplies to growing cities such as Philadelphia, New York, and Baltimore, all of which engaged in huge municipal reservoir projects in the mid-nineteenth century (see 1-027, 1-034, 1-040).

Well-capitalized canal companies, investing in the infrastructure for commercial navigation, constituted a third source of large-scale and systematic dam-building programs. Before the dawn of the railroad era, watercourses were virtually the only way to transport heavy loads long distances over land, and the developing nation saw the value of creating a network of canals linking navigable rivers. The Erie Canal—completed in 1825 and the first all-water route across the Appalachian Mountains—is perhaps the best known of these nineteenth-century projects, but a number of other canals were built over the course of the century, spurred on by Treasury Secretary Albert Gallatin's transportation plan of 1808. Not to be outdone in taking advantage of the resources of the North American continent, the British crown built the impressive Rideau Canal system in 1832, consisting of fifty-two solidly built dams that ran from Kingston, at the head of Lake Ontario, to Ottawa. The goal was to provide an all-Canadian route to the Great Lakes following the hostilities of the War of 1812 (see 2-006). Construction of the James and Kanawha Canal system, which began as early as 1785,

IN-008

continued piece by piece for fifty-five years until it stretched 200 miles inland, yet it never reached the Kanawha River, a tributary of the Ohio (see 1-045). Other significant canals were the Lehigh Canal system, built in 1829 to access the coal fields of Pennsylvania (see 1-049); the Muskingum River system of 1841 that extended over a hundred miles through southeastern Ohio (see 2-011); and the many dams and locks that facilitated traffic between the Great Lakes and their connected rivers.

IN-008. Keane Valley Mill, East Branch of the Ausable River, Keane Valley, Essex County, New York, ca. 1900. DETR, ca. 1900. P&P,LC-D4-33207.

Early inventions for navigational dams included the bear trap gates of the Lehigh Canal, patented in 1819 by Josiah White (see 1-052), and the movable gates of the Erie Canal, which could pivot to allow high water to pass and close to raise river levels to a navigable depth. As dams grew larger and their builders used masonry with expectations that the dams would be lasting structures, a more developed engineering science was required, one capable of resolving the problems posed by heavy structures on natural foundations. Some these problems included seepage through rock foundations and the tendency of water flowing over a dam's crest to undercut its foundations on the downstream side.

IN-009. Grand Coulee Dam under construction, Columbia Basin Reclamation Project, Columbia River, between Douglas and Grant Counties, Washington, 1933–1942. Bureau of Reclamation, 1936. P&P,LC-dig-ppmsca-17334.

IN-009

WATER TRANSFORMS THE WEST

With the westward movement of settlement, the character of dam building changed dramatically. Whereas the eastern landscapes had provided steadily flowing rivers in valleys and the relatively gentle terrain of the Appalachian Mountains, landscapes of the trans-Mississippi West presented new difficulties and required new ways of thinking about dam building. In the East, dams were built for mills, canals, and water storage for municipalities. But in the American West of the mid-nineteenth century, dams served one purpose above all others—mining. California's gold mines, Colorado's silver mines, and Montana's copper mines generated instant boom towns. This landscape was vertiginously mountainous with rapidly flowing streams and rivers that cut canyons through foothills and disappeared into desert and prairie. Water descended from the sky in sudden thunderstorms as clouds released their burdens over the mountain ranges, and it ran down the steep mountainsides in torrents before it could be captured or used.

Western miners needed water in their search for metals, and they used huge quantities of it to wash through mineral-bearing soils. Early miners became experts at moving earth with water. To store and direct the water, they developed technologies such as dams, flumes, sluiceways, trestles, and pipes. They welcomed the tremendous force generated by the large drops in elevation that occurred when they diverted mountain streams to mining sites, channeling these with high-pressure nozzles, developed in 1857. They got used to seeing vast fields of water-soaked earth slowly drain away and solidify into silts and sediments. Hydraulic mining techniques facilitated a number of dam-building projects in the West, and they worked in water distribution systems for power generation and irrigation as well.

Mining also used hydropower, lots of it—to operate stamp mills for crushing ore, grinding machines, sawmills, and forges. The precisely engineered turbines and raceways developed in the New England textile towns did not suit the rough-and-ready, do-it-yourself requirements of the early miners, who turned to easily built water wheels like the hurdy gurdy and, later, the Pelton impulse wheel. Such technical developments, along with the high head conditions presented by western streams—and above all, the immediately available and well-paying market for power—set the stage for the extremely rapid development of hydroelectricity in the West, leading the way for a number of national firsts in hydroelectric power generation, particularly in California, Idaho, and Oregon.

The race for western resources was an enterprise that occurred on an industrial scale and with epic speed. Steam shovels began to play a role in earth-moving operations like the construction of the Panama Canal, an enormous undertaking that had the weight of the federal government behind it and the personal endorsement of President Theodore Roosevelt. Such techniques were used to move soil and rock for early zoned earthfill dams, such as Milner Dam in Idaho (see 7-021), Avalon and McMillan dams in New Mexico (see 5-024), and Keelchus and Tieton dams in Washington State (see 7-057, 7-068), all from the first decades of the twentieth century. For some earthfill dams, hydraulic earth-laying techniques were used to build sections of the dam or all of it. This involved pumping a slurry of earth and water into a trough in order to build up an embankment layer by layer. As the water gradually drained off, the fill would compact and the process could be repeated. An early example in the West is Conconully Dam in

Washington State; toward the East, builders of Michigan's Croton Dam and the Miami River dams in Ohio applied this technique (see 2-062, 2-078–2-080). As the century progressed, earthfill construction, either hydraulic or zoned, was being used for truly massive projects like Fort Peck Dam in Montana (see 4-048), Anderson Ranch Dam in Idaho (see 7-049), and Oroville Dam in California, which still holds the record as the country's tallest dam.

For dams in remote areas, it required painstaking efforts to bring materials to locations that had no roads or navigable rivers. This led to new methods of dam construction that relied on engineering science to develop designs that used less material in a more efficient way. Examples include steel dams like Redridge in the Upper Peninsula of Michigan (see 2-128), Hauser in Idaho, and the reinforced concrete, multiple arch dams of John Eastwood (1857–1924), built in California and the West in the early twentieth century (IN-010).

If miners were leading the edge of settlement, leaping across the prairies hoping to strike it rich in the far West, they beat a path that farmers followed—helped by the federal government, which aimed to consolidate its control over vast tracts of sparsely populated land. But farmers also required water, and by 1902 the government agreed to support irrigation in the arid trans-Mississippi West, much as it had supported the construction of canals and railroads in the East over the previous century.

HYDROELECTRICITY AND DAM BUILDING

The great era of dam building is inseparable from the electric age. The first electrical power plants in the nation were hydroelectric plants. These early plants, built in the 1880s, were direct current stations that powered arc and incandescent lights. The first central station plant that served a network of customers was a hydroelectric plant built in Appleton, Wisconsin, in 1882. Seven years later, the American Electrical Directory listed two hundred electric companies that used waterpower for some or all of their power needs.

Because it was not possible to transmit direct current any great distance, early users of electricity had to be near the hydroelectric plants that generated power, as was the case with the southern textile industries that lined the Chattahoochee River in Georgia (see 1-075) and the copper mills of Great Falls, Montana (see 4-015). The development of alternating current changed all that, allowing electricity to travel miles from its source of production. The close of the nineteenth century saw tremendous advances in transmitting electricity over long distances. In 1889, power was sent 14 miles from Willamette Falls to

IN-010

IN-010. Anyox Dam, Anyox Creek, Alice Arm, British Columbia, Canada, 1922–1924. Unidentified photographer, 1924. (Water Resources Center Archives) TC547 J33 1995,page 217.

Portland, Oregon; six years later, California's Folsom Plant No. 1 sent 11,000 volts of alternating current 22 miles to Sacramento to power a number of businesses. In 1895, the country's first large-scale hydroelectric plant began operating in Niagara Falls, New York, transmitting power to Buffalo, 20 miles away. By 1900, transmission lines up to

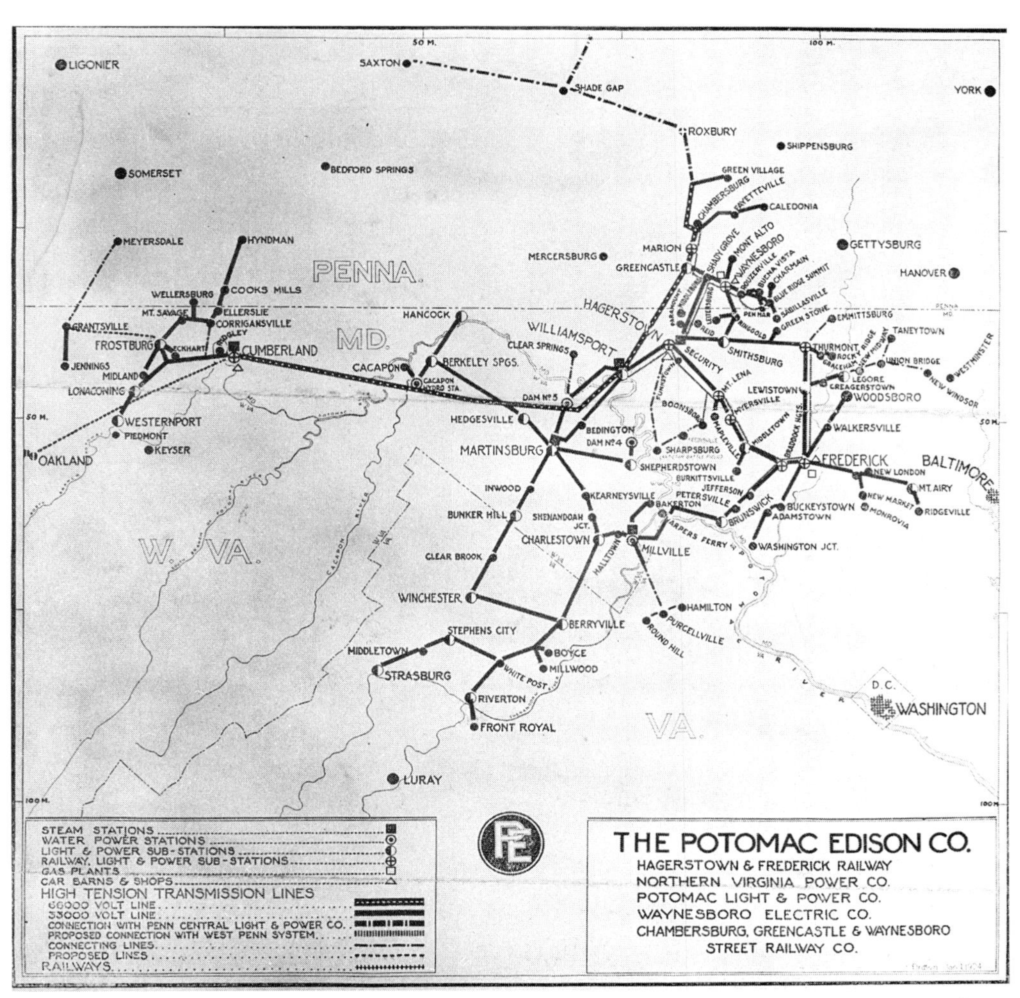

IN-011. Map of system, Potomac Edison Company. P&P,HAER, WVA,2-HEDVI.V,1-63.

IN-011

140 miles long were under construction.[5] By that date, hydroelectric power accounted for more than 40 percent of the nation's supply of electricity.[6]

The demand for power was insatiable, and companies sprang up to meet the opportunity, building plants and hanging transmission wires across the urban and industrial regions of the country (IN-011). In its history of electric utility regulation, the League of Women Voters of California describes the results: "a disorderly, sometimes dangerous, and always needless multiplication of pipe lines and overhead pole lines, [that] . . . demonstrated to the satisfaction of most that . . . utilities were 'natural monopolies.' Legislatures granted them exclusive service territories. Recognizing that this protection from competition could be abused, [legislatures] also set up regulatory bodies to control utility rates and ensure that they operated in the public interest."[7]

The economic historian Werner Troesken elaborates: "Between 1907 and 1924, nearly thirty states created statewide regulatory commissions to govern the behavior of private utility companies. At the time, just about all observers—government officials, consumers, the managers of public utility companies, academic economists, and

RECORD SETTING DAMS FOR HEIGHT

DATE	HEIGHT	DAM
1904	212 ft	Cheesman Dam, South Platte River, CO (concrete arched gravity; see 4-013)
1909	214 ft	Pathfinder Dam, North Platte River, WY (masonry arched gravity; see 4-041)
1910	350 ft	Shoshone Dam, Shoshone River, WY (concrete thin arch; see 4-042)
1911	356 ft	Roosevelt Dam, Salt River, AZ (masonry arched gravity; in-006)
1929	389 ft	Diablo Dam, Skagit River, WA (concrete arched gravity; see 7-098)
1932	417 ft	Owyhee Dam, Owyhee River, OR (concrete arched gravity; see 7-074)
1936	726 ft	Hoover Dam, Colorado River, NV (concrete arched gravity; see 6-105)
1968	770 ft	Oroville Dam, Feather River, CA (earthfill)

political scientists—viewed the creation of state utility commissions as a positive development that would promote both equity and economic efficiency."[8]

In the process of power generation, the head (the distance the water drops from reservoir to turbine) was a critical factor in turbine design, since turbines require a significant amount of force to spin. While hydroelectric plants in the state of New York took advantage of the Niagara escarpment to obtain the necessary head (see 2-058, 2-059), western plants sited in mountainous terrain had the "head start" in this regard, examples being San Joaquin Electric Light Company's Plant No. 1, built in California in 1896; Shoshone Falls Hydroelectric Plant of 1907, built in Idaho (see 7-019); and the similarly named Shoshone Dam and Hydroelectric Plant of 1909 in Colorado (see 6-009). All plants had to be able to transmit their electric power to consumer markets, whether those were in California cities or Montana mining industries.

And, of course, as engineers became more familiar with handling high head conditions, dam heights grew. This led to another set of engineering firsts, as varying site conditions required different structural solutions. Sites with solid rock foundations were good locations for thin-shell arch structures. Sites with plenty of clay, sand, or gravel suggested an earthfill dam, while inaccessible sites where both manpower and building materials were at a premium were likely to take advantage of a multiple-arch

IN-012. Keokuk Lock and Dam No. 19 under construction, Mississippi River, Keokuk, Lee County, Iowa, 1910–1914. Anschutz, 1911. P&P,LC-DIG-ggbain-50130.

or buttressed structure. For sheer height, it was difficult to improve on the concrete gravity dam in an arched shape; see the table on page 24 for progressive heights and dam types.

At the other end of the scale, low-head plants presented other opportunities in terms of dam design, hydroelectric power generation, and ecological considerations. The low-head hydroelectric plants in Wisconsin and Michigan did not require the impoundment of a reservoir to generate power, because they could count on a continuous and reliable flow of water to run the turbines, albeit at slower speeds. Also known as run-of-the-river dams, early examples like the Fox River dams in Appleton, Wisconsin, built in 1882, did not generate much power compared to contemporary high-head plants like the ones at Niagara Falls. But by the time the Sault Sainte Marie hydroelectric facility opened in 1902 on the canals that drain Lake Superior into Lake Huron, it was the first low-head plant with directly connected turbines and generators, and the largest low-head plant in the United States (a distinction it still holds today). It generated electricity for 50 percent of Michigan's Upper Peninsula (see 2-061). Low-head hydroelectric generation was also used at the American Falls Dam in Idaho, built in 1902 (see 7-016); Michigan's Croton Dam of 1908 (see 2-062); and the massive Keokuk Dam across the Mississippi River, built in 1914 (IN-012).

THE FEDERAL GOVERNMENT AND THE MULTIPURPOSE DAM

Dam building in the twentieth-century United States received a huge push from federal involvement in water management—promoting navigation and controlling flood damage in the rivers of the East, and promoting irrigation in the arid West. Three agencies played a central role in federal dam building: (1) the U.S. Army Corps of Engineers, focusing on navigation and flood control in the eastern seaboard and the Mississippi River drainage basin; (2) the Tennessee Valley Authority, engaged in a program of regional planning and modernization exclusively in the Tennessee River drainage basin; and (3) the Bureau of Reclamation, the federal dam-building giant that began its institutional life by supporting irrigation projects in the western states. Each agency has realized huge, coordinated strategies that involved systematic dam building across whole watersheds, accomplishing multiple goals in the process. As federal agencies, they have been vulnerable to political pressures and changes in government; it is not surprising, then, that they pioneered the concept of the multipurpose dam—the dam that serves many functions and pays for itself along the way.

U.S. Army Corps of Engineers

The federal government's first foray into dam building was through the U.S. Army Corps of Engineers. The Corps' involvement with waterways was minimal before the middle of the nineteenth century, restricted to the charting of rivers and the occasional clearing of obstructions that hindered traffic. But during the Civil War, its activities scaled rapidly upward with the military engineering of dams, bridges, defenses, and road systems. By the 1870s, the Corps had begun to develop expertise in the construction of navigation channels and the specialized dams that went with them. In the late nineteenth century, it assumed responsibility for canals privately built on the Monongahela and Allegheny rivers and began to modernize them, putting lessons from these waterways into practice on the Ohio and then the Mississippi rivers.

In 1875, the Corps completed a systematic survey of movable dam designs from Europe, focusing on developments in France, England, and Germany. These countries, with their extensive canal systems and abundant river traffic, had invented many kinds of dams to raise river depths so they would be easier to navigate. In 1878, the Corps used lessons from this study to complete its first design for a dam on the Ohio River. Ohio No. 1 Lock and Dam used two types of movable dam system: a weir of chanoine wickets over the navigation channel, and a bear trap dam over the rest of the riverbed. The wickets, modeled after the design of French engineer Jacques Chanoine, were heavy wooden

IN-013. Sections through navigable pass and weir no. 1, Davis Island Dam (Lock & Dam No. 1), Ohio River, Pittsburgh Vic., Allegheny County, Pennsylvania, 1878–1885. U.S. Army Corps of Engineers, delineator, 1889. TC549 U53 Folio, plate no. 2, detail.

IN-014 Section through bear trap, Davis Island Dam (Lock & Dam No. 1), Ohio River, Pittsburgh Vic., Allegheny County, Pennsylvania, 1878–1885. U.S. Army Corps of Engineers, delineator, 1889. TC549 U53 Folio, plate no. 18, detail.

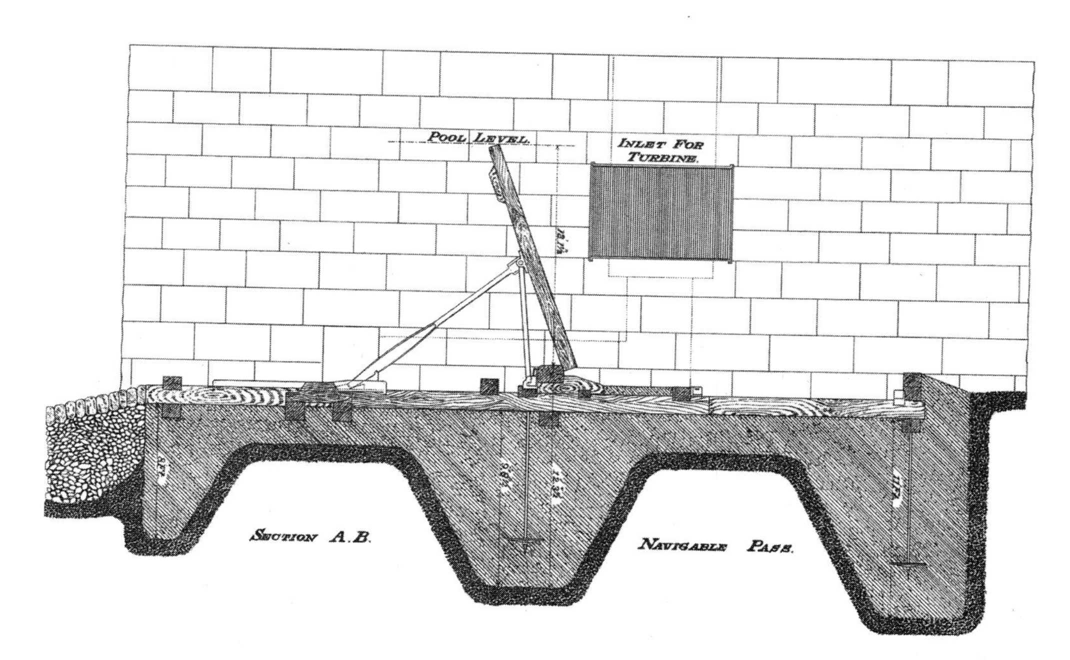

IN-013

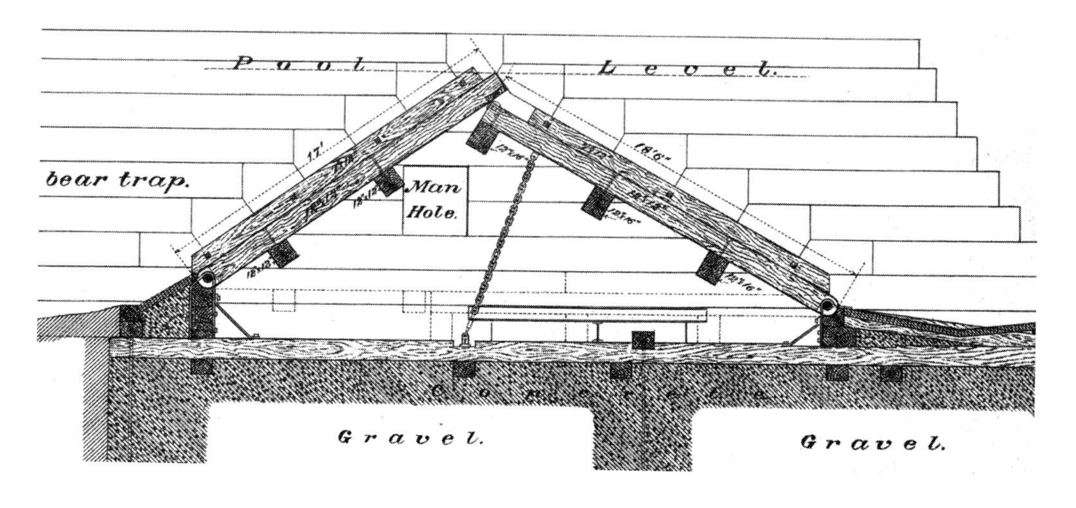

IN-014

shutters that could be raised and lowered manually. When the river was low, the wickets were raised to increase navigation depths; during floods, they were lowered so that water and boats passed freely (IN-013). The bear trap design was an earlier American invention used on the Lehigh Canal (see I-049–I-051). It employed two leaves hinged at their bases to form a triangle in cross section, and meeting at the top, or crest, of the gate. When lowered, the leaves slide over each other to form a flat bed for the passage of boats or river debris. The gate is raised by admitting water from the upstream side into the space beneath the leaves, lifting the buoyant lower leaf, which in turn lifts the upper leaf. The Corps was so satisfied with the success of its first dam on the Ohio that it extended the

IN-015

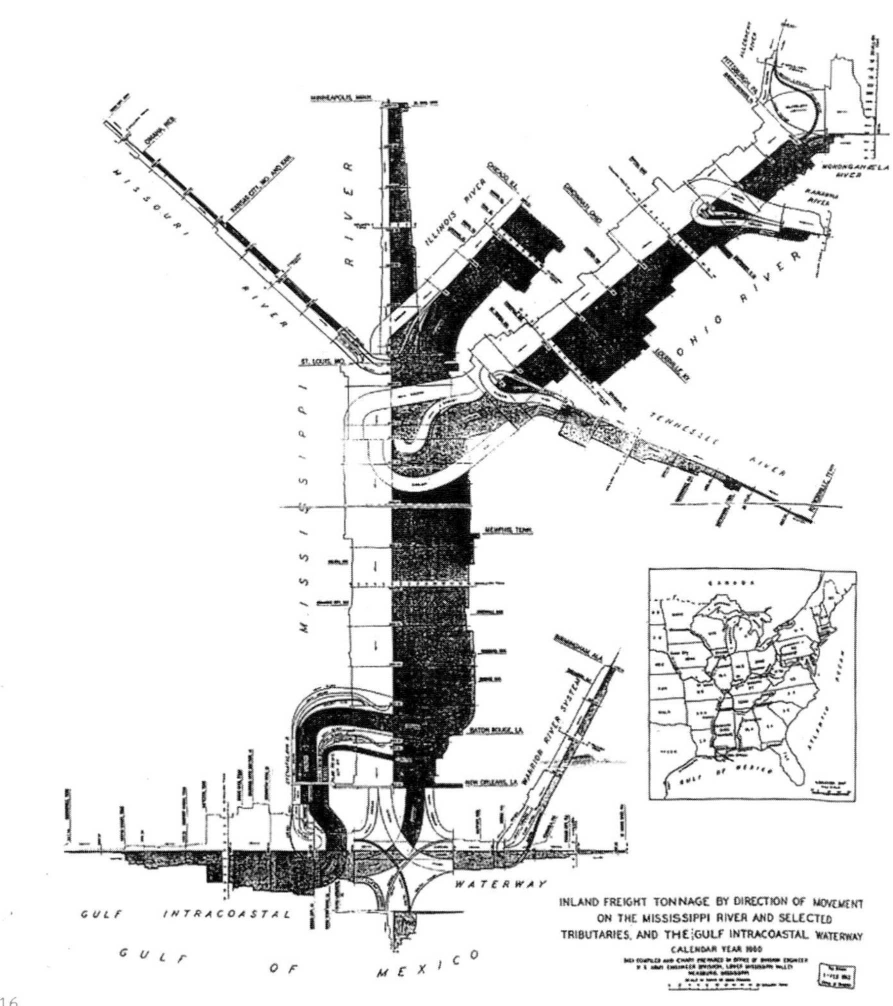

IN-016

IN-015. Lock and Dam No. 9., Mississippi River Nine-Foot Channel Project, Lynxville, Crawford County, Wisconsin, 1936–1938. U.S. Army Corps of Engineers, 1938. P&P,HAER,WIS,12-LYNX.V,1-67.

IN-016. Inland freight tonnage by direction of movement on the Mississippi River and selected tributaries. U.S. Army Corps of Engineers, 1960. G&M,G3701,P5 1960. U48, p. 283.

system the length of the river, building fifty-one additional dams using a similar design. By 1929, the Corps' canalization of the Ohio was complete, and barges could count on a 9-foot navigation depth year-round between the Mississippi River and Pittsburgh.

From 1927 to 1940, the Corps developed a similar 9-foot project on the Mississippi, turning it into a series of lakes with twenty-six dams between Minneapolis and Saint Louis. As on the Ohio, these were movable dams that could be raised in floods and closed the rest of the time to guarantee reliable navigation on the river. Army engineers continued their improvements on the movable dam, in this case making refinements to the (pivoting) tainter gate and (hoisted) roller gate. All of the Corps' channel projects were built to promote interstate commerce, in accordance with the mandate of the U.S. Congress (IN-015–IN-016).

Navigation was not the only motivation for the Corps' participation in dam building. In the 1920s, the natural cycles of flooding on the Mississippi and Ohio rivers began to create political pressure for federal involvement in flood control. With passage of the Flood Control Act of 1928, the Corps abandoned its policy of containing the Mississippi in its banks and began to construct overflow channels that directed floodwater away from populated centers. The Bonnet Carré Spillway upstream from New Orleans is a good example (see 3-084).

An interesting parallel development in the Ohio River basin was the Miami Conservancy District, a regional initiative created after the disastrous floods of 1913 to protect people living in the floodplain of the Miami River, one of the Ohio's tributaries. The engineer Arthur Morgan (1878–1975) designed a valley-wide system of five earthfill dams that, when they were completed in 1922, represented the most comprehensive flood protection system in the nation. The unique characteristic of the Miami River dams is that they hold back water only at flood times, remaining dry embankments the remainder of the year (see 2-077–2-080).

Tennessee Valley Authority

Morgan went on to become the first director of the Tennessee Valley Authority (TVA), a flagship project of Franklin Roosevelt's New Deal. Established in 1933 as a dam-building project on the Tennessee River and its tributaries, the TVA expanded to encompass a much wider agenda, including flood control, power production, regional development, soil conservation, and rural resettlement. A triumvirate of directors ran the agency: the engineer Arthur Morgan, the lawyer and public power advocate David Lilienthal, and the scientist Harcourt Morgan. In the TVA, the concept of the multi-

IN-017

IN-017. Chickamauga Lock and Dam, Tennessee Valley Authority, Tennessee River, Chattanooga, Hamilton County, Tennessee, 1936–1940. Tennessee Valley Authority, ca. 1940. P&P,LC-dig-fsa-8e07982.

purpose dam found its fullest development—as a means to prevent floods, provide irrigation, support navigation, and generate hydroelectricity for domestic, agricultural, and industrial purposes. In short, this project used dams as instruments of modernization and national integration.

From an engineering perspective, TVA dams were not particularly large or innovative since, at least at its outset, the Authority relied on existing designs from the U.S. Army Corps of Engineers and consultants from the Bureau of Reclamation. Rather, its key innovation was to conceive of dams as public facilities that deserved significant architectural treatment. And with the adoption of a forward-looking modernist aesthetic, the architectural quality of TVA projects—not only its dams but its locks, public facilities, outbuildings, and even roads and landscaped parks—was of an extraordinarily high level. There is no doubt this helped to contribute to the positive press coverage these structures enjoyed across the country.

In the nine years that elapsed between its establishment and the country's entry into World War II, the TVA replanted 40,000 square miles of land in seven southern states with trees and seeded the land with demonstration farms, nurseries, and fish hatcheries.

IN-018

IN-018. Hiwassee Dam, Tennessee Valley Authority, Apalachia River, Murphy Vic., Cherokee County, Tennessee, 1936–1940. Tennessee Valley Authority, ca. 1940. P&P,LC-dig-fsa-8e00544.

It built seven mainstream and eleven tributary dams, transforming 650 miles of river into a chain of lakes that allowed goods to circulate throughout the region. It strung transmission wires far beyond the reach of the river, bringing electricity for the first time to farms and households in one of the most rural regions of the country. It relocated tens of thousands of people from low-lying farms and resettled them in existing towns and new model communities. New freeways threading through the valley brought in visitors to witness the incredible transformation of the Appalachian hill country into a productive modern landscape. In short, by the eve of World War II the TVA had expanded—to use the words of Franklin Roosevelt—to touch "all manners of human concerns."[9] It had its contradictions, to be sure: while its aim in the early years was to re-establish a natural equilibrium in the region, it ultimately became a center for wartime atomic research and a significant producer of nuclear power. But even with its limitations and contradictions, the TVA remains an example of a regional plan that conceived of people, technology, and nature as part of one interdependent and productive landscape—and for that, it has earned a lasting place in the imagination of governments worldwide (IN-002, IN-017–IN-018).

Bureau of Reclamation

The third major federal agency involved in dam building was the Bureau of Reclamation. Pre-dating the TVA by thirty years, the Bureau was created after a good deal of political agitation by western senators and representatives, who wanted the federal government to play as important a role in fostering development in their region as it had in supporting railroads, roads, canals, and harbor improvements in the eastern parts of the country. In the West, this meant irrigation. By 1900, pro-irrigation planks formed parts of both Democratic and Republican platforms, and congressional opposition to western irrigation projects waned when a filibuster by angry western representatives killed a bill for improvements to rivers and harbors.[10] With the election of Theodore Roosevelt to the presidency in 1901, executive support for irrigation projects was ensured. The following year saw the passage of the Reclamation Act, which established a role for the federal government in developing and financing water projects.

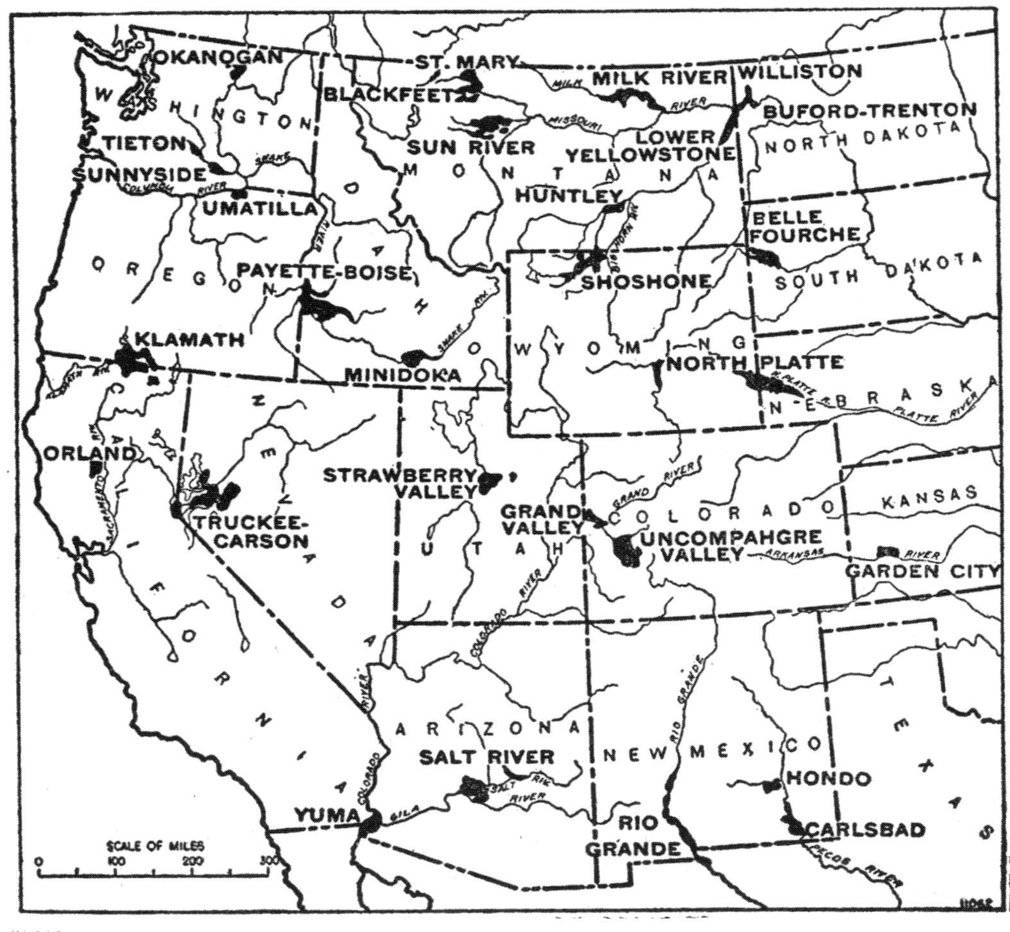

IN-019

IN-019. Principal United States Reclamati Service projects in the western United States. 1919. TC823.6 B7 1974,p. 25.

The new United States Reclamation Service identified potential projects in each western state that had federal lands, the sale of which would provide revenue for the program. For obvious reasons, the first five projects of the Service were widely distributed across several states: from Nebraska (North Platte) and Montana (Milk River), to Colorado (Uncompahgre), Arizona (Salt River), and Nevada (Truckee-Carson) (IN-019). Numerous other projects followed, and in 1923 the agency was renamed the Bureau of Reclamation.

The Bureau of Reclamation employed a wide range of dam designs, consistently learning from earlier projects as it began the next. In the first decade of the century, it set a number of national records for masonry and thin-shell concrete arched dams, including Pathfinder and Shoshone dams in Wyoming (see 4-025–4-026) and Roosevelt

IN-020. Powerhouse construction at Boise River Diversion Dam, Boise Irrigation Project, Boise River, Boise Vic., Ada County, Idaho, 1908–1910. United States Reclamation Service, 1912. P&P,HAER, ID,1-BOISE.V,1-A-37.

IN-020

IN-021

Dam in Arizona (IN-006), followed by Arrowrock Dam in Idaho (see 7-047) and Hoover Dam on the Nevada-Arizona border (see 6-93). It developed innovative techniques for cooling concrete as it set, controlling cracking, and perfecting outlet and intake works. It also built earthfill dams on a large scale, pioneering the use of steam shovels, the hydraulic fill technique, and the construction of cores in these types of dams, including all of the record-setting earthfill dams discussed earlier.

In a few cases, Reclamation went with less proven designs, when these were indicated by difficult soil conditions, restricted budgets, or remote locations. The agency decided, for example, to adopt the floating weir technique pioneered by British engineers in India for the difficult desert conditions posed in the Granite Reef, Ashurst-Hayden, Sacaton, and Laguna Diversion dams in Arizona (see 6-058, 6-083, 6-087). This overflow-type dam was called "floating" because it was designed to rest on silt rather than requiring excavation to bedrock. Also in Arizona, the agency decided on a material-efficient, thin-shell multiple dome design for Coolidge Dam (see 6-080) and on a multiple arch design for Bartlett Dam for similar reasons (see 6-071). Reclamation's only slab and buttress dam was Stony Gorge Dam in California (see 8-093).

IN-021. Grand Coulee Dam, Columbia Basin Reclamation Project, Columbia River, between Douglas and Grant Counties, Washington, 1933–1942. Bureau of Reclamation, 1941. P&P,LC-USZ62-13341.

IN-022

IN-022. Shasta Dam, Central Valley Reclamation Project, Sacramento River, Redding Vic., Shasta County, California, 1938–1945. Unidentified photographer, 1950. HD1695 W4 R68 2006.

It did not take long for the Bureau of Reclamation to realize the financial benefits of including hydroelectric plants in its irrigation projects, and this undoubtedly contributed to its longevity and vitality as a government agency. By the 1930s, it had outstripped the U.S. Army Corps of Engineers as the chief dam-building agency in the country, and its specialty was the multipurpose dam. Hoover, Grand Coulee, and Shasta dams all stem from this period (see 6-092; IN-021, IN-022). Reclamation's success led to a western-style showdown over dams on the Missouri River, as the Bureau's preeminence in dam building for western irrigation projects ran up against the Corps of Engineers' historic dominance in building navigational dams in the Mississippi drainage basin. The result was a standoff that led to both agencies having a hand in the project, with the Bureau's head, William Sloan, sharing principal billing with the Corps' Brigadier General Lewis Pick in the congressional authorization of the massive Pick-Sloan Project. After a late start, the Bureau of Reclamation also became involved in California's dam building, constructing most of that state's Central Water Project. Similarly, it built the lion's share of the large dams on the Columbia River—and consequently played a key role in power generation for the entire Northwest.

IN-023

IN-023. Protest against Tuttle Creek Dam, Blue River Valley, Stockdale Vic., Kansas, 1955. United Press, 1955. P&P,LC-dig-ppmsca-17288.

ECOLOGICAL COMPLEXITY AND FRAGILE BEHEMOTHS

From the beginning of reservoir dam building, the flooding of settled areas has been a problem, particularly since river basins tend to be the oldest inhabited parts of any region. The larger the dam, the larger the problem. Damage from flooding was the first and most visible ecological impact of dam building. It has led to the appropriation of long-settled lands through eminent domain, the removal and relocation of towns and settlements, the flooding of agricultural valleys, the exhumation of graveyards, and the inundation of archaeologically significant sites (IN-023). San Francisco's construction of O'Shaughnessy Dam for its municipal water supply led to widespread concern over the dam's impact on the Hetch Hetchy Valley, a scenic resource comparable (and adjacent) to Yosemite Valley, the country's first and best-known national park (IN-024). With time and a deeper understanding of natural systems, problems other than flooding have come to light, such as the impact of large dams on the movement of anadromous fish that travel each year from the ocean to their spawning grounds hundreds of miles inland. Proposed solutions have included fish ladders, parallel fish canals, "fish-friendly" turbines, crest gates, and weirs—and even busing the fish around the impediment! Dam-building agencies have countered such criticisms by promoting the recreational uses of new reservoirs, with the Tennessee Valley Authority going so far as to promote its "scenic resources" in pamphlets and to develop picturesque boat launches, automobile "parkways," and tourist cabins in conjunction with its dams.

Dams built in the mining landscapes of the Far West presented new ecological challenges, as toxic mine tailings accumulated behind dams, creating some of the most

IN-024

IN-024. Hetch Hetchy reservoir at draw-down, Tuolumne River, Yosemite National Park, Tuolumne County, California. Acme, 1934. P&P,LC-dig-ppmsca-17258.

polluted sites in the country. One such example was Thompson Falls Hydroelectric Dam in Montana, built in 1917 downstream from the copper fields of Butte and Anaconda (see 7-029). With the 1962 publication of Rachel Carson's book *Silent Spring*, criticisms arose regarding the relentless pursuit of technological development without due regard for its effects on nature. The ecological movement gained additional force with the transmission of photographs of the earth taken by astronauts circling the planet. It became clear that the natural world had a value beyond its utility for humans. While nineteenth-century conservationists spoke of the wonderful and sublime effects produced on the citizen by awesome displays of nature, these late-twentieth-century ecologists were concerned not with the impact of nature on people, but with the integrity of natural systems themselves. The very notion of an ecological system soon developed.

One political outcome of this new way of thinking involved a number of acts of Congress. A series of Endangered Species Acts began to offer protection for habitat, the Wild and Scenic River Preservation Act of 1968 protected scenic rivers from hydroelectric projects, and the subsequent National Environmental Policy Act required federal agencies to take environmental considerations into account in decision making. In 1970, the government established the Environmental Protection Agency to protect human health and to safeguard the environment. Two dams built in this period encountered significant opposition because of their impact on the ecological environment: the Tennessee Valley Authority's Tellico Dam, which damaged habitat for the tiny snail darter fish; and the Bureau of Reclamation's Glen Canyon Dam of 1964, which flooded a magnificent stretch of the Colorado River canyon (see 6-117). Canada was experiencing a similar outcry over Hydro Quebec's dams on native lands in the far northern reaches of Quebec.

These battles led to a radical shift in the political landscape that had previously supported dam building in the United States—a shift that has ultimately led to a moratorium on the construction of large-scale dams and a transformation in the management of existing dams. While formerly irrigation, municipal water use, and hydropower generation were the principal factors determining releases of water, now habitat management, the simulation of natural riverine conditions, and maintaining the health of a river must all be taken into account in the management plans regulating the operation of every large dam. Ecological management of dams received an additional boost in 1986, when the Electric Consumers Protection Act required federally regulated power producers to give equal consideration to values other than power production—such as energy conservation, fish and wildlife conservation, and recreation—when making licensing decisions. Pollution clean-up acts such as the Resource Recovery Act of 1970 and the "Superfund" Act a decade later involved the federal government in removing dams that had become repositories of toxic mine tailings.

Power generation, too, has been subject to the shifting winds of political ideology. Indeed, the late 1990s saw a reversal of arguments made early in the century, for the regulation of public utilities and for public power. With the deregulation of power markets, utilities found themselves pressed by power producers and dealers, leading to rapacious price gouging by companies such as Enron, as voters flip-flopped on the value of regulation. Hydroelectricity production, like dam building, requires enormous investments and constant oversight and is consequently an industry well suited to government involvement and control. Issues of safety and security, the mitigation of environmental effects, the large expanses of land involved, and even the continuity of responsibility all point toward a central role for governments in such enterprises. Yet even with government involvement, when the pendulum swings away from careful stewardship, disasters can happen, as was evident with the failure of the levee system in New Orleans during Hurricane Katrina in 2005.

Hydroelectric power production still has much to offer as one pillar of a national energy policy. Existing large dams and plants need continued management to minimize their negative impacts on the environment and to distribute their benefits equitably and for the public interest. Micro-hydroelectric production (like other renewable sources) is still a relatively untapped area—although after the energy crisis of the 1970s, a number of laws simplified licensing for small-scale hydropower facilities and required utilities to purchase power produced from renewable resources.[11] In parts of the country with an active hydrological cycle, run-of-the-river plants offer a vision for non-polluting sustainable energy production. In short, dams could have a future as promising as their history is inspiring, but for this to happen, some radical rethinking of their potentials and their effects needs to take place.

STRUCTURE OF THE BOOK

The book is organized by watershed. Water management in the continental United States falls into eighteen administrative districts, corresponding to the country's major drainage basins. The eight sections in this book roughly follow this division, but I cluster them into larger regions. For example, the three administrative districts of New England, Middle Atlantic, and South Atlantic–Gulf have been clustered into the Atlantic section. Similarly the Great Lakes, the Ohio River basin, and the Tennessee River basin are collected into the section Great Lakes–Appalachia.

BOOK SECTION	WATER MANAGEMENT DISTRICT
Atlantic	New England, Mid-Atlantic, and South Atlantic–Gulf Regions
Great Lakes–Appalachia	Great Lakes, Ohio, and Tennessee regions
Mississippi	Upper Mississippi and Lower Mississippi regions
Missouri	Souris-Red-Rainy and Missouri regions
Arkansas–Texas Gulf–Rio Grande	Arkansas-White-Red, Texas Gulf, and Rio Grande regions
Colorado–Great Basin	Upper Colorado, Lower Colorado, and Great Basin regions
Columbia	Pacific Northwest region
California	California region

Movement through the book proceeds geographically from east to west. Presenting a history of dam building by region rather than chronologically gives a real sense of how these engineering works were intimately bound up with local geographies—not only the physical geography of a place, its hydrology and geology, but also the cultural and social practices of its inhabitants and their relation to the land.

Within each section, subheadings emerge out of cultural practices specific to that region. For example, mill dams were a characteristic feature of the eastern seaboard and the first wave of settlement westward, while navigational dams were widespread from the eastern part of the country to the Mississippi—but no further west, since the landscape was far too dry to sustain inland navigation except in rare instances. Conversely, irrigation dams exist only in the western states. The Great Lakes feature logging dams and run-of-the-river hydroelectric projects, and levees are distinctive to the flooding lowlands of the Mississippi. Finally, municipal reservoir dams figure prominently on both coasts. I have aimed to strike a balance between large institutional dam-building efforts carried out by federal agencies, and the anonymous vernacular constructions that were much more widespread and arguably as influential in the development of the country.

Last, in preparing this book, I have had the pleasure to wander in my imagination down the rivers of the country. Each is distinctive as it winds along from its particular source to its inevitable mouth, carving canyons out of precipitous mountain valleys or depositing alluvial silts as it breaks through meanders on a spreading plain. The rapidly flowing streams of New England rushing over rock and gravel beds toward a cold Atlantic could hardly be more different from the interlaced networks of southern waterways that scarcely seem to move at all in the broad lowlands of their courses. In the Great Lakes, the entire landscape is aquatic, and the cultures that have thrived there have had to adapt appropriately, creating networks of trade that have continued from hand-paddled canoes to diesel barges. The Mississippi, more than any other river, has formed the identity of this continent, with its seasonal flooding forcing its original inhabitants into cities on mounds, and its vast reach bringing cultures into contact and communication with each other, forming the distinctive hybrids and patois that have become American. The naturalist Loren Eisley described western rivers unforgettably when he imagined he *was* the Platte River, floating effortlessly down its placid current with his head pointed toward the Rocky Mountains and his toes toward the Mississippi.[12] And the Colorado and Columbia rivers both are giants, twin descendants of the Continental Divide that flow thunderously across the short distances between their headwaters and the Pacific. The distinctive character of each of these rivers has contributed to the unique quality of its watershed, as people have maneuvered its rapids, tapped it for agriculture, harnessed it for work, and settled near its banks to trade, farm, mine, and manufacture. Following these rivers, at least in my imagination, has been a lesson in the history and culture of this land.

NOTES

1. National Research Council, Committee on Restoration of Aquatic Ecosystems: Science, Technology, and Public Policy, *Restoration of Aquatic Ecosystems: Science, Technology, and Public Policy*, (Washington, DC: National Academy Press, 1992), 26.
2. U.S. Army Corps of Engineers, *National Inventory of Dams*, revised 2001, crunch.tec.army.mil/nid/webpages/nid.cfm.
3. Lewis Mumford, "The Architecture of Power," *New Yorker* 17, no. 2 (June 7, 1941): 58.
4. David E. Nye, *American Technological Sublime* (Cambridge, MA: MIT Press, 1994).
5. The Nevada City–Oakland line built by Bay Counties Power Company stretched 140 miles.
6. PPL Corporation, "The History of Hydropower Development," in *A Study in Hydropower*, Curriculum for Grades 4–8 (Allentown, PA: PPL Corporation, 2000), 14–20.
7. League of Women Voters of California, "A World of Its Own: Electric Utility Regulation in California," in *Keeping California's Lights On: The League and Energy* (Sacramento: League of Women Voters of California, Energy Position, updated 2004), ca.lwv.org/lwvc/edfund/citizened/natres/energy/lights_on-1.pdf.
8. Werner Troesken, "Regime Change and Corruption: A History of Public Utility Regulation," draft paper (Cambridge, MA: National Bureau of Economic Research, 2004), www.nber.org/books/corruption/CR04/Troesken.pdf.
9. Franklin D. Roosevelt, Message to Congress on Muscle Shoals Development, House Doc. 15, 73rd Cong., 1st sess., April 10, 1933, 7. Cited in numerous popular publications of the time, including Paul Hutchinson, "Revolution by Electricity: The Significance of the Tennessee Valley Experiment," *Scribner's Magazine* 96, no. 4 (October 1934): 195.
10. U.S. Department the Interior, Bureau of Reclamation, *The Bureau of Reclamation: A Very Brief History*, Office of Program and Policy Service, History Program, www.usbr.gov/history/borhist.htm.
11. Examples include the Congressional Acts: Public Utility Regulatory Policies Act (1978), Energy Security Act (1980), Crude Oil Windfall Profit Tax (1980).
12. Loren Eiseley, *The Immense Journey* (New York: Vintage Books, 1957), 19.

ATLANTIC

NEW ENGLAND

St. John, St. Croix, Penobscot, Kennebec, Merrimack, Connecticut, and Housatonic Rivers

While larger rivers like the St. John and Connecticut were major thoroughfares to the interior, most of New England's rivers were fast-flowing streams in granite beds. Early colonists floated crops, lumber, and goods downstream, building wooden crib dams (gravity structures made of timber boxes filled with earth or

1-001. Sawmill at Glen Falls on the Hudson River, New York, 1829. Jacques G. Milbert, artist, 1829. P&P,LC-USZ62-1052(LOT 4397-R).

1-002. Sawmills on the Penobscot River, Oldtown, Penobscot County, Maine. 1854. P&P,LC-USZ62-39456.

1-002

rock) to keep water levels deep enough for navigation. They also dammed streams to power wood and grain mills, using technologies brought from Europe. By the middle of the nineteenth century, the potential power in the Merrimack and Houstonic rivers had attracted many industries to their riverbanks, which were transformed into mill ponds and power canals lined with textile mills, paper mills, and mineral-processing factories.

MID-ATLANTIC

Delaware, Susquehanna, Potomac, Rappahannock, and James Rivers

The easily navigable waterways of the mid-Atlantic states were vital arteries in the development of cities, farms, and industries in this prosperous region. Linked to the great protected harbors of Chesapeake and Delaware bays, and reaching far into the interior of the productive farms and coal mines of Virginia, Pennsylvania, and New York, the rivers of the mid-Atlantic region were developed equally to aid navigation, provide reliable water supplies for the burgeoning metropolises, and generate hydroelectric power for industry and municipalities.

1-003. Barclay's Iron Works on the Susquehanna River, Ulster, Bradford County, Pennsylvania, ca. 1830. G. Wall, delineator, 1851. P&P,LC-USZ62-32569.

1-003

SOUTH ATLANTIC–GULF SOUTH

Roanoke, Cape Fear, PeeDee, Santee, Savannah, Apalachicola, Mobile, and Pearl Rivers

The rivers of the Carolinas rise as lively streams in the Appalachian Mountains and become slow and navigable as they flow through flat coastal plains, although they are prone to flooding. Over the past fifty years, flood-control and hydroelectric schemes have altered the flows of these rivers, creating some of the largest reservoirs in the United States. Dam building in this region has a longer history, though, particularly for navigation and hydropower. The southern states pioneered the development of hydroelectricity in an outgrowth of the cotton mill economy, which gained additional impetus with the Tennessee Valley Authority in the 1930s. Water-produced electricity has contributed to the industrialization of the South—particularly in aluminum production and the military and aerospace industries.

1-004. Group of young doffers and superintendent, Catawba cotton mill, Newton, Catawba County, North Carolina. Lewis Hine, photographer, 1908. P&P,LC-USZ62-91170.

1-004

NEW ENGLAND MILL DAMS

From the earliest colonization of New England, farmers powered their mills with the region's rapidly flowing rivers. The timber-crib weirs or wing dams, built to divert water into mill races, required frequent replacing when washed out by spring floods or undercut by the rapid currents. They were made out of local materials and by rules of thumb. In the late eighteenth century, mills grew up around falls or rapids, or along canals that bypassed them. Paterson, New Jersey, for example, took advantage of the great falls of the Passaic River to supply its raceways and power canals. With the patronage of Alexander Hamilton, Pierre-Charles L'Enfant (1754–1852) designed the beginnings of an elaborate canal system there in 1792–1794. By the 1840s, it was supplying power to over one hundred mills, producing locomotives and textile machines.

As mill owners organized themselves to realize larger-scale projects, they began to construct power canals for their operations, building more durable masonry dams over sizable rivers. The success of Lowell, Massachusetts, as a mill town spurred Lawrence, Holyoke, and many other towns to develop hydropower for industry. The result was the first large-scale industrialization in the United States, and a lasting transformation of New England's landscape from farming to industry.

1-005

1-005. Bellows Falls Dam, Connecticut River, Bellows Falls, Windham County, Vermont, 1791–1802. DETR, 1907. P&P,LC-D4-70086.

The Connecticut was the first major river in the country to be improved for travel, with about 250 miles open to navigation by 1810. The Bellows Falls canal, located to bypass the Great Falls of the Connecticut River, was a major influence on the growth of the village because of the power it provided for mills. Produce and lumber traveled downriver on flat-bottomed boats, descending 50 feet through nine locks. By the 1870s, paper mills in Bellows Falls were among the first in the country to use wood pulp instead of rags for raw material. This photograph shows the rocks of the Connecticut River gorge, not the canal. The wooden crib dam maintains water levels to feed the power canal.

1-006

1-006. Bellows Falls Dam and Hydropower Canal, Connecticut River, Bellows Falls, Windham County, Vermont, 1791–1802. Lucien R. Burleigh, delineator, 1886. G&M,G3754.B4A3 1886. B8,detail.

1-007. Dewatered downstream face of Burlington Woolen Mill Dam, Winooski River, Burlington, Chittenden County, Vermont, 1876. Wayne Fleming, photographer, 1992. P&P,HAER,VT,4-BURL.V,1A, no. 5.

A good example of timber crib construction, using concrete, rather than masonry, as fill.

1-007

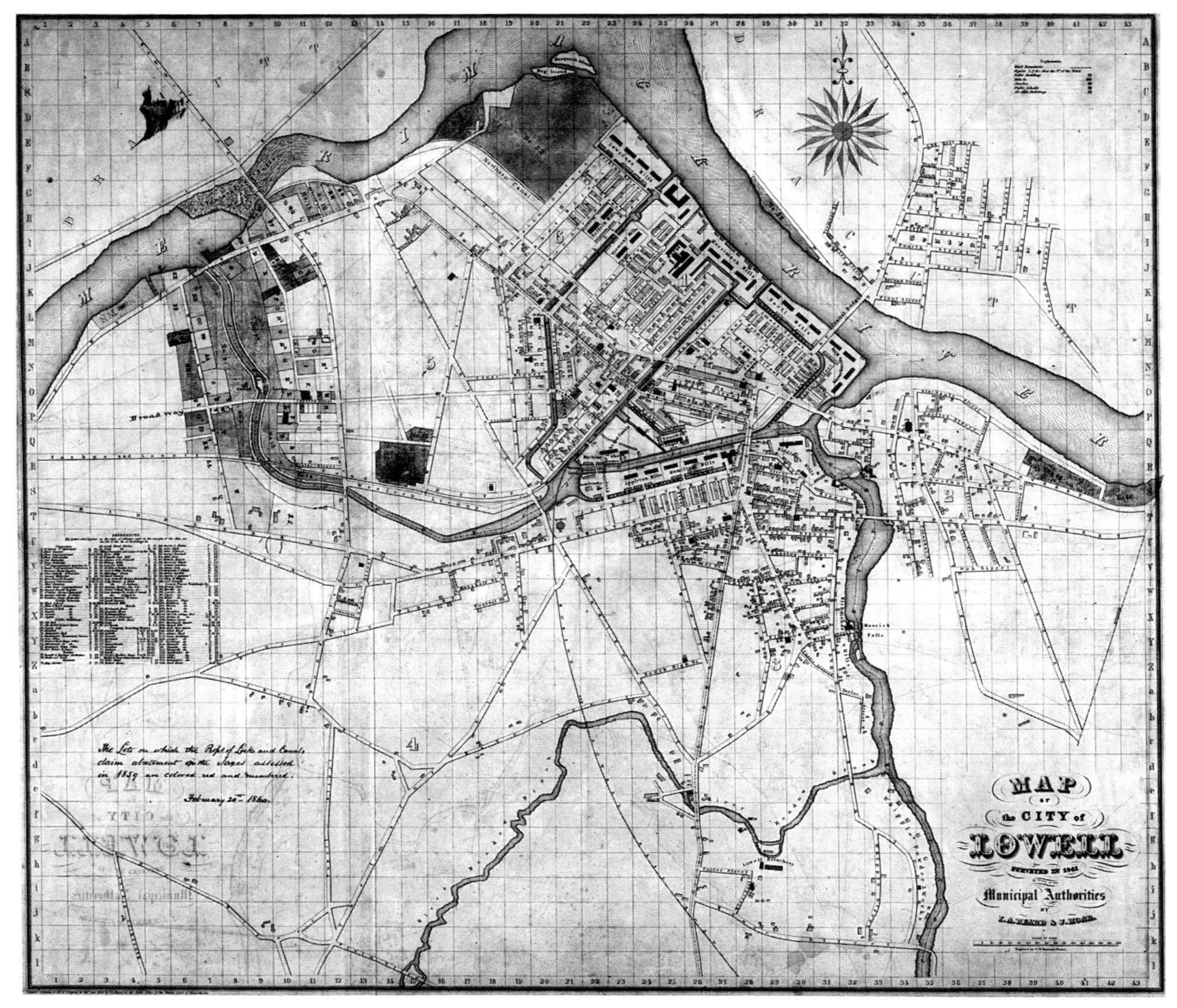

1-008

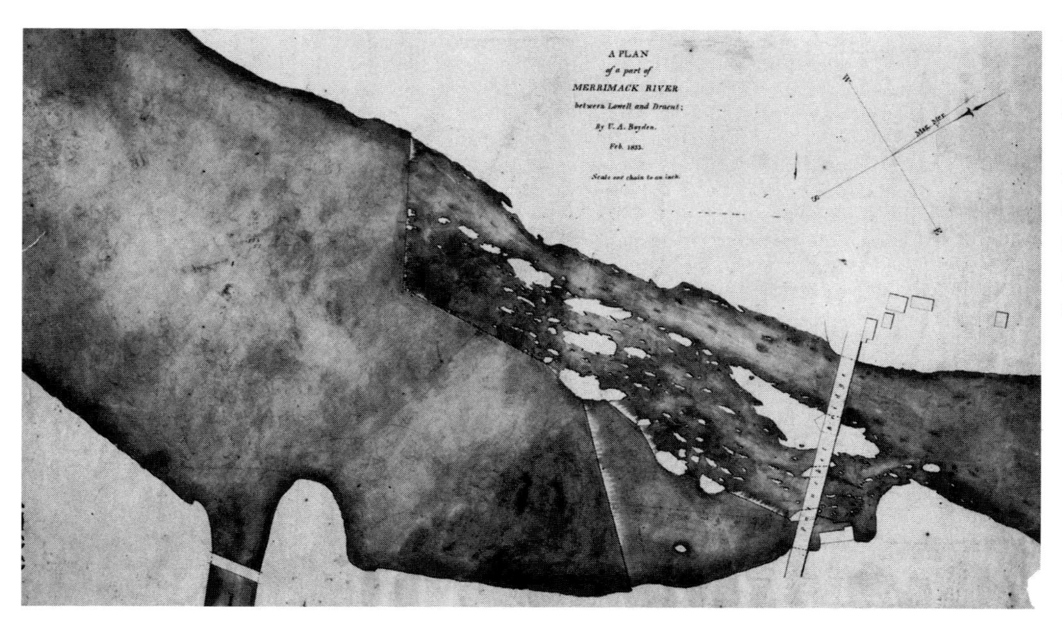

1-009

1-008. City of Lowell, Massachusetts, 1841. L. A. Beard and J. Hoar, delineators, 1841. P&P,HAER,MASS,9-LOW,8, no.13.

1-009. Plan of Pawtucket Dam, Merrimack River, Lowell, Middlesex County, Massachusetts, 1826, 1833. U. A. Boyden, delineator, 1833. P&P,HAER,MASS,9-LOW,8A, no. 8.

1-010. Pawtucket Dam, Merrimack River, Lowell, Middlesex County, Massachusetts, 1826, 1833, 1847, 1875. R. E. Westcott, photographer, 1896. P&P,LC-dig-ppmsca-17295.

1-011. Old dam visible at reconstruction, Pawtucket Dam, Merrimack River, Lowell, Middlesex County, Massachusetts, 1847, 1875. M. Sanborn, photographer, 1875. P&P,HAER,MASS,9-LOW,8A, no. 3.

1-012. Section of the old and new dam, Pawtucket Dam, Merrimack River, Lowell, Middlesex County, Massachusetts, 1847, 1875. James Francis, delineator, 1875. P&P,HAER,MASS,9-LOW,8A, no. 9,detail.

Lowell, Massachusetts, was one of the earliest centers of industrial manufacturing in the country. In 1820, a group of Boston industrialists purchased a transportation canal that was built to bypass Pawtucket Falls in the Merrimack River, and they adapted it to supply waterpower for a number of textile mills. Over the nineteenth century, the Lowell canal system developed to become one of the largest hydropower sites in the country. Pawtucket Dam, first built in 1826 as a wooden crib dam, was replaced twice before it was rebuilt as a masonry gravity overflow dam in 1875.

1-010

1-011

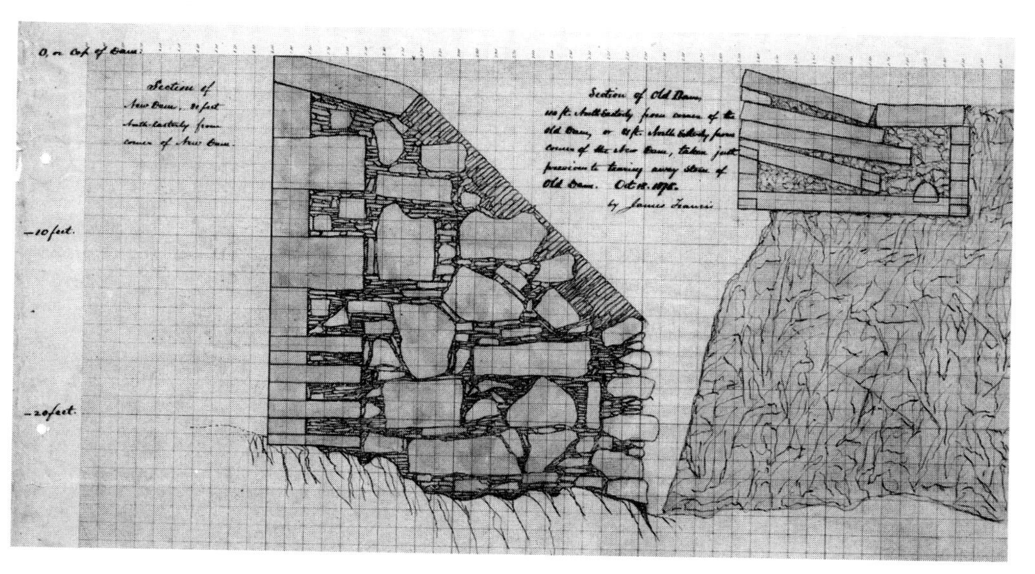

1-012

1-013

1-013. Great Stone Dam, Merrimack River, Lawrence, Essex County, Massachusetts, 1831, 1845–1848. Keystone View Company, 1928. P&P,LC-dig-stereo-1s01722.

After the success of Lowell, industrialists looked for other sites where waterpower could be harnessed to power industry. This site, several miles downstream from Lowell, was first dammed in 1831 to serve the power canal at Lawrence Mill. By 1845, work began on an enormous straight-crested stone gravity dam that was capable of raising water levels 30 feet. On its completion in 1848, it was the largest dam in the country for many years, serving textile mills in Lawrence. It still provides hydroelectricity today.

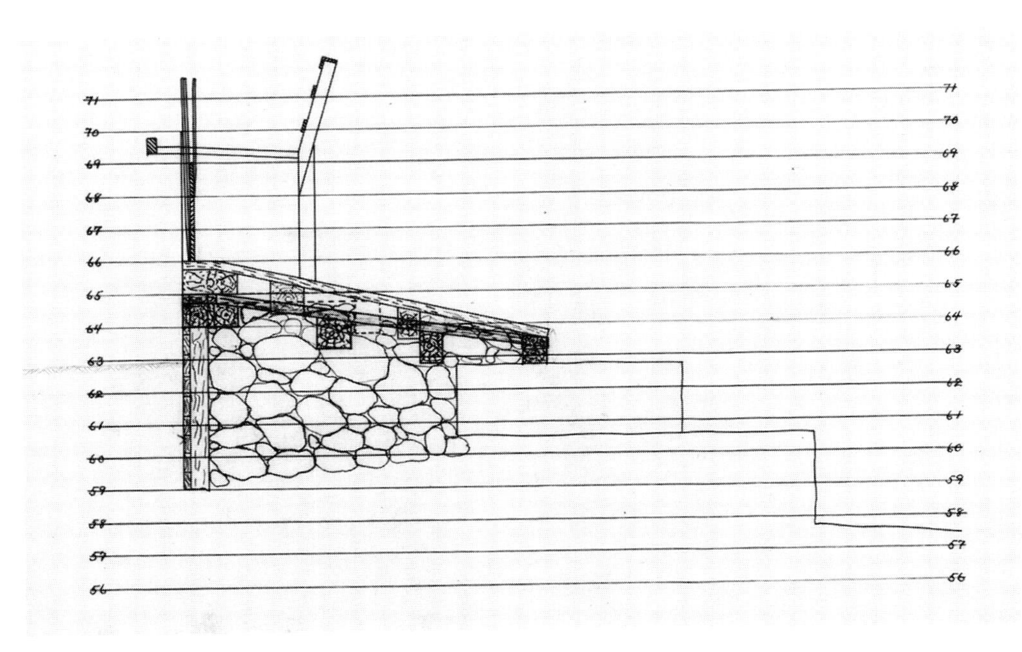

1-014

1-014. Section, Great Stone Dam, Merrimack River, Lawrence, Essex County, Massachusetts, 1831, 1845–1848. 1913. P&P,HAER,MASS,9-LOW,13A, no. 1.

1-015

1-015. Bird's-eye view of Holyoke, Massachusetts, 1881. A. F. Poole, delineator, 1881. G&M,G3764.H7A3 1881 .P6,detail.

In order to power their paper mills, in 1848 a consortium of wealthy mill owners incorporated with a capital stock of $4 million to build the largest of all timber and rubble dams across the Connecticut River. After the failure of the first attempt in 1848, work on a second dam began immediately. This massive crib dam, containing four million board feet of lumber, was pinned to the rock riverbed with iron bolts and protected at its crest by sheets of iron. The dam was built in two sections, one on each side of the natural ledge of the falls. The mill owners also purchased eleven hundred acres of land and constructed canals, mills, streets, and dwellings. The project transformed Holyoke from a farming village into a manufacturing center. Because the dam discharged so much water over its crest, it suffered undermining of its foundations. The timber and stone apron added in 1870 shifted the problem downstream. It was replaced in 1899 by a masonry gravity dam with a curved downstream face.

1-016

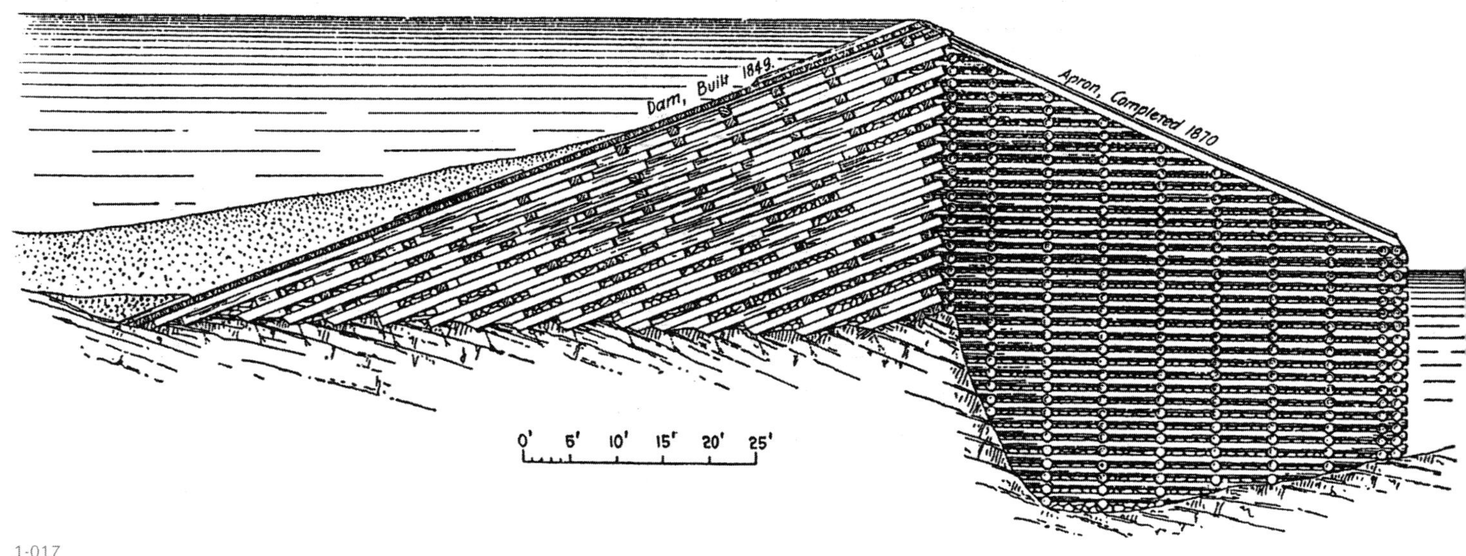

1-017

1-016. Holyoke Dam, Connecticut River, Holyoke, Hampden County, Massachusetts, 1848–1849. J. Bowker, delineator, 1868. P&P,LC-USZ62-107308.

1-017. Section, Holyoke Dam, Connecticut River, Holyoke, Hampden County, Massachusetts, 1848–1849, modified 1868–1870. Edward Wegmann, delineator, 1911. TC547 J33 1995,p. 16.

1-018. Derby Dam and Hydropower Canal, Housatonic River, Derby, New Haven and Fairfield Counties, Connecticut, 1870, rebuilt 1892. Hughes & Bailey, delineator, 1920. G&M,G3784.D7A3 1920 .H8 Fow 81,detail.

Built by the Ousatonic Water Power Company to supply water to Shelton mills, this project includes a dam and canals, and a later hydroelectric plant and fishway.

1-019. Derby Dam and Hydropower Canal, Housatonic River, Derby, New Haven and Fairfield Counties, Connecticut, 1870, rebuilt 1892. Cindy L. Newberg, photographer, 1987. P&P,HAER,CONN,5-DERB,1, no. 5.

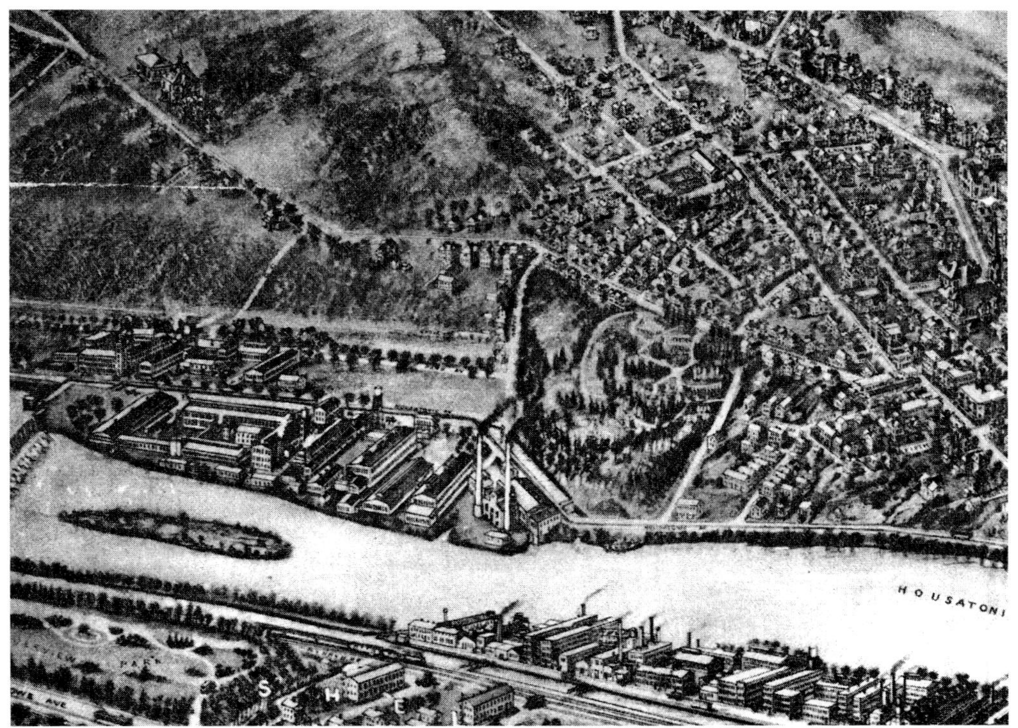

1-018

1-019

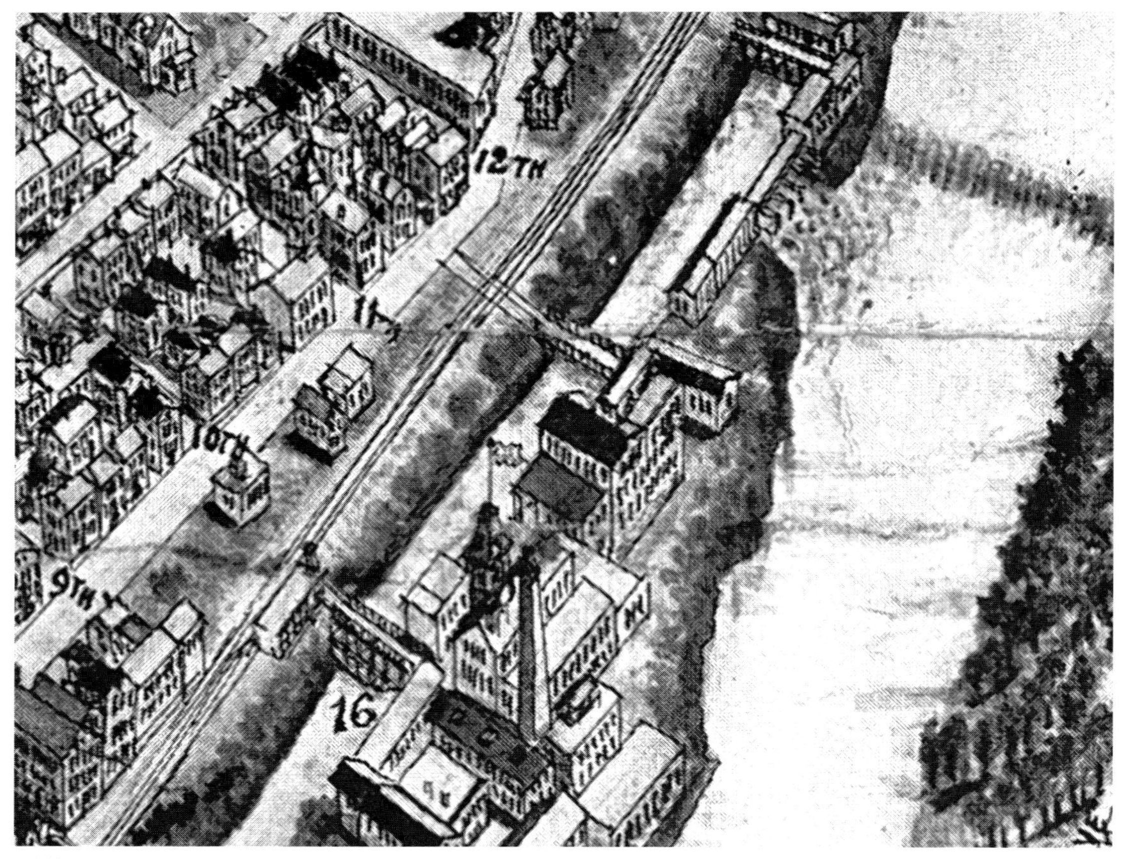

1-020

1-021

1-020. Norwich Water Power Company Dam, Shetucket River, Norwich, New London County, Connecticut, 1882. Hughes & Bailey, delineators, 1912. G&M,G3784.N8A3 1912.H8, detail.

Built in 1882 as a replacement for an earlier structure, this dam increased the flow of water to manufacturers along the canal.

1-021. Norwich Water Power Company Dam, Shetucket River, Norwich, New London County, Connecticut, 1882. Wayne Fleming, photographer, 1994. P&P,HAER,CONN,6-NOR,21A, no. 2.

The dam is of massive stone masonry and timber-crib construction, designed by Hiram Cook (ca.1833–?), who was also president of the Norwich Water Power Company. Most dams built this way failed due to erosion under the downstream face. Nineteenth-century engineers tried various designs for dam crests to reduce the energy of the falling water, and for aprons to protect the riverbed at the toe of the dam.

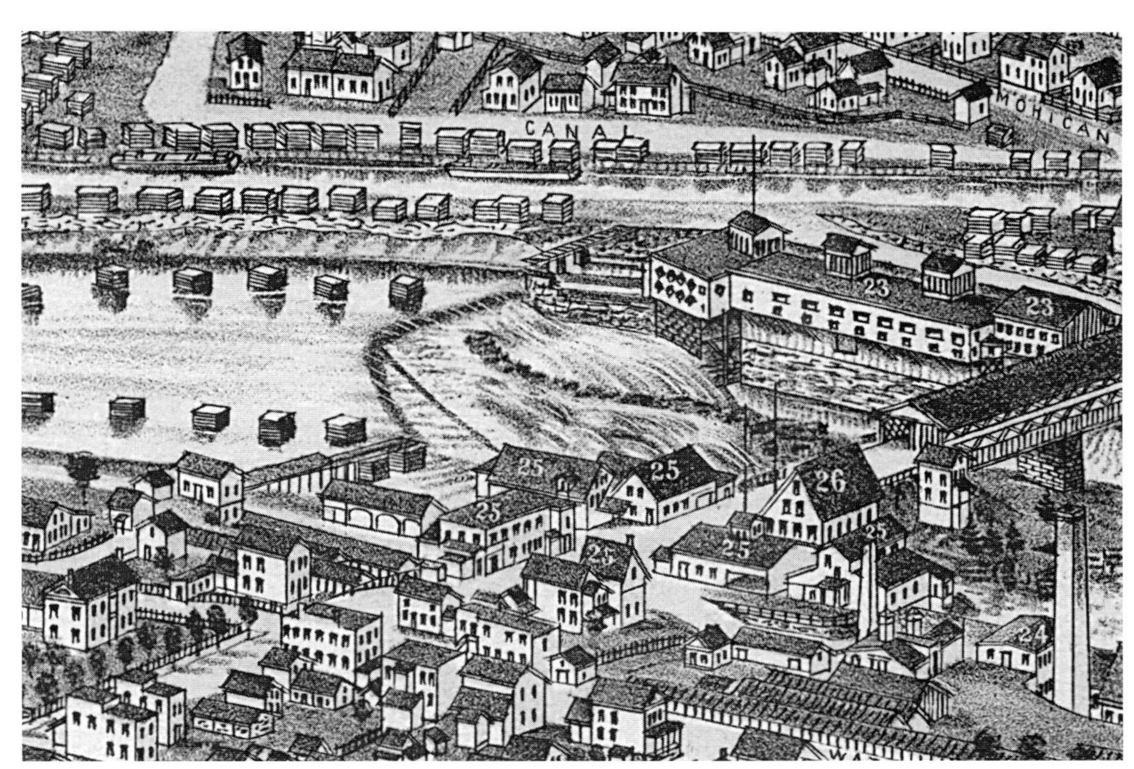

1-022

1-022. Aerial view of Glen Falls Dam, Hudson River, Glen Falls, Warren and Saratoga Counties, New York, 1884. Lucien R. Burleigh, delineator, 1884. G&M,G3804.G5A3 1884. B8,detail.

This dam is one of three structures to have spanned the Hudson River at Glen Falls from the late eighteenth to the twentieth centuries. Glen Falls dams provided waterpower for numerous industries, including sawmills, grist mills, limestone sawmills, cotton factories, planing and turning mills, and paper mills. A partnership of a large local paper mill and International Paper Company built the dam.

1-023

1-023. Glen Falls Dam, Hudson River, Glen Falls, Warren and Saratoga Counties, New York, 1913–1915. Greg Troup, photographer, 1993. P&P,HAER, NY,57-GLEFA,1, no. 5.

The dam is a concrete structure, built in three sections to follow the arc of the highest part of the limestone riverbed, with a height that varies from 9 to 13 feet. A concrete log chute traverses the dam. Its unusual design borrows from both the buttress and gravity dam types, with oversized and heavily angled buttresses that compensate for the lack of reinforcing in the concrete.

1-024

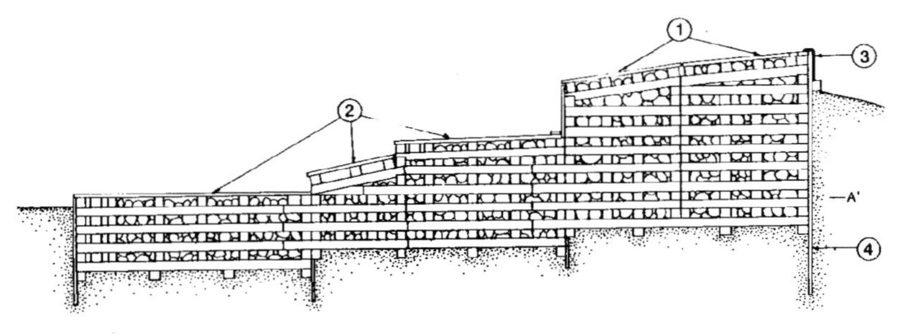

1-025

1-026

1-024. Ruins of Sewall's Falls Dam and Hydroelectric Plant, Merrimack River, Concord, Merrimack County, New Hampshire, 1892–1894, destroyed 1984. Dennis E. Howe, photographer, 1992. P&P,HAER,NH,7-CON,11, no. 14.

These falls on the Merrimack were a fishing site for the Penacook Indians and a European trading post in the 1650s. Although the state legislature authorized a canal at this site as early as 1833, when the dam was finally built in 1894, it was used for hydroelectricity.

1-025. Section, Sewall's Falls Dam and Hydroelectric Plant, Merrimack River, Concord, Merrimack County, New Hampshire, 1892–1894, destroyed 1984. David R. Starbuck, delineator, 1992. P&P,HAER,NH,7-CON,11, sheet 14.

This was one of the longest timber-crib dams in the eastern United States and an early application of the three-phase system, servicing the city of Concord.

1-026. Ripogenus Dam, Penobscot River, Millinocket Vic., Penobscot County, Maine, 1920. Unidentified photographer, 1936. P&P,LC-dig-ppmsca-17325.

The Penobscot is Maine's most difficult river, dropping 70 feet per mile as it passes through Ripogenus Gorge. The Great Northern Paper Company built the dam across this gorge to impound sufficient water so logs could be transported 35 miles downstream to the mill in Millinocket. In the 1950s, a tunnel was drilled in the rock to supply a hydroelectric station. In 1972, pulp drives were banned on the river.

MUNICIPAL RESERVOIR DAMS

From the late eighteenth century, the growing demand for water in cities on the eastern seaboard was largely met by chartered companies that invested in infrastructure and charged users, as was the case with canals or turnpikes. Limited service, unreliability, and an inability to keep up with growing populations were some of the problems that dogged cities that relied on private water supply. By the early nineteenth century, Philadelphia, Boston, and New York had problems with the quantity and quality of their water, suffering from polluted wells and cisterns, water-borne epidemics, and shortages of water for firefighting, yet they were reluctant to take on municipal debt. Arguments raged over the merits of chartering private companies (thereby fostering competition in water supply) versus municipal ownership, which would remove the profit motive and ensure a broader mandate to supply good-quality water to all neighborhoods. In Boston, the debate lasted twenty years, pitting the business view of water as a commodity against the popular view of water as a common resource, with urban reformers playing a key role in supporting a municipal system, for its effect on health and morality. The eventual success of public water supply in the major cities of the eastern seaboard was the result of overlapping interests between a working-class electorate and public health advocates, social reformers, and temperance activists among the elites.

1-027. Fairmount Dam and Waterworks, Philadelphia, Pennsylvania, 1814–1815, 1821–1822. Thomas Birch, delineator, 1824. P&P,LC-USZ62-2587.

1-027

The nation's first capital, and one of its most important cities, Philadelphia established the country's first large-scale municipal water supply in 1801, in direct response to the yellow fever epidemics that struck the city during the previous decade. The scheme, developed by Benjamin Latrobe (1764–1820), involved pumping water from the Schuylkill River to the Center Square Water Works and then up to tanks on a hill, where it would be gravity-fed to distribution pipes. As the demand quickly outgrew the supply, Latrobe's former assistant, Frederick Graff, and John Davis completely redesigned the system in the neoclassical Fairmount plant (1812–1817).

In 1822, the city joined forces with Schuylkill Navigation Company to build a dam across the river so that the more reliable waterpower, instead of steam, could drive the pumps. The 1,204-foot-long Fairmount Dam also fed the navigation company's canals. A timber-crib dam with earth and stone infill, it sits obliquely across the Schuylkill River to cope with the river's flood flow. By 1835, the water works, set behind Greek Revival facades in an extensive park, were a major tourist attraction. Graff went on to advise New York City on its water system of 1842, and Boston on its of 1848.

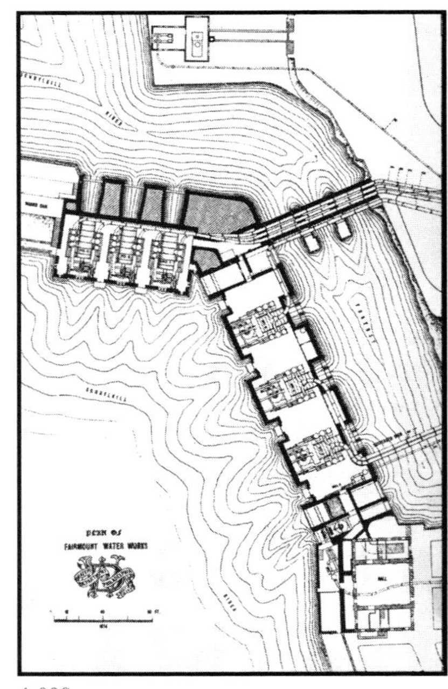

1-028

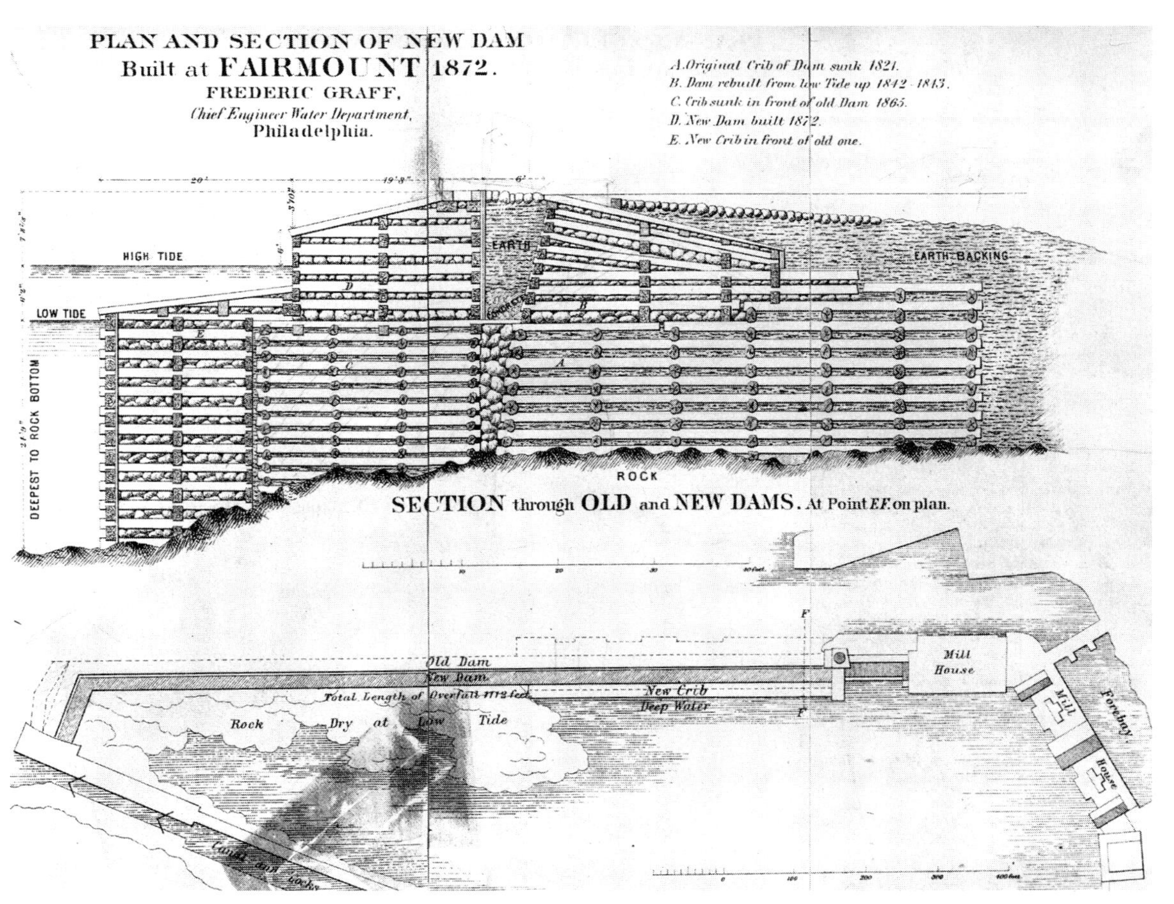

1-029

1-028. Plan of waterworks, Fairmount Dam, Schuylkill River, Philadelphia, Pennsylvania, 1814–1815, 1821–1822, 1860–1862, 1867–1872. Water Department of Philadelphia, 1874. P&P,HAER,PA, 51-PHILA, 328-141.

1-029. Section through new and old Fairmount dams, Schuylkill River, Philadelphia, Pennsylvania, 1814–1815, 1821–1822, 1860–1862, 1867–1872. Water Department of Philadelphia, 1872. P&P,HAER,PA,51-PHILA, 328, no. 140.

1-030. Section of engine house showing Boulton and Watts engine, Fairmount Dam, Schuylkill River, Philadelphia, Pennsylvania, 1814–1815. Courtesy of Franklin Institute, 1815. P&P,HAER,PA,51-PHILA,328-109.

1-031. Section through turbine wheel and flume, Fairmount Dam, Schuylkill River, Philadelphia, Pennsylvania, 1814–1815, 1821–1822, 1860–1862, 1867–1872. Water Department of Philadelphia, 1868. P&P,HAER,PA,51-PHILA,328-139.

1-032. Plan of turbine wheel and flumes, Fairmount Dam, Schuylkill River, Philadelphia, Pennsylvania, 1814–1815, 1821–1822, 1860–1862, 1867–1872. Water Department of Philadelphia, 1868. P&P,HAER,PA,51-PHILA,328-137.

1-033. Detail view of turbine gears, Fairmount Dam, Schuylkill River, Philadelphia, Pennsylvania, 1814–1815, 1821–1822, 1860–1862, 1867–1872. Philadelphia City Archives. P&P,HAER,PA,51-PHILA,328-144.

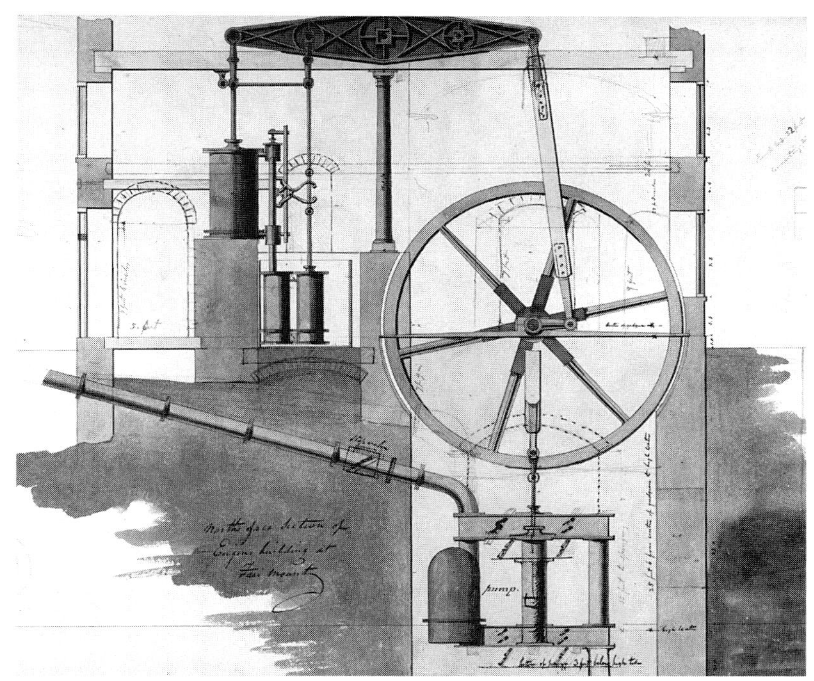

1-030

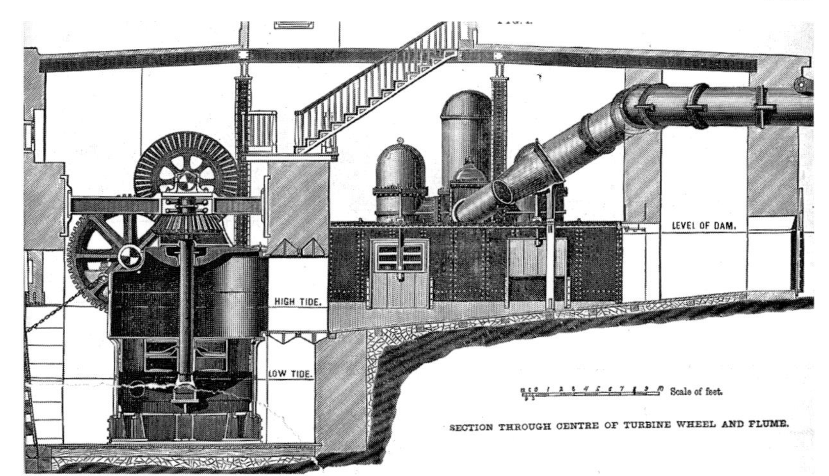

1-031

FIG. 2.

GROUND-PLAN OF TURBINE WHEEL, PUMPS, FLUME, AND MAINS.

1-032

1-033

From the early days of its settlement, New York City relied on wells for its fresh water. As the city grew, this led to sewage contamination, cholera, and water shortages that affected firefighting. After considerable study, the city decided to supply its needs from the Croton River, making the Croton Dam and Aqueduct, designed by city engineer John B. Jervis (1795–1885), the second major municipal water supply project in the United States.

Jervis's first design for an earthen dam failed at its foundation while still under construction. For his next attempt, he stabilized the weak northern foundation with an abutment built of wood and masonry. A masonry dam was built on top, creating a total structure 430 feet long and 50 feet high. A "stilling basin" protected the dam's water face—this was a massive earthen bank that sloped gradually to the crest of the dam. Jervis also designed a wooden crib dam downstream that kept the timber foundation saturated and the lower portion of the main dam under water, counteracting the horizontal force of reservoir water against the principal dam.

1-034. Croton Dam, Croton River, Croton-on-Hudson, Westchester County, New York, 1837–1842. George Hayward, delineator, ca. 1842. P&P,LC-dig-ppmsca-17374.

1-034

The 41-mile-long Croton Aqueduct traversed numerous tunnels and bridges (including the High Bridge over the Harlem River) on its way to a receiving reservoir at Yorkhill (in Central Park) and a distributing reservoir on 42nd Street (now the site of the New York Public Library). In 1860, an additional pipe was added to the High Bridge. By 1900, the city had to further increase its supply, completing the New Croton Aqueduct system and submerging the original Croton Dam.

1-035. Croton Aqueduct High Bridge crossing the Harlem River to Yorkhill reservoir, New York, New York, 1862. Ca. 1862. P&P,LC-dig-ppmsca-17375.

1-035

1-036

1-036. Spillway, New Croton Dam, Croton River, Croton-on-Hudson, Westchester County, New York, 1892–1907. Jack Boucher, photographer, 1978. P&P,HAER,NY,60-CROTOH.V,1, no. 10.

Designed by the New York City Aqueduct Commission, this dam was the tallest in the world when it was built. It is a masonry gravity dam, almost 300 feet high and equally as wide, and 2,200 feet long (including its 1,000-00 spillway). It still supplies water to New York City.

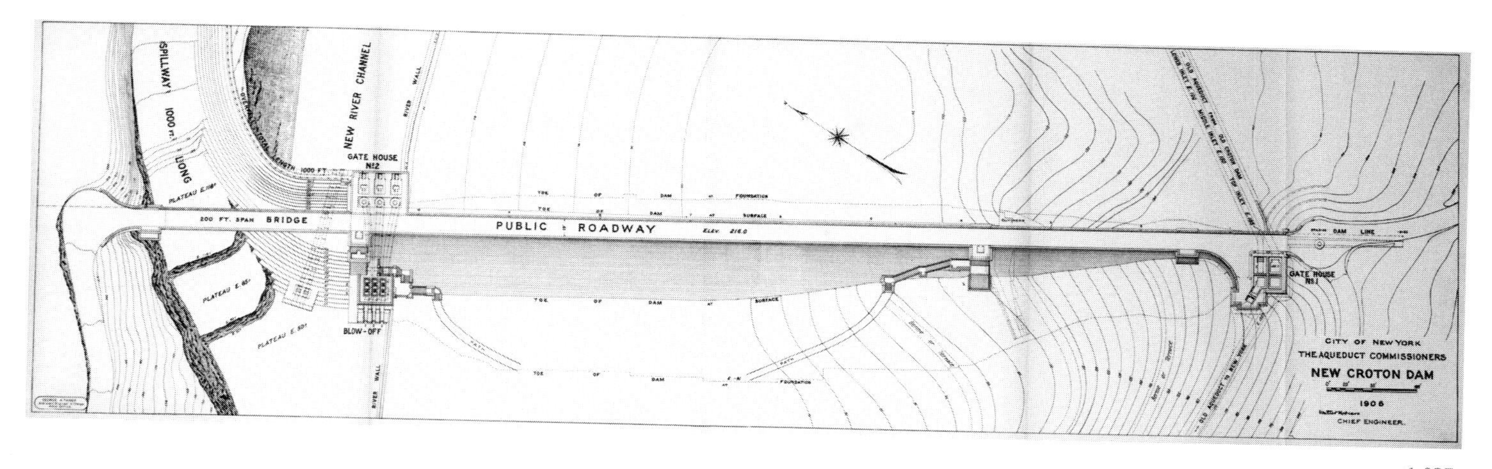

1-037

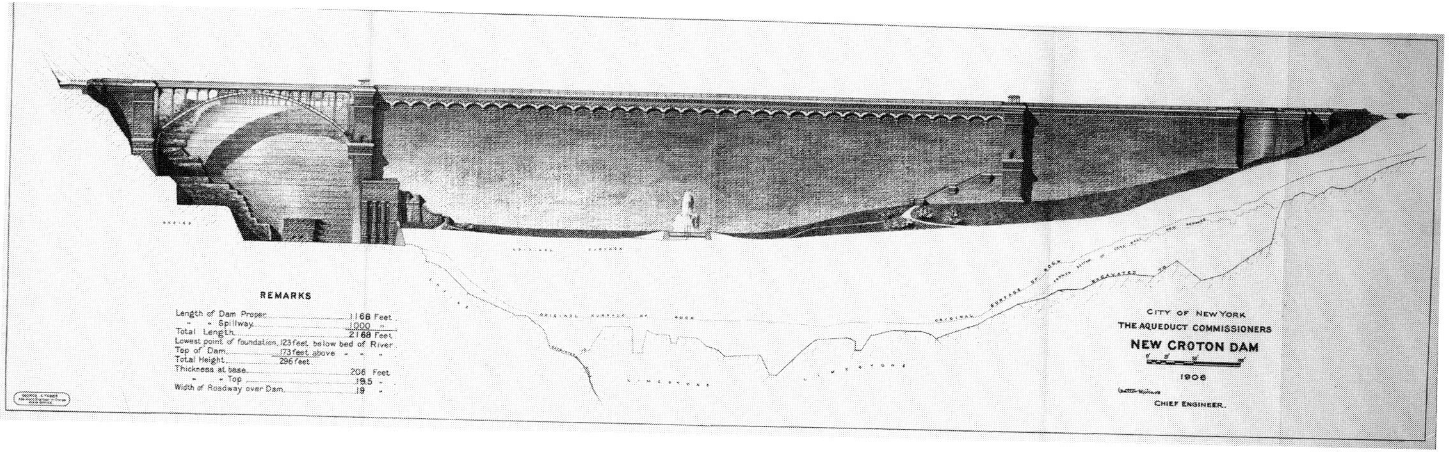

1-038

1-037. Plan of New Croton Dam, Croton River, Croton-on-Hudson, Westchester County, New York, 1906. New York City Aqueduct Commission, 1906. P&P,HAER,NY,60-CROTOH.V,1, no. 21.

1-038. Downstream elevation of New Croton Dam, Croton River, Croton-on-Hudson, Westchester County, New York, 1906. New York City Aqueduct Commission, 1906. P&P,HAER,NY,60-CROTOH.V,1, no. 22.

1-039. New Croton Dam, Croton River, Croton-on-Hudson, Westchester County, New York, 1892–1907. Jack Boucher, photographer, 1978. P&P,HAER,NY,60-CROTOH.V,1, no. 28.

1-039

BALTIMORE

The city of Baltimore had difficulty securing a steady and sufficient supply of water for its population. In the early nineteenth century, a private water company constructed a reservoir in the city to receive water from the Jones Falls, building additional higher reservoirs and a distribution system later. Frustrated with limited service, contaminated water, and a number of cholera outbreaks, the city purchased the system in 1854 and began a series of improvements, including Lake Roland and Druid Lake dams. Lake Roland Dam is a masonry gravity dam with a rubble core. The massive Druid Lake Dam, 800 feet long and 119 feet high, with a base width of 640 feet, was the largest earthen dam in the United States at the time of its completion.

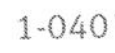

1-040

1-041

1-040. Lake Roland Dam, Jones Falls River, Towson Vic., Baltimore County, Maryland, 1858–1861, abandoned 1915. Marge Green, photographer, 1991. P&P,HAER,MD,3-TOW.V,2, no. 2.

1-041. Druid Lake, Baltimore, Maryland, 1870–1871. DETR. P&P,LC-D4-16541.

1-042. Wachusett Dam under construction, South Branch of Nashua River, Clinton, Worcester County, Massachusetts, 1901–1905. Unidentified photographer, 1904. P&P,LC-dig-ppmsca-17372.

This straight-crested masonry gravity dam, built for Boston's municipal water supply, serves a 12-mile-long tunnel that feeds the Sudbury Reservoir. It was designed and built under the direction of Frederic P. Stearns (1851–1919), chief engineer of Boston's Metropolitan Water Board.

1-043. Montreat Dam, Lake Susan, Montreat, Buncombe County, North Carolina, 1924. Unidentified photographer, ca. 1941. WC. P&P,LC-dig-ppmsca-17322.

Montreat, for "mountain retreat," was established in 1897 as a Christian camp meeting grounds. While retreat participants lived in tents in the first years, houses and a hotel soon followed. In 1907, the Presbyterian Church of the United States obtained control of Montreat, building an auditorium and inn and, a decade later, a school and college. In 1916, the original wooden dam collapsed in a flood. It was replaced by the present dam in 1924.

1-042

1-043

For eighteenth-century colonists along the eastern seaboard, the Appalachian Mountains presented a formidable barrier to westward settlement. In this era before the railroads, canal building seemed an ideal solution to promoting commerce, although it was a challenging engineering enterprise.

Spurred on by Treasury Secretary Albert Gallatin's transportation plan of 1808, federal and state governments subsidized canal building into the hinterlands and across the four major eastern capes, creating a network of waterways in the developing country. The Hudson River, navigable by ocean vessels for 150 miles of its length, rose to preeminence with the successful completion of the Erie Canal in 1825. This canal linked the eastern seaboard with the Great Lakes, providing the first all-water route across the Appalachian mountain range, while other canals connected the Hudson to the Saint Lawrence and Delaware rivers. Likewise, the head of the Chesapeake Bay was linked by canal to the Delaware, and further canals were built along the Potomac and James rivers in the hope of reaching the Ohio River valley on the other side of the Appalachian mountain range. Other significant canals included the Chesapeake & Ohio and the Lehigh systems, built to reach the coal fields of Pennsylvania. Canals required dams, and the dams on these rivers serve as illustrations of both vernacular and engineered waterworks, revealing lessons learned and improvements made over the course of the nineteenth century.

1-044. James River & Kanawha Canal Dam, Richmond, Virginia, 1785–1851, abandoned ca. 1880. DETR, ca. 1910. P&P,LC-D4 72498.

1-045. Workmen picnicking at dam, Maury River Navigation Canal, Lexington Vic., Rockbridge County, Virginia, 1851–1860, part of James River & Kanawha Canal system. Unidentified photographer, ca. 1860. P&P,LC-USZ62-43804.

1-045

1-044

James River & Kanawha Canal

Navigation of the James River played an important role in early Virginia and the settlement of the interior; in fact, as early as the mid-eighteenth century, plans were afoot to link the James River to the Ohio River basin. The idea was to link the headwaters of the James with the New River, which was the only watercourse that cut across the Appalachian Mountains, and from there to descend to the fertile Kanawha River valley. Dams and locks would be required across the entire length of the watercourse.

Work on the canal system began in 1785. The first segment, built to circumvent the falls near Richmond, was complete by 1822. By 1840, the navigable waterway ran from Richmond to Lynchburg, and by 1851 it had reached its farthest extent—197 miles inland, in Botetourt County. At this time, it was the primary commercial route in the state. Freight boats brought tobacco and wheat from western Virginia to market and returned with finished goods from the city. Passenger boats ran on a regular basis between Lynchburg and Richmond. The canal fell into decline during the Civil War, and by 1880 the Richmond and Allegheny Railroad was laying its tracks on the towpath of the canal. Canal construction never reached the Kanawha River. Built at the twilight of the canal era in 1860, the Maury Canal was an extension of the James River system, bringing traffic up the Maury River in western Virginia.

1-046. Dam, Maury River Navigation Canal, Lexington Vic., Rockbridge County, Virginia, 1851–1860, part of James River & Kanawha Canal system. Unidentified photographer, ca. 1860. P&P,LC-dig-cph-3a44044.

1-046

Schuylkill Navigation Canal

Shortly after the discovery of coal in 1814 in Schuylkill County, in the central Appalachian Mountains, the Schuylkill Navigation Company was formed to develop a reliable shipping route on the Schuylkill River. The canal system (1817–1825), one of the first in the United States, extended over 100 miles from the Fairmount Dam in Philadelphia (1-027) to the anthracite coalfields near Pottsville, Pennsylvania, rising nearly 600 feet. It included 32 dams on the Schuykill of varying heights, 62 miles of canals, 92 locks, and a tunnel. The combination of hydropower and transport guaranteed the industrial development of the towns of Reading, Norristown, and Pottsville, as well as Manayunk, the site of Flat Rock Dam. As the only way of shipping coal from the Schuylkill anthracite fields to Philadelphia (and, via the Delaware and Raritan canal, to New York City), the canal was highly profitable, carrying nearly a million tons of coal per year from 1830 to 1840. With the advance of the railroad, the canal company attempted expansion but could no longer compete by 1870. The canal last carried freight in 1916. Manayunk was incorporated into the city of Philadelphia in 1854.

1-047

1-047. Flat Rock Dam, Schuylkill Navigation Canal, Schuylkill River, Philadelphia, Pennsylvania, 1817–1825. Coleman Sellers, photographer, 1862. P&P,LC-DIG-stereo-1s01484.

Lehigh Canal

The Lehigh Canal system was built to transport coal from the Lehigh River valley to Easton, Pennsylvania, on the Delaware River. The earliest coal transportation on the Lehigh began in 1792, the year after coal was discovered in Mauch Chunk. These shipments floated downriver to Philadelphia on wooden rafts, but rapids and shallows on the Lehigh made navigation difficult, and the company foundered. In 1818, the industrialist Josiah White, interested in securing a reliable supply of coal for his iron works, obtained permission to improve the river. Having established the Lehigh Navigation Company, he proceeded to construct a series of timber-crib dams with a lock system of his own design, which he called the bear trap.

1-048

1-048. Panoramic view of Bear Mountain and the Lehigh Valley, Mauch Chunk, Carbon County, Pennsylvania, Charles F. Schneur, photographer, 1879. P&P,LC-dig-pan-6a19686.

1-049. Lehigh Canal Dam, Lehigh River, Mauch Chunk, Carbon County, Pennsylvania, 1827–1829. Charles F. Schneur, photographer, 1879. P&P,PAN US GEOG-Pennsylvania no. 67 (E size),detail.

1-049

The bear trap gate was hinged to the floor and opened by the pressure of water delivered from a reservoir via a flume. In the raised position, the gate formed part of the dam. When water was discharged into the bear trap, the raft was raised several feet and propelled out of the gate on a surge that swept it over the rapids. When the gate was closed, the bear trap again formed part of the dam. Although bear trap gates enabled a raft to ride on a rush of water from one slack-water pool to the next, they worked in only one direction (see 2-025).

In 1827, the waterway was rebuilt as a conventional canal with lift locks that allowed traffic in both directions. The enlarged canal was 46 miles long, with eight dams and numerous locks and aqueducts that allowed it to climb 355 feet. Within five years of its completion, the canal had been linked to both Philadelphia and New York by means of the Delaware Division Canal and the Morris Canal, respectively, ensuring the industrial preeminence of this part of Pennsylvania for many years. The system was again extended in 1835–1838 and reached its peak volume in 1855, when it carried more than a million tons of cargo. The railroad eclipsed it shortly afterward.

1-050. Lehigh Canal Dam No. 7 (Hamilton Street Bridge Dam), Lehigh River, Allentown, Lehigh County, Pennsylvania, 1827–1829. Unidentified 1979. P&P,HAER,PA,48-ALLEN,3A, no. 15.

1-050

1-051

1-051. Detail of rock crib dam and planking on Lehigh Canal Dam No. 7, Lehigh River, Allentown, Lehigh County, Pennsylvania, 1827–1829. Robert G. Miller, photographer, 1983. P&P,HAER,PA,48-ALLEN,3A, no. 10.

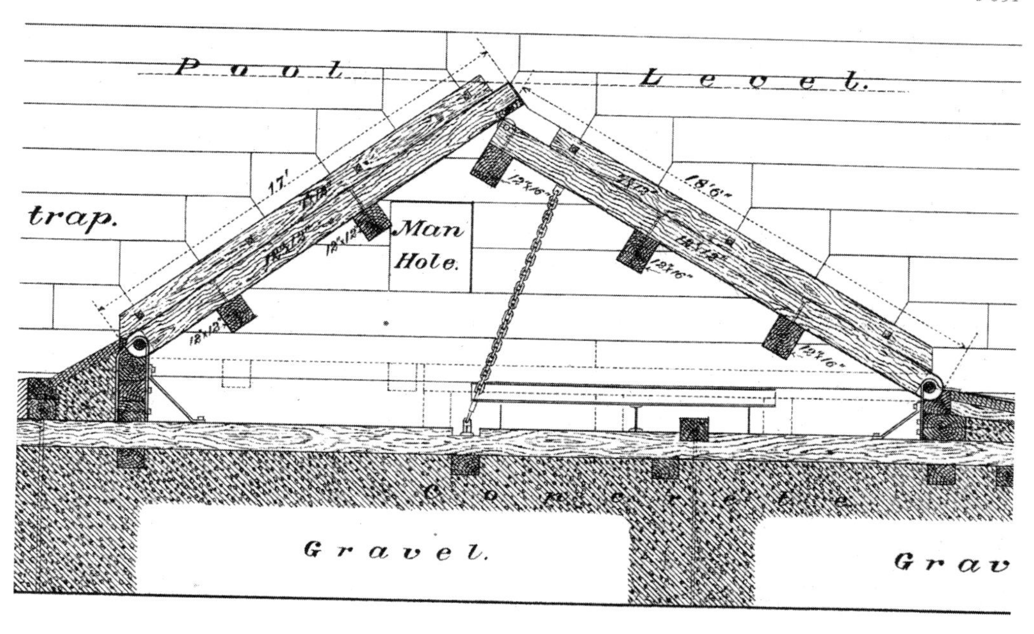

1-052

1-052. Section through bear trap, Davis Island Dam (Lock & Dam No. 1), Ohio River, Pittsburgh Vic., Allegheny County, Pennsylvania, 1878–1885. U.S. Army Corps of Engineers, delineator, 1889. TC549 U53 Folio,plate no. 18,detail.

Chesapeake & Ohio Canal

George Washington's Potowmac Company of 1785 was the earliest effort to improve navigation on the Potomac, involving the construction of a number of small canals to bypass the 80-foot drop at Great Falls and allow passage of flatboats downstream. In 1824, the company's holdings were acquired by the Chesapeake & Ohio Company, which aimed to link the Potomac to the Ohio River valley. Construction on the canal began in 1828, and sections opened for navigation as they were completed: Georgetown to Seneca in 1830, Harpers Ferry in 1833, Hancock in 1839, and finally Cumberland, Maryland, in 1850. The 184-mile canal was dug and blasted through rock, with a number of carefully built masonry features—seventy-four lift locks to accommodate the 600-foot climb to Cumberland, eleven aqueducts to cross tributary streams, and a 3,120-foot tunnel. Seven timber-crib dams with feeder and guard locks diverted the Potomac River into the canal. The canal's chief engineer, Benjamin Wright (1770–1842), had been the lead engineer on the Erie Canal.

From the outset, the canal faced stiff competition from the Baltimore & Ohio Railroad, which had begun construction the same year as the canal but completed its track to Cumberland in 1842. From there, passengers could continue to the Ohio Valley by stage-

1-053. C&O Canal, Potomac River, Great Falls, Fairfax County, Virginia, 1828-50. Robert Latou Dickinson, artist, 1918. P&P,LC-USZC4-11100.

1-053

coach on the National Road. The railroad eventually continued westward, but the canal never advanced past Cumberland. Even so, coal and timber freight allowed the canal to operate profitably by the 1870s, until a flood in 1889 bankrupted the company. Its long-time rival, the Baltimore & Ohio Railroad, bought it out; following a second flood in 1924, the canal closed down.

The canal's original timber and crib dams leaked heavily, failing to provide sufficient water to maintain navigable levels. They were replaced by more substantial masonry dams in the 1860s. The hydropower potential at the dam sites attracted mills and other industries early on, particularly at the canal terminus Cumberland (Dam No. 8), which became the second-largest city in Maryland, after Baltimore. In 1902, the Martinsburg Electric Light Company developed a hydroelectric plant at Dam No. 5, reusing the power plant and turbines from a mill; four years later, the company leased water rights on Dam No. 4 to build a new powerhouse. Because of frequent flooding, the generators were located three stories above river level and connected to the submerged turbines with a rope drive.

1-054. C&O Canal Dam No. 4, Potomac River, Shepherdstown Vic., Jefferson County, West Virginia, 1832–1835, rebuilt 1860–1869, powerhouse 1906–1909. Photograph courtesy Potomac Edison Company, ca. 1910. P&P,HAER,WVA, 2-SHEP.V,1, no. 17.

1-054

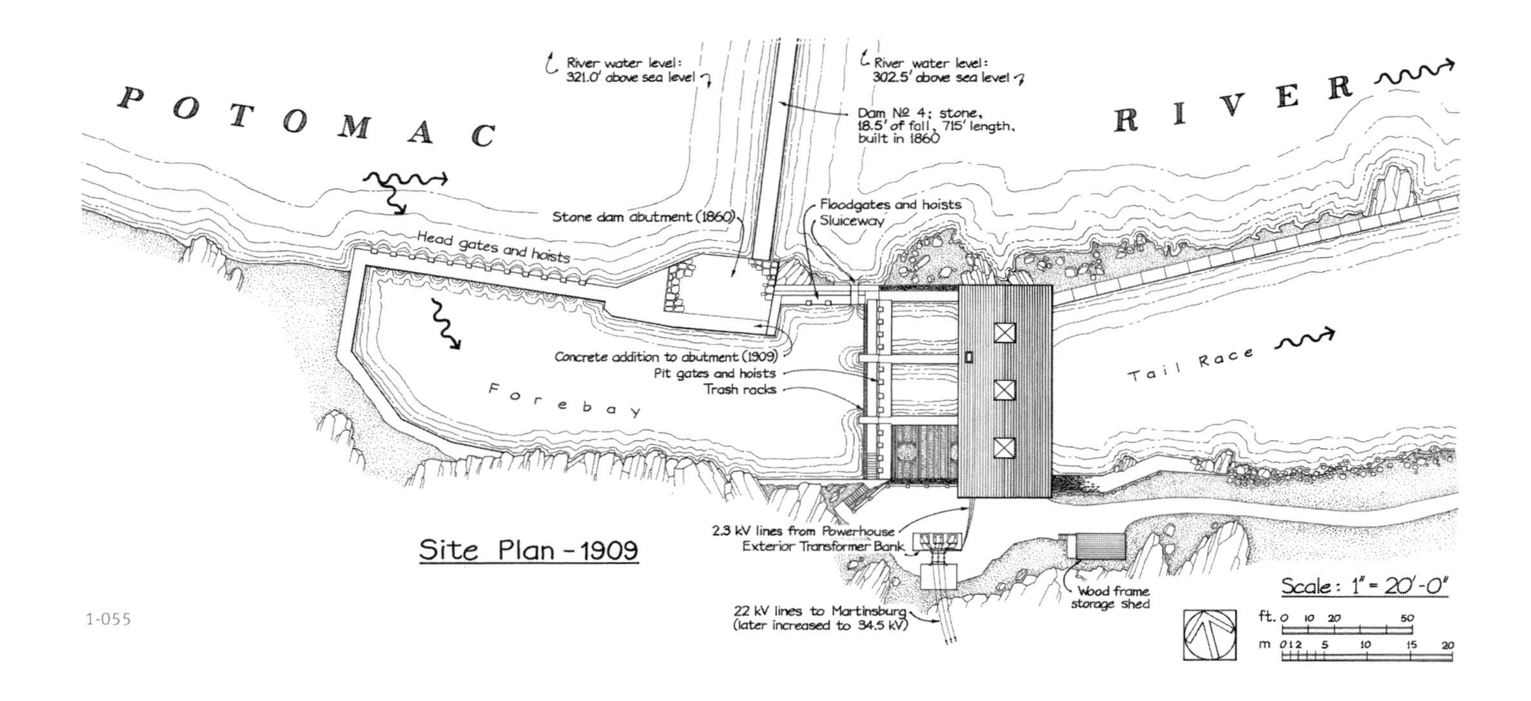

1-055

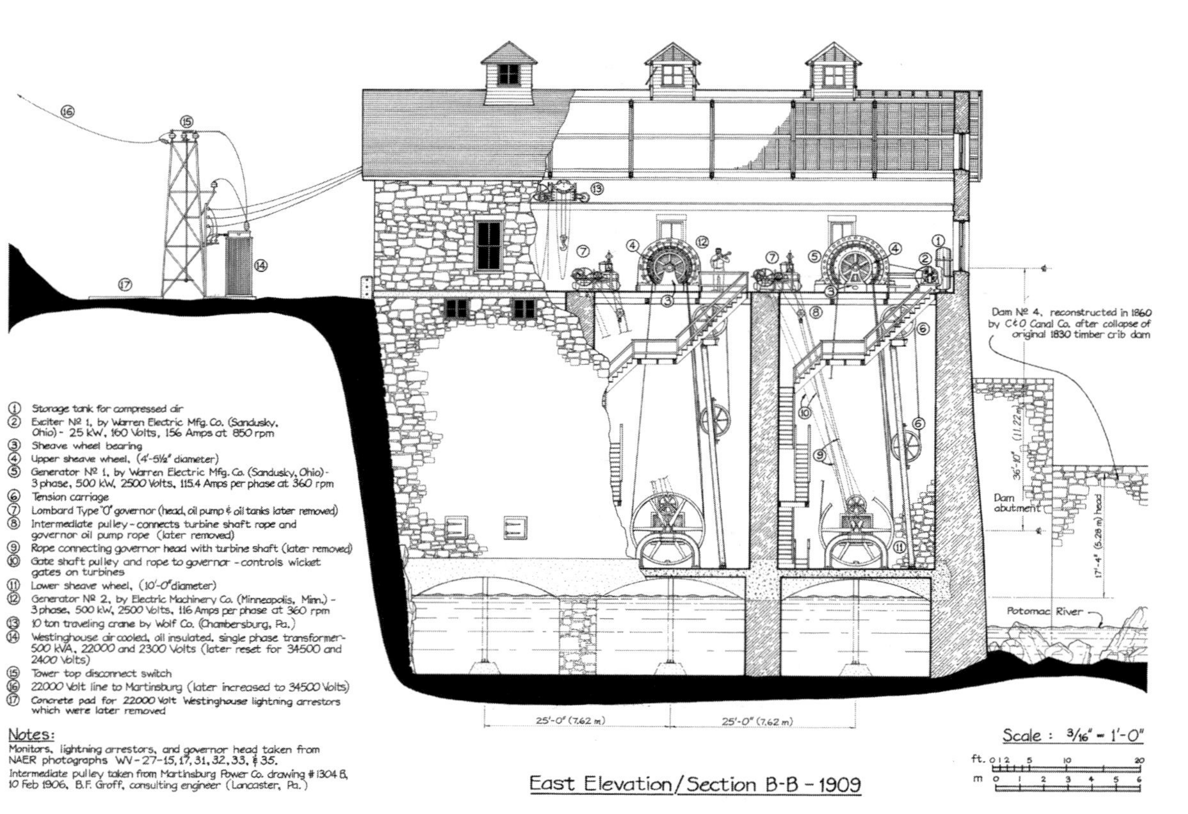

1-056

1-055. Site plan, C&O Canal Dam No. 4, Potomac River, Shepherdstown Vic., Jefferson County, West Virginia, 1832–1835, rebuilt 1860–1869, powerhouse 1906–1909. John R. Bowie and Neil Richardson, delineators, 1980. P&P, HAER,WVA,2-SHEP.V.1.

1-056. East elevation, C&O Canal Dam No. 4, Potomac River, Shepherdstown Vic., Jefferson County, West Virginia, 1832–1835, rebuilt 1860–1869, powerhouse 1906–1909. John R. Bowie, delineator, 1980. P&P,HAER,WVA,2-SHEP.V, 1-, sheet no. 6.

1-057. C&O Canal Dam No. 5. Potomac River, Hedgesville Vic., Berkeley County, West Virginia, 1832–1835, rebuilt 1860–1869, powerhouse 1902–1905. Photograph courtesy Potomac Edison Company, ca. 1904. P&P,HAER,WVA,2-HED-VI.V,1, no. 1.

1-058. Power house, forebay, and river, C&O Canal Dam No. 5. Potomac River, Hedgesville Vic., Berkeley County, West Virginia, 1860–1869, powerhouse 1902–1905, hydro plant rebuilt 1917–1918. Photograph courtesy Potomac Edison Company, 1933. P&P,HAER,WVA,2-HEDVI.V,1, no. 49.

1-059. Power house and substation during flood, C&O Canal Dam No. 5. Potomac River, Hedgesville Vic., Berkeley County, West Virginia, 1860–1869, powerhouse 1902–1905, hydro plant rebuilt 1917–1918. Photograph courtesy Potomac Edison Company, 1937. P&P,HAER,WVA,2-HEDVI.V,1, no. 50.

1-057

1-059

1-058

1-060

1-060. C&O Canal Dam No. 8, Potomac River, Cumberland, Allegany County, Maryland, 1850. Robert Shriver, photographer, 1862. P&P,LC-DIG-stereo-1s01490.

Warrior River Navigation System

In 1871, Congress authorized construction of a navigational channel along the Warrior and Black Warrior rivers in Alabama. The 455-mile system, built from 1888 to 1917, was the result of federal and commercial interests working together to facilitate traffic on the shoal-prone rivers to the state capital, Tuscaloosa. The six modern locks and dams, completed from 1954 to 1991, flooded the nineteen original locks and dams.

1-061

1-061. Holt Lock & Dam, Warrior River, Tuscaloosa County, Alabama, 1966. David Delting, photographer, 1992. P&P,HAER,ALA,63-HOLT,3, no. 1.

In the Piedmont region of Appalachia, it had long been customary to dam streams and rivers to power mills both small and large. When mercantile blockades during the Civil War prevented southern cotton from reaching textile manufacturers in England, local textile industries boomed. Hydroelectric plants soon replaced the mechanical mills. Generally, local settlements sprang up near the new hydroelectric plants, as the development of long-distance transmission lines took some time to work out. Throughout the Piedmont South, textile mills and their villages dot the landscape. This industry was integral to the creation of a "new South" after Reconstruction and transformed the lifestyles of the white rural poor through wage labor.

The first two dams illustrated in this section, Womack's Mill Dam and Pearle Mill Dam, are typical of a small-scale use of hydropower for manufacturing. Located miles upstream from the mills, these dams directed water into a headrace to feed turbines that drove the machinery at the mills.

1-062. Womack's Mill Dam, County Line Creek, Yanceyville, Caswell County, North Carolina, 1826. Jet Lowe, photographer, 1979. P&P,HAER,NC,17-YANV.V,3, no. 1.

1-062

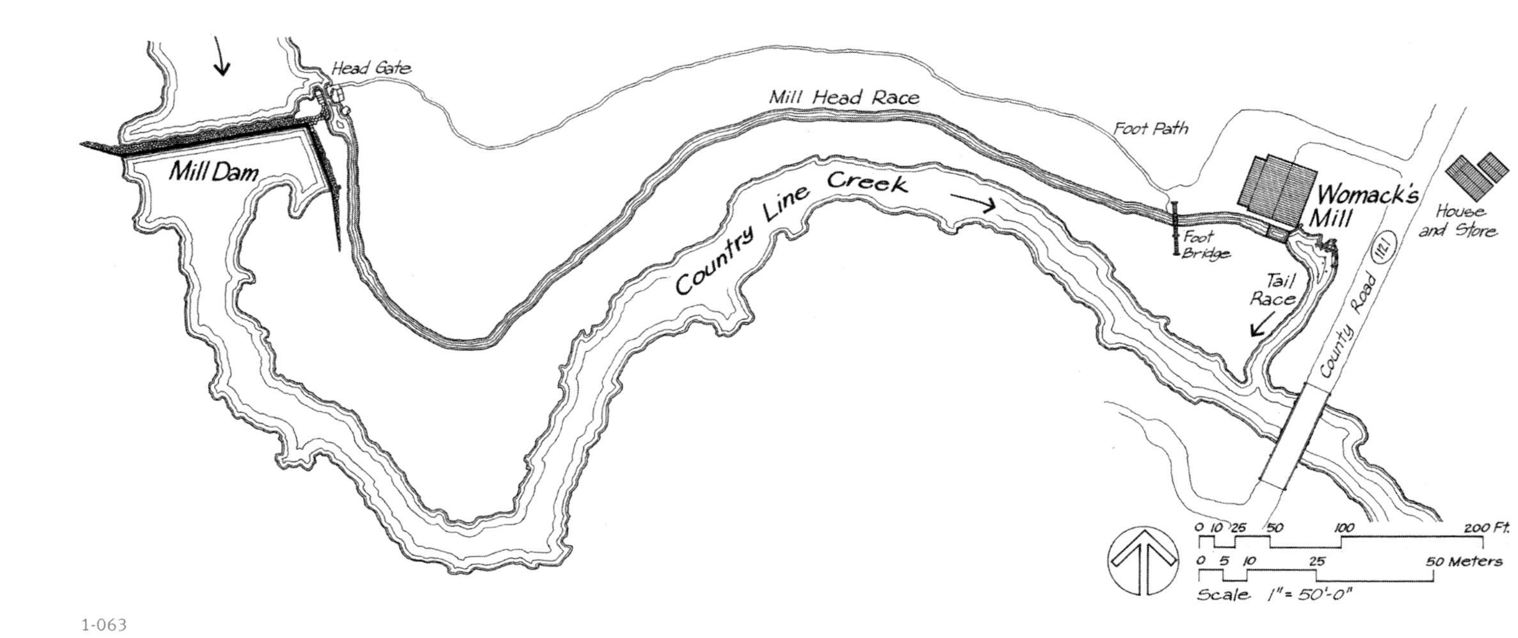

1-063

1-064

1-063. Site plan of Womack's Mill, County Line Creek, Yanceyville, Caswell County, North Carolina, 1826. A. Kokoris and K. Savoie, delineators, 1979. P&P,HAER,NC,17-YANV.V,3,sheet no. 1.

1-064. Mill tailrace at Womack's Mill Dam, County Line Creek, Yanceyville, Caswell County, North Carolina, 1826. Jet Lowe, photographer, 1979. P&P,HAER,NC,17-YANV.V,3, no. 5.

1-065. First floor plan of Womack's Mill, County Line Creek, Yanceyville, Caswell County, North Carolina, 1826. A. Kokoris, delineator, 1979. P&P,HAER,NC,17-YANV.V,3.

1-066. Cellar plan of Womack's Mill, County Line Creek, Yanceyville, Caswell County, North Carolina, 1826. A. Kokoris, delineator, 1979. P&P,HAER,NC,17-YANV.V,3.

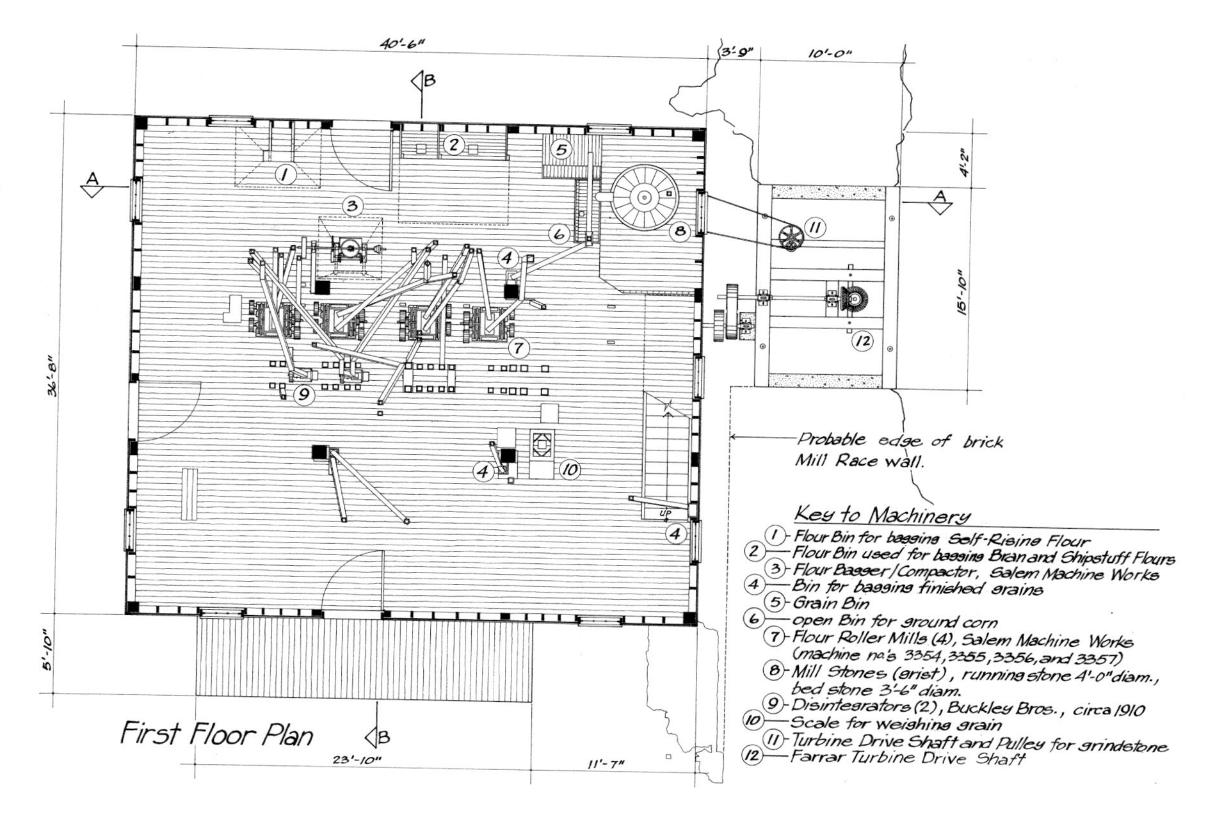

1-065

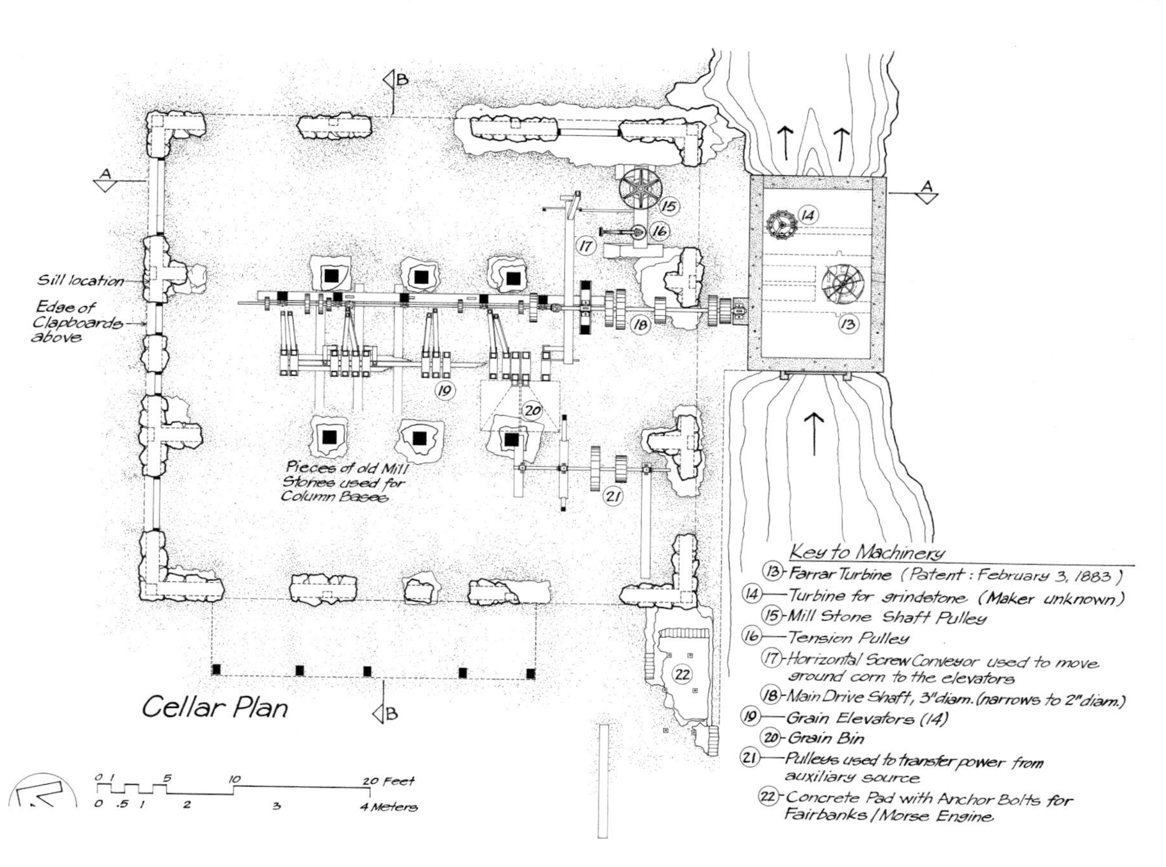

1-066

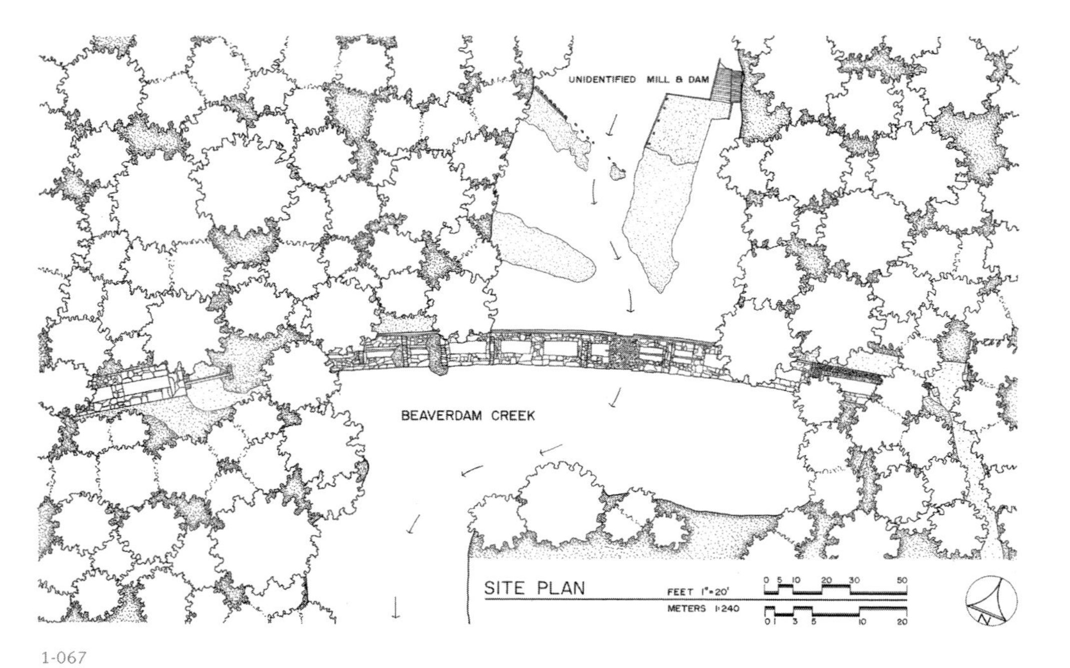

1-067

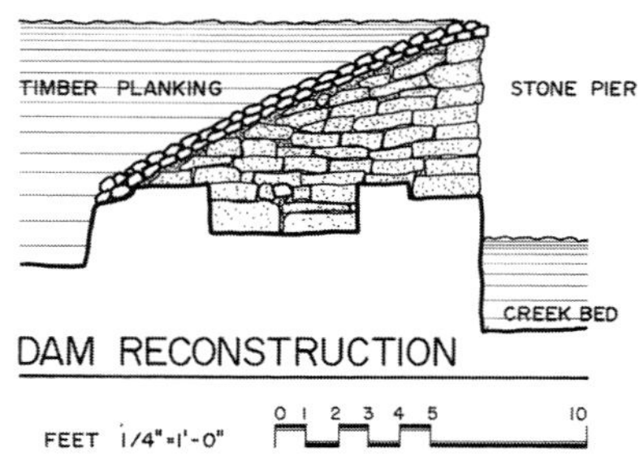

1-068

1-069

1-067. Site plan of Pearle Mill Dam, Beaverdam Creek, Elberton Vic., Elbert County, Georgia, 1890–1895. Richard J. Cronenberger, delineator, 1980. P&P, HAER,GA,53-ELBE.V,1-,sheet no. 5.

1-068. Reconstructed section of Pearle Mill Dam, Beaverdam Creek, Elberton Vic., Elbert County, Georgia, 1890–1895. Richard J. Cronenberger, delineator, 1980. P&P, HAER,GA,53-ELBE.V,1-,sheet no. 5.

1-069. Ruins of Pearle Mill Dam, Beaverdam Creek, Elberton Vic., Elbert County, Georgia, 1890–1895. Dennis O'Kain, photographer, 1980. P&P,HAER,GA,53-ELBE.V,1, no. 15.

1-070. South elevation, Pearle Cotton Mill, Elbert County, Georgia, 1895. Richard J. Cronenberger, delineator, 1980. P&P, HAER,GA,53-ELBE.V,1-sheet no. 4.

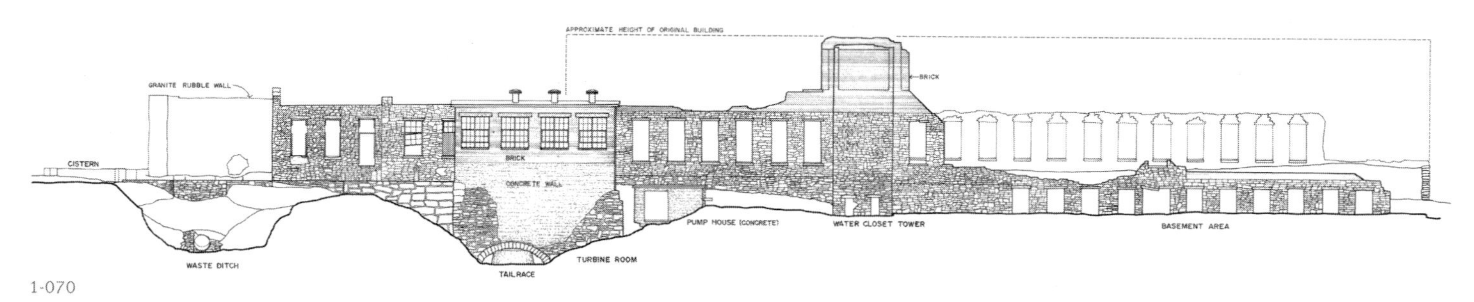

1-070

1-071. Spinning room, Pearle Cotton Mill, Elbert County, Georgia, 1895. From the collection of Mrs. O. W. Davis. P&P,HAER,GA,53-ELBE.V,1, no. 26.

1-072. Riverdale Cotton Mill Dam, Chattahoochee River, Langdale Vic., Chambers County, Alabama, 1866. Jet Lowe, photographer, 1997. P&P,Unprocessed items,survey no. HAER AL-166-D, no. 2.

Riverdale and Langdale Mills, built in the 1860s, were the first two cotton mills to use Chattahoochee River hydroelectric power. Their success led others to copy them, until each little town in Chambers County had its own cotton mill; these towns have all merged now into Valley, Alabama, named for the Chattahoochee River valley.

1-073. Langdale Cotton Mill Dam and Powerhouse, Langdale, Chambers County, Alabama, ca. 1860. Jet Lowe, photographer, 1997. P&P,Unprocessed items,survey no. HAER AL-167-B, no. 1.

1-071

1-072

1-073

The growth of Columbus as a major center in the textile industry depended largely on water power developed along the Chattahoochee River. The falls of the Chattahoochee had been harnessed in Columbus as early as 1828, supplying power for textile mills by the 1840s. Hydroelectric generation on the river began in 1880, supplying power for the city's streetlights and transit system as well as its industries and residences. The first dam, City Mills (1828), was a wooden dam that straddled islands for support. Several other wooden wing, or partial, dams followed, diverting water into mill races serving factories (Clapp's Factory Dam, 1832; Coweta Falls Dam,

1-074. Falls of the Chattahoochee. Bowen & Company, delineator, 1827. P&P,HAER,GA,108-COLM,16, no. 1.

1-075. Map of Chattahoochee River through Columbus, Muscogee County, Georgia. Donald F. Stevenson, delineator, 1977. P&P,HAER,GA,108-COLM,16,sheet no. 1.

1-074

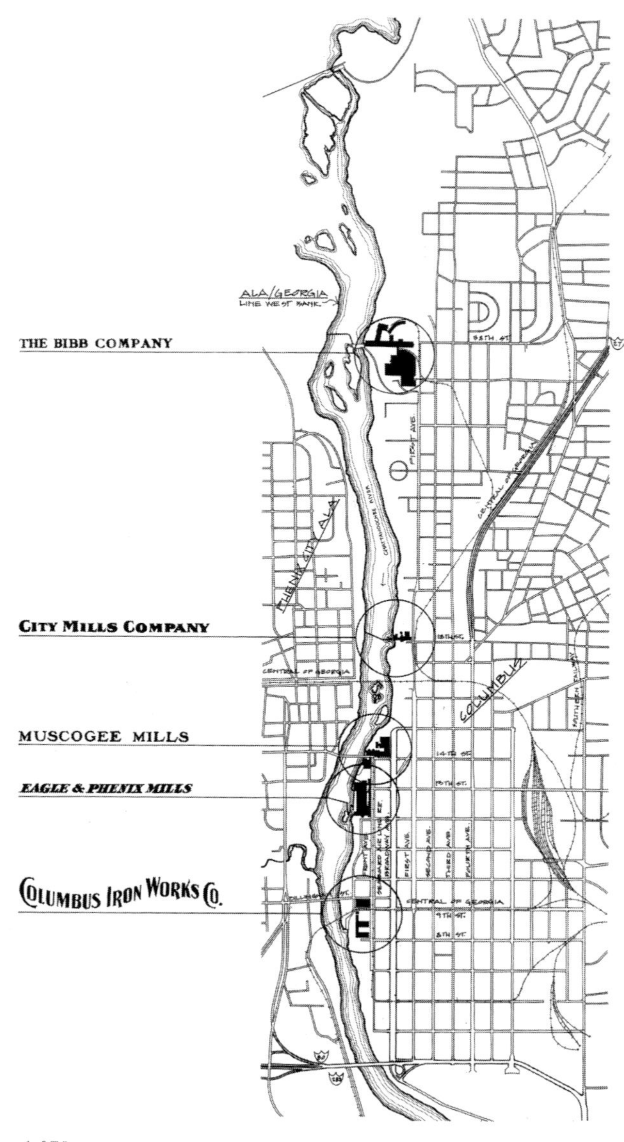

1-075

1-076

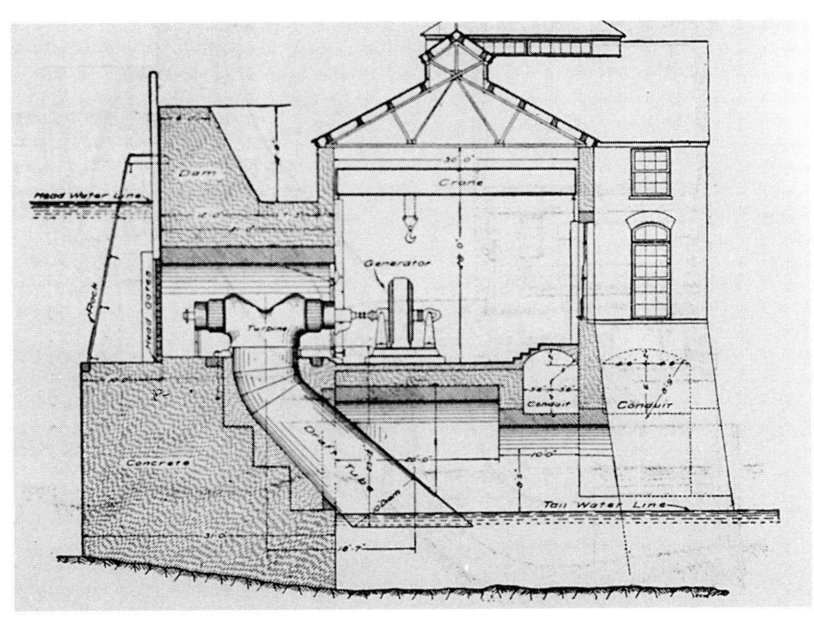

1-077

1-076. North Highlands hydroelectric dam, Chattahoochee River, Columbus, Muscogee County, Georgia, 1901–1902. P&P,HAER,GA,108-COLM,16, no. 7.

1-077. Cross-section of power plant, North Highlands Dam, Chattahoochee River, Columbus, Muscogee County, Georgia, 1901–1902. P&P,HAER,GA,108-COLM,25, no. 1.

1-078

1-079

1-080

1-078. Hamberger Cotton Mills, Electric Light Plant, and power canal, Columbus, Muscogee County, Georgia. Havens. P&P,HAER,GA,108-COLM,16, no. 6.

1-079. Goat Rock Dam, Chattahoochee River, Columbus, Muscogee County, Georgia, 1911–1912. H. J. Harvey, photographer, 1921. P&P,HAER,GA,108-COLM, 16, no. 13.

1-080. Bartlett's Ferry Dam, Chattahoochee River, Columbus, Muscogee County, Georgia, 1927. P&P,HAER,GA,108-COLM, 16, no. 16.

1844). In 1882, the masonry Eagle and Phenix Dam replaced the smaller wooden structures; within twenty years, powerhouses were progressively electrified.

The North Highlands Dam of 1902 provided water for mechanical and electrical powerhouses. The simultaneous supply of both kinds of power shows that electricity served as a mere substitute for mechanical power, useful when transmission over a distance was necessary. Early users of electric power used line shafts and rope drives to reach several machines from a central motor, barely modifying the layout of mechanically powered factories for the new technology. Additional hydroelectric plants were built at Goat Rock Dam (1911–1912), Bartlett's Ferry (1927), and Oliver Dam (1959–1960).

1-081

1-081. Oliver Dam, Chattahoochee River, Columbus, Muscogee County, Georgia, 1959–1960. Unidentified photographer, ca. 1959. P&P,HAER,GA,108-COLM,16, no. 18.

1-082

1-082. Whitney Dam, Yadkin River, Albemarle Vic., Stanly County, North Carolina, 1904–1910. Unidentified photographer, ca. 1910. WC. P&P,LC-dig-ppmsca-17323.

The original purpose of this masonry dam made of local granite was to supply hydroelectricity to the mills and cities of the North Carolina Piedmont region. After the first owner went insolvent in 1910, l'Aluminum Française purchased the dam and built the adjacent town of Badin. When that company pulled out with the advent of World War I, the Aluminum Company of America bought the town and the half-finished plant, completing a new dam known as the Narrows Hydroelectric Plant in 1917.

In the early years of the twentieth century, numerous small-scale hydroelectric plants were constructed throughout the Carolinas and Georgia. In some cases, these involved upgrading the masonry or timber-crib dams of the hydropower era; in others, they entailed the construction of new concrete dams, headraces, and powerhouses. The facilities served growing power markets in cities and mills. Gregg Shoals is one such early low-head hydroelectric plant.

In the mid-Atlantic region, early hydroelectric projects were built at a much larger scale, to supply consumer markets in the large cities of the eastern seaboard and industry in the Alleghany Mountains.

1-083. Gregg Shoals Dam and Hydroelectric Plant, Savannah River, Lowndesville Vic., Abbeville County, South Carolina, 1907. P&P,HAER,SC,4-SAVRI,2, no. 8.

1-084. Horizontal turbines, Gregg Shoals Dam and Hydroelectric Plant, Savannah River, Lowndesville Vic., Abbeville County, South Carolina, 1907. P&P,HAER,SC,4-SAVRI,2, no. 15.

1-083

1-084

1-085

1-085. Holtwood (McCalls Ferry) Dam, Susquehanna River, Martock, Lancaster County, Pennsylvania, 1906–1910. C. F. Havercamp, photographer, 1911. P&P,LC-dig-pan-6a09344.

Located on the lower Susquehanna, Holtwood Dam is a 2,392-foot-long, concrete overflow gravity dam. When it was built, it was the longest dam and largest hydroelectric plant in the country. It was also the first to use the Kingsley thrust bearing, which employed a thin film of oil to support the massive weight of the turbine shaft, avoiding wear on the moving parts. While earlier bearings of this size would have required replacement every two months, the Kingsley bearing was still in excellent condition when rebuilt for 60-cycle service forty years after its installation.

1-086. Warrior Ridge Dam, Juniata River, Petersburg, Huntingdon County, Pennsylvania, 1906. Jet Lowe, photographer, 1990. P&P,HAER,PA,31-PETBU.V,1, no. 1.

The Ambursen Hydraulic Construction Company, a major dam builder in the early twentieth century, designed and built this 400-foot-long concrete dam. An early example of a buttressed hollow-core dam, it received significant attention in the press and was patented by Ambursen in 1904. The plant distributed power to the nearby cities of Huntingdon, Altoona, and Tyrone in the Alleghany Mountains.

1-087. Conowingo Dam and Hydroelectric Plant, Susquehanna River, head of Chesapeake Bay, Maryland, 1926–1928. Theodor Horydczak, photographer, ca. 1930. P&P,LC-H812-1184-010.

Near the mouth of the Susquehanna, the river drops off 90 feet, giving this site its Indian name meaning "at the rapids." The Philadelphia Electric Company built this dam to take advantage of the hydropower potential and supply nearby consumer markets in eastern Pennsylvania and New Jersey. With a length of one mile, it is one of the largest privately financed hydroelectric power plants in the United States. When it was completed, it had a generating capacity surpassed only by the hydroelectric plants at Niagara Falls. To accommodate floods, a spillway runs along half of its length. Conowingo and the other massive Susquehanna dams had a serious impact on the shad fishery. After a trial period of trucking fish upstream in the 1970s, fish lifts and ladders were built at all four dams.

1-086

1-087

1-088

1-088. Turbine Hall of Conowingo Hydroelectric Plant, Susquehanna River, Maryland, 1926–1928. Theodor Horydczak, photographer, ca. 1930. P&P, LC-H824-T01-1184-006-B.

1-089. Safe Harbor Dam and Hydroelectric Plant, Susquehanna River, Lancaster County, Pennsylvania. 1929–1931. AP, 1931. P&P,LC-dig-ppmsca-17287.

This dam was a joint investment of Pennsylvania and Baltimore power companies. Power was first delivered a mere twenty months after work began in the river. The original seven generating units were increased to twelve in 1985.

1-089

By mid-century the federal government had become involved in large scale hydroelectric dam building on the slow-moving rivers of the south Atlantic states.

Although the Santee and Cooper rivers of South Carolina were first connected in the nineteenth century, it wasn't until 1926 that the federal government awarded a license for a dam and canal system. With the onset of the Great Depression, the government agreed to finance the project, provided that a state agency managed it. The resulting Santee Cooper Public Service Authority, established along the lines of the Tennessee Valley Authority, was empowered to develop the Santee, Congaree, and Cooper rivers for navigation, power generation, and watershed conservation.

Over the opposition of private utilities, work on the Santee and Pinopolis dams began in 1939, linking the rivers to form the Santee Cooper Lake System. As with the TVA projects, the Santee Cooper Project required a significant amount of resettlement and forest clearing—the latter left incomplete due to the accelerated pace of work brought on by the war. Economic benefits have included rural electrification and recreational development.

1-090. Aerial view of Santee Dam, Santee River, Clarendon County, South Carolina, 1939–1941. Hamilton Wright, 1941. P&P,LC-dig-ppmsca-17283.

The 8-mile-long earthen Santee Dam was a huge undertaking, employing 12,000 workers under the Works Progress Administration. The spillway alone is 3,400 feet long and has 62 tainter gates.

1-090

1-091

1-091. Construction of spillway, Santee Dam, Santee River, Clarendon County, South Carolina, 1939–1941. Acme, 1941. P&P, LC-dig-ppmsca-17284.

1-092. Clark Hill Dam, Savannah River, McCormick County, South Carolina, 1946–1954. Acme, 1950. P&P,LC-dig-ppmsca-17239.

Built by the U.S. Army Corps of Engineers as a multipurpose dam for flood control and power generation, Clark Hill contributed to the development of the Savannah River Atomic Energy Reservation, which manufactures nuclear weapons. During its construction, the Corps of Engineers faced controversies over forced resettlements, forest clearing, cemeteries and road relocations, and damage to archaeological sites. When completed, Clark Hill Reservoir extended 40 miles up the Savannah River, creating one of the largest inland bodies of water in the South.

1-092

A DAM DISASTER

As part of its paper mill complex in north-central Pennsylvania, the Bayless Pulp and Paper Company built a 50-foot-high, 550-foot-long, concrete gravity storage dam. Because the foundation conditions at the site failed to provide a solid, non-porous support for the structure, leakage under the dam eroded the foundations. In 1911, the dam failed by sliding along its base, killing seventy-eight people and causing $14 million in damage. This failure is a classic example of water penetrating under a gravity dam and pushing up the bottom of a structure. The Austin flood served as the impetus for the first federal and state regulations and inspection laws for dams.

1-093. Austin Dam, Freemans Run, Austin, Potter County, Pennsylvania, 1909, failed 1911. Almeron Newman, photographer, 1911. P&P,LC-dig-pan-6a09293.

1-093

GREAT LAKES-APPALACHIA

THE GREAT LAKES

Lakes Superior, Michigan, Huron, Erie, and Ontario; and the Saint Lawrence River

The five Great Lakes and their tributaries form the world's largest freshwater system. Linked to one another through the Saint Clair, Saint Mary's, and Niagara rivers, the Great Lakes flow into the Saint Lawrence River to reach the Atlantic Ocean. The lakes were the main passage into French North America, leading to Lake Winnipeg and the Saskatchewan River to the north and the Mississippi River basin to the south. French fur traders and missionaries have left place names all through these regions.

The historical importance of the Great Lakes region lies in its central location as a communication hub for North America. Its local resources of timber and iron ore fed the development of industry, and its location on the major transportation routes led to the development of great cities like Milwaukee, Chicago, Detroit, Cleveland, Buffalo, Toronto, and Montreal. The earliest dams in the region were built to facilitate traffic among the lakes and waterways such as the Erie, Welland,

2-001. Workers installing spiral case, Cherokee Dam, Tennessee Valley Authority, Holston River, Rutledge Vic., Jefferson County, Tennessee, 1940-1941. Tennessee Valley Authority, 1941. P&P,LC-dig-ppmsca-17357.

and Soo canals. The limestone escarpment of Canadian Shield forms dramatic gorges in southern Ontario and northwestern New York. High-head hydroelectric plants, including the famous drop at Niagara Falls, influenced dam development in this region.

The generally slow-flowing rivers of Michigan's Lower Peninsula feed Lakes Huron and Superior, and thousands of waterways that served as easy routes for transporting logs to regional centers dot the neighboring states of Minnesota and Wisconsin. In this watery environment, mill dams were a fixture from the earliest days of settlement—to power sawmills, paper mills, and, later, many other kinds of industry. While Michigan's low-lying landscape would not seem to be well suited to the development of hydroelectric plants,

2-002

2-002. Niagara Falls between Lakes Erie and Ontario. Unidentified photographer, 1911. P&P,LC-dig-cph-3c16318.

its heavy rainfall and reliably flowing streams made the construction of low-head power plants a good investment. These facilities, fed by a constant flow of water, were a dependable source of electricity, and they played a prominent role in the growth of regional electric power systems.

OHIO RIVER BASIN

Allegheny, Monongahela, Muskingum, Kanawha, Scioto, Great Miami, Kentucky, Green, and Wabash Rivers

The Ohio River forms at the confluence of the Allegheny and the Monongahela rivers and flows nearly a thousand miles until it reaches the Mississippi. The Ohio River valley was a major route for westward settlement and commerce. Because the river was prone to flooding in the spring and too shallow for navigation in the summer, the creation of a year-round channel required enormous efforts. Many innovations in dam design were further developed in this region—two of the most notable being movable dams (i.e., with operable gates) of the type invented for coal transport on the Lehigh canal, and the use of dry dams for flood control—the latter a design of Arthur Morgan, who later became a director of the Tennessee Valley Authority.

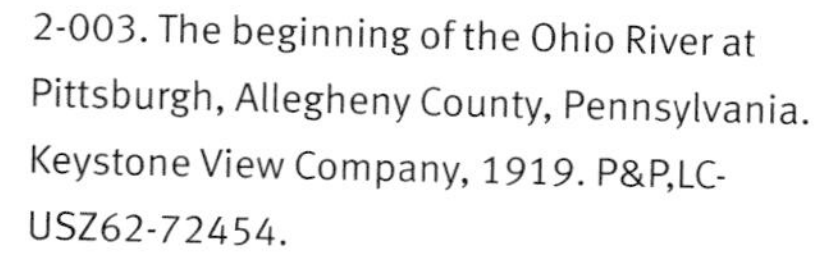

2-003. The beginning of the Ohio River at Pittsburgh, Allegheny County, Pennsylvania. Keystone View Company, 1919. P&P,LC-USZ62-72454.

2-003

2-004

2-004. Inundation of the Ohio River at Cincinnati, Hamilton County, Ohio, 1862. George M. Finch, delineator, 1862. P&P,LC-USZ61-1856.

TENNESSEE RIVER BASIN

Cumberland, Elk, Apalachia, Little Tennessee, French Broad, and Clinch Rivers

The Tennessee River drains a basin of 41,000 square miles before it joins the Ohio shortly before its junction with the Mississippi. Its shoals and rapids made navigation difficult, and, like the Ohio, it was prone to severe flooding. Since the establishment of the Tennessee Valley Authority in 1933, nine main-stem and many more tributary dams have transformed the river into a chain of lakes. The idea of the multipurpose dam, with its functions of navigation, power generation, and flood control, was further extended in the TVA—with irrigation, recreation, aesthetics, and regional and economic development. A federally funded experiment in regional planning, the TVA helped to promote the industrial development of the region, its recreational resources, and soil conservation. Today the Tennessee River is one of the world's great irrigation and hydropower systems.

DAMS BUILT FOR NAVIGATION

The earliest dams in the Great Lakes region were built to ease navigation among the lakes and major continental waterways. In 1783, the British government built four locks on the north shore of the Saint Lawrence, and a fur company constructed the first Soo Canal in 1797 to bypass the Saint Mary's River between Lakes Superior and Michigan. After the War of 1812, the Canadians built the Rideau Canal as a military supply route from Montreal to Lake Ontario, which would bypass the Saint Lawrence River on the U.S. border.

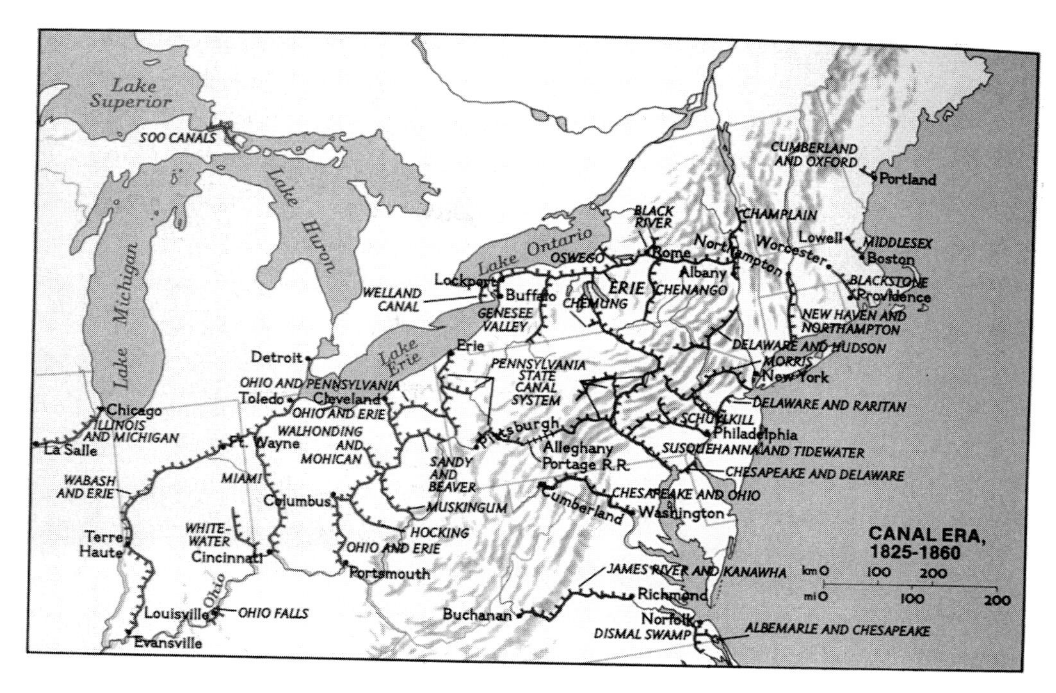

2-005

2-005. Diagram of canals in the United States, 1825–1860. 1985. G&M,G1201.S1 N3 1988 fol.,copy 1.

2-006

2-006. Rideau Canal Lock & Dam, from Ottawa (Ottawa River) to Kingston (Lake Ontario), Ontario, 1826–1832. DETR, ca. 1900. P&P,LC-D4-12746.

The 202-kilometer (125-mile) waterway has forty seven locks and fifty two dams built of stone. Although it never served its original purpose as a military supply route, it supported commercial traffic until the early twentieth century, when it began primarily to serve pleasure craft. This photograph shows the first lock in the system, below Parliament Hill in Ottawa.

Their equally impressive masonry Welland Ship Canal of 1833 bypasses Niagara Falls between Lakes Erie and Ontario.

In the United States, the Erie Canal was the first waterway to link the Atlantic seaboard to the Great Lakes region, ensuring New York City's preeminence as a commercial center. After the completion of the Erie Canal, canal building picked up in speed, and the years 1825–1860 saw many new canals connecting the Erie Canal and the Great Lakes with the Ohio, Muskingum, Miami, Wabash, and Mississippi rivers, linking them into a large transportation network.

2-007

2-007. Erie Canal at Little Falls, Herkimer County, New York, 1817–1825. William Norris, delineator, 1834. G&M,G3701.P3 1834.N6 Vault: RR 2a,detail.

Erie Canal Dams

Extending 360 miles from Troy on the Hudson River to Tonawanda on the Niagara River, the Erie Canal connects New York City with the Great Lakes. With its three branch canals—the Champlain (1817–1826), the Oswego (1829), and the Cayuga and Seneca (1829)—it forms the New York State barge canal. Built in sections, it was complete by 1825. The Erie Canal contributed to New York City's financial development, opened eastern markets to midwestern farm products and encouraged immigration to that region, and helped to create many large cities. Its initial success started a wave of canal building in the United States. After the 1850s, railroads began to compete with the canal; from 1904 to 1918 the canal was modernized. The opening of the New York State Thruway and the Saint Lawrence Seaway ensured the canal's commercial demise.

Several types of dam were built for the Erie Canal. Reservoir dams supply water to the canal. Fixed and movable dams span the channeled streambeds, turning the rivers into a series of pools connected by locks. The fixed dams are overflow weir dams, first built out of timber cribbing and later rebuilt in masonry. The Mohawk movable dams look like truss bridges on concrete piers. Vertical steel channels are anchored between the truss and a concrete sill on the riverbed. Sliding gates lowered between the channels impound the water. When raised during a flood, they let the waters pass freely. There are eight of these dams between Schenectady and Little Falls. Lower Oswego Dam (shown here) is fixed, while Lock No. 11 is a movable dam.

2-008

2-008. Erie Canal at Rome, Oneida County, New York, 1817–1825. Lucien R. Burleigh, delineator, 1886. G&M,G3804.R7 A3 1886.B8,detail.

2-009. Lower Oswego Dam, Oswego River and Canal, Oswego, Oswego County, New York, 1827–1829, rebuilt 1864. DETR, ca. 1900. P&P,LC-D4-12158.

2-009

2-010

2-010. Movable dam at Lock No. 11, Erie Canal, Amsterdam, Montgomery County, New York, 1817–1825, 1862, 1904–1918. John Collier, photographer, 1941. P&P,LC-USF34-081150-C.

2-011. Muskingum River Navigation System Lock & Dam No. 1, Muskingum River, Marietta, Washington County, Ohio, 1836, 1889, 1913. Unidentified photographer, ca. 1915. P&P,LC-dig-ppmsca-17321.

The Muskingum River system consisted of ten dams, five bypass canals, and eleven locks on the Muskingum River to create a navigation channel from the Ohio and Erie Canal at Dresden, to the city of Marietta on the Ohio River. Completed in 1841, the system extended over 100 miles through southeastern Ohio, opening it up to trade and development.

2-011

Monongahela River Navigation System Dams

Flowing from the mountains of West Virginia to the coal country of Pittsburgh, Pennsylvania, the Monongahela was one of the first rivers in the United States to be converted into a slack water navigation system. Because the river flowed through the famed Pittsburgh coal seam and could bring an enormous tonnage of coal to markets further downstream, investment in navigational improvements on the Monongahela was commercially attractive. In 1841, the Monongahela River Navigation Company began work on seven dams between the West Virginia border and Pittsburgh. In 1897, the federal government purchased the system and assigned further improvements to the U.S. Army Corps of Engineers. As the locks and dams were replaced, building materials changed, operations became mechanized, and the number of locks and dams on the river were reduced, providing larger slack-water pools, higher dams, and larger locks. The late-nineteenth-

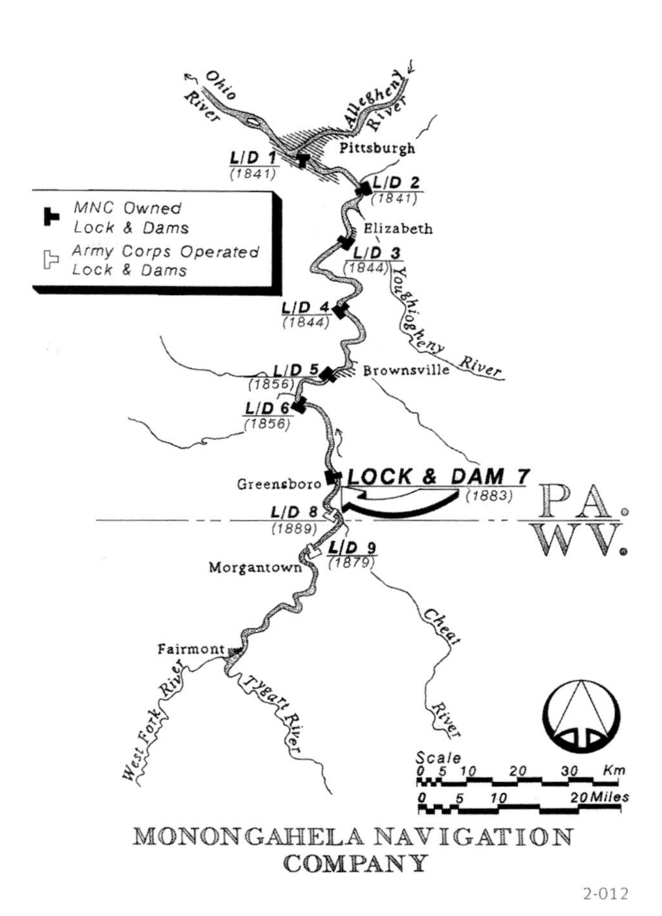

2-012

2-012. Monongahela River Navigation Company dams on the Monongahela River, Pennsylvania and West Virginia, 1841–1883. Curtis Burlbaw, delineator, 1994. P&P,HAER,PA,30-GREE,1-,sheet no. 2,detail.

2-013. Plan and section of Lock & Dam No. 7, Monongahela River Navigation System, Monongahela River, 1883. Jonathan Gill, delineator, 1994. P&P,HAER,PA, 30-GREE,1-,sheet no. 4,details of plan and section A-A.

2-014. Plan and section of Lock & Dam No. 7, Monongahela River Navigation System, Monongahela River, 1926. Jonathan Gill, delineator, 1994. P&P,HAER,PA, 30-GREE,1-,sheet no. 4,details of plan and section B-B.

2-013

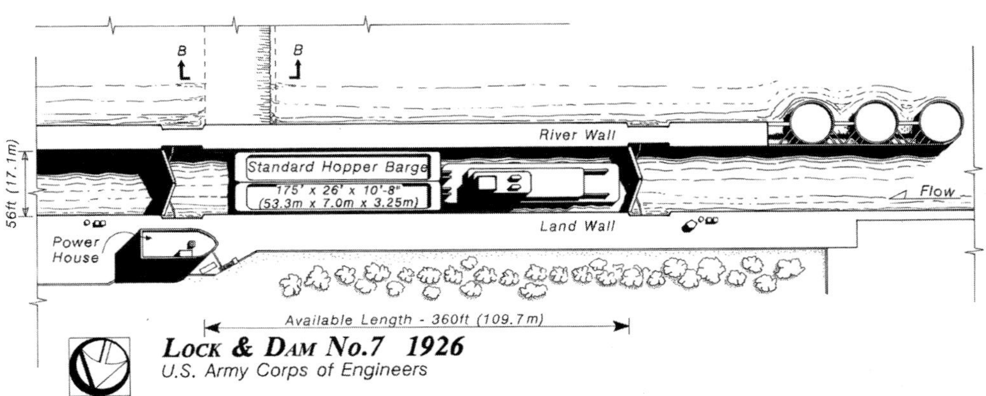

2-014

2-015

2-015. Lock & Dam No. 7, Monongahela River Navigation System, Monongahela River, Greensboro, Greene County, Pennsylvania, 1883, replaced 1926. Jet Lowe, photographer, 1994. P&P,HAER,PA,30-GREE,1, no. 1.

century dams were timber-crib and rubble fill structures, with quarried stone lock chambers that were hand operated. By the 1920s, higher dams were built in concrete and the hydraulic mechanisms allowed one person to operate the chambers.

Soo Canals

These canals bypass the rapids on the Saint Mary's River between Lakes Superior and Huron. The first canal was built on the Canadian side by the Northwest Fur Company and destroyed in the War of 1812. Forty years later, the State of Michigan completed the first canal on the American side (the State Locks). Placed under the control of the U.S. Army

2-016

2-016. Ojibwe fishing for whitefish in the Sault rapids, St. Mary's River between Lakes Superior and Huron, 1900. DETR, 1900. P&P,LC-D4-13076.

2-017. Old State lock, U.S. Sault Sainte Marie Canal, St. Mary's River between Lakes Superior and Huron, Sault Sainte Marie, Chippewa County, Michigan, 1853–1855. 1905. P&P,LC-dig-ppsmca-17297.

2-017

2-018. Weitzel lock with whaleback freighter, U.S. Sault Sainte Marie Canal, Saint Mary's River between Lakes Superior and Huron, Sault Sainte Marie, Chippewa County, Michigan, 1881. 1905. P&P,LC-dig-ppsmca-17298.

2-018

2-019

2-019. U.S. Sault Sainte Marie Canal, Saint Mary's River between Lakes Superior and Huron, Sault Sainte Marie, Chippewa County, Michigan, 1853–1855, 1881–1919, hydroelectric plant 1902. DETR, 1905. P&P,LC-D4-21873.

Corps of Engineers, the locks were modernized with the addition of the Weitzel (1881), Poe (1896), and the twin Davis and Sabin locks (1919). The U.S. canal prospered with the westward expansion of the railroads, as it facilitated the transport of iron ore from Michigan's Upper Peninsula to steel mills in the southern Great Lakes. In 1895, the Canadians rebuilt their canal with a 900-foot electrically operated lock; but after the Americans embarked on a second modernization following World War II, the Canadian canal lost tonnage. It now operates for pleasure craft only. The waterways are popularly called the Soo Canals.

Ohio River Navigation System Dams

The U.S. Army Corps of Engineers began improving the Ohio River in 1824 by dredging sandbars and removing snags, and in 1830 it built a canal and three locks to circumvent the Falls of the Ohio at Louisville, Kentucky. In 1875, Congress authorized the first dam on the Ohio at Davis Island; the structure was completed ten years later. By 1910, twelve lock and dam complexes had been built, and by 1929, canalization of the river was complete, with fifty-two locks and dams that ensured year-round depths of nine feet between the Mississippi River and Pittsburgh.

The dam portion of each complex comprised piers connecting two types of movable dam systems: (1) a weir of chanoine wickets over the navigable channel, and (2) a bear trap dam over the rest of the riverbed. The wickets were heavy wooden shutters 4 feet wide and 12 feet long that pivoted from their foundation on the riverbed. They were raised when the river ran low, damming a pool of water deep enough to allow boats to navigate from lock to lock. During high water or floods, they could be lowered to the riverbed to allow water and boats to pass freely. It required a large work force to raise and lower the wickets one at a time by boat. The bear trap dam allowed water, but not boats, to pass through. From the 1920s onward, the lock and dam complexes were rebuilt with larger locks, concrete dams with hydraulically operated roller gates, and hydroelectric plants, leaving today twenty dams and forty-nine power-generating facilities on the Ohio River.

The first lock and dam on the flood-prone and highly traveled Ohio River, Davis Island Dam was an experimental project that tested the skills of nineteenth-century engineers.

2-020

2-020. U.S. Army Engineers dredging at the outer end of Rising Sun Dike, Ohio River, Rising Sun Vic., Ohio County, Indiana, 1884. E. J. Carpenter, photographer, 1884. P&P,LC-USZ62-52696.

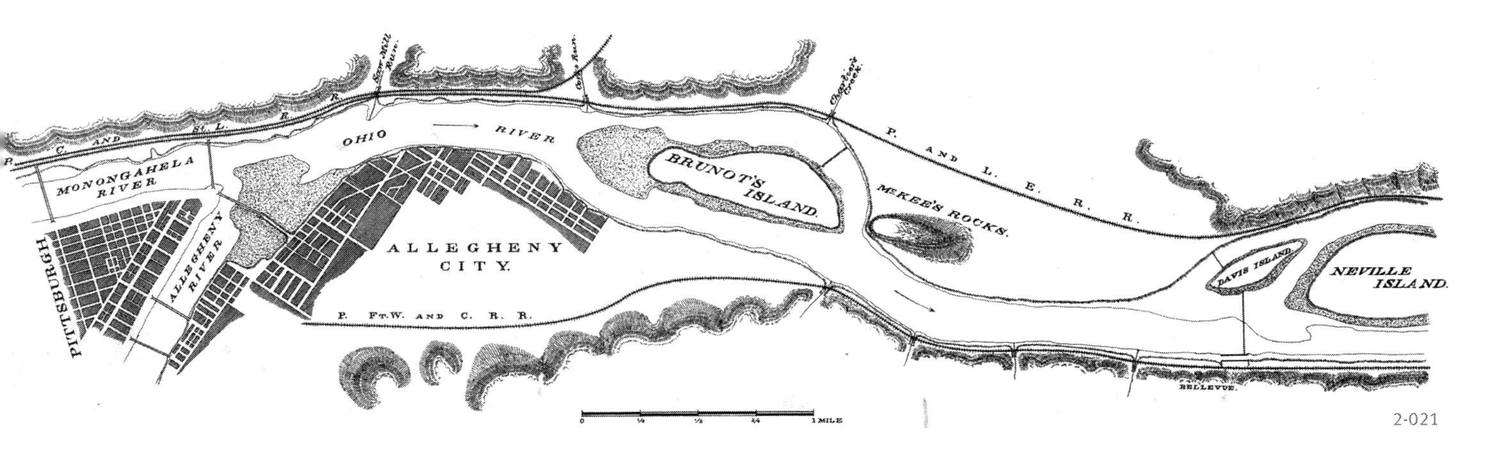

2-021

2-021. Location of Davis Island Dam (Lock & Dam No. 1), Ohio River, Pittsburgh Vic., Allegheny County, Pennsylvania, 1878–1885. U.S. Army Corps of Engineers, delineator, 1889. TC549 U53 Folio,plate no. 1,detail.

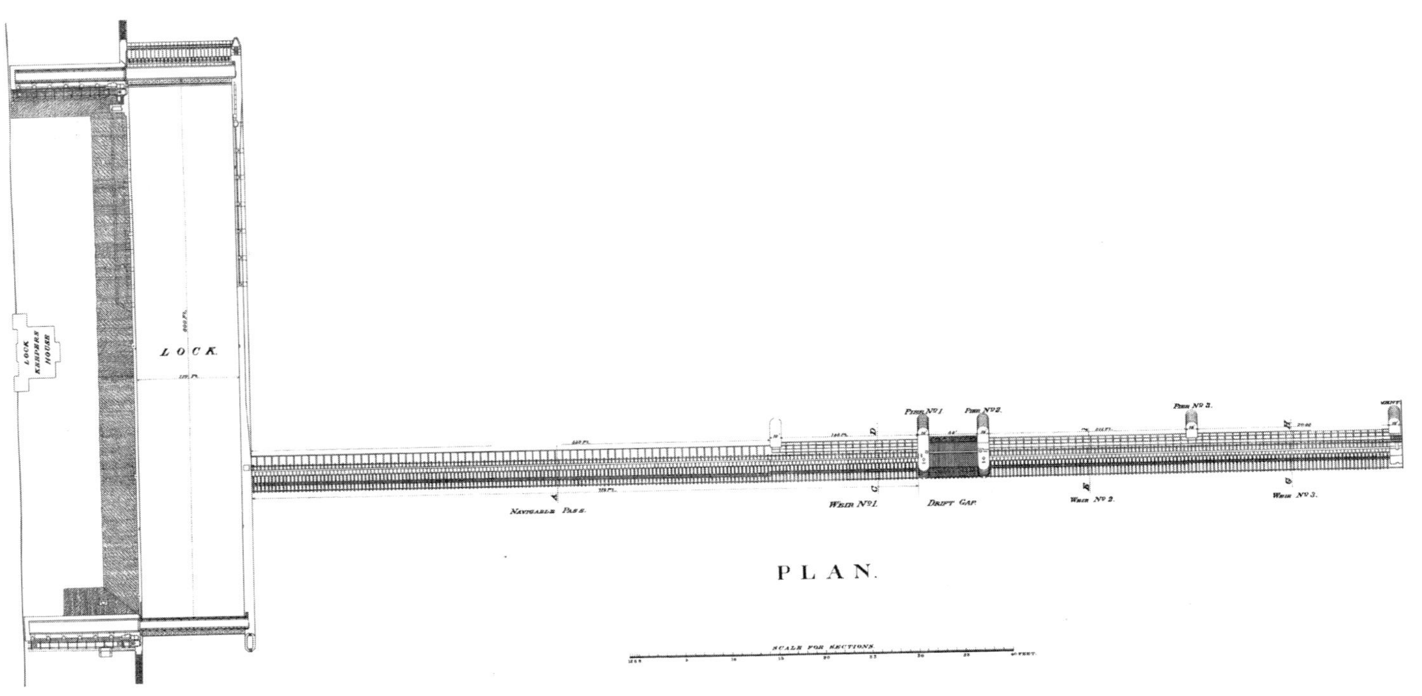

2-022

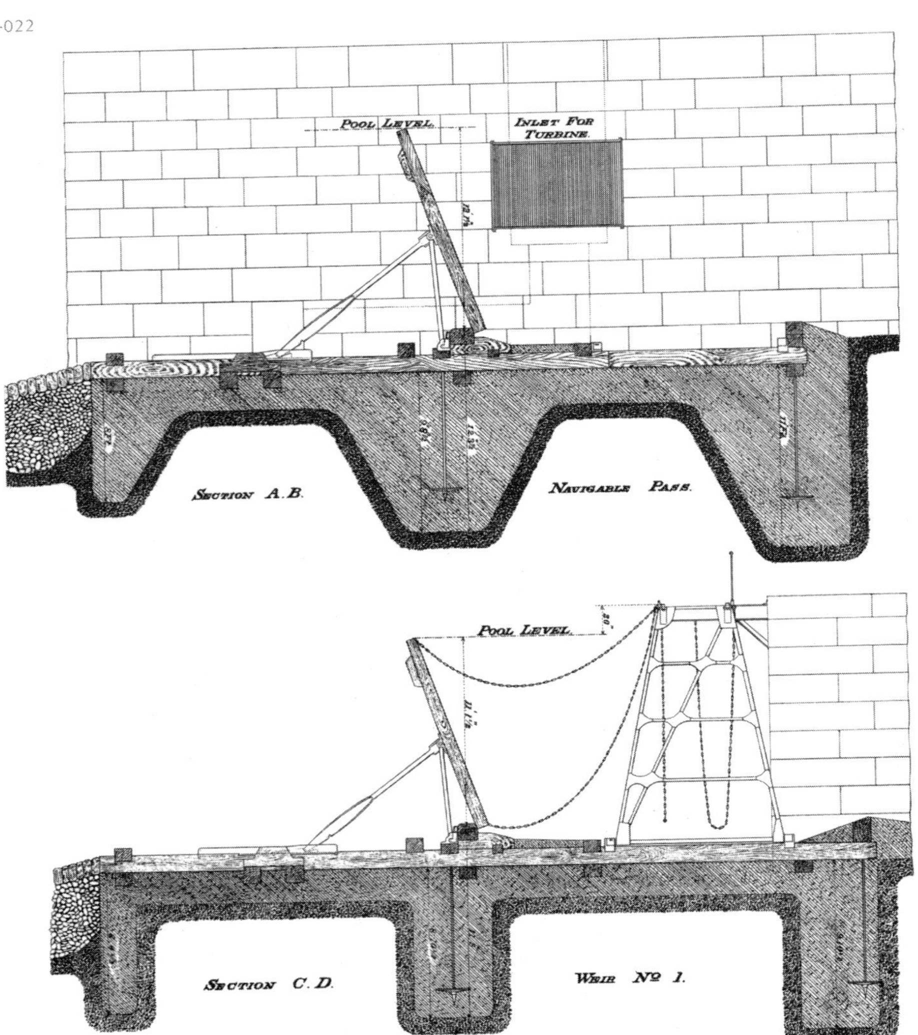

2-023

2-022. Plan, Davis Island Dam (Lock & Dam No. 1), Ohio River, Pittsburgh Vic., Allegheny County, Pennsylvania, 1878–1885. U.S. Army Corps of Engineers, delineator, 1889. TC549 U53 Folio,plate no. 2,detail.

2-023. Sections through navigable pass and weir no. 1, Davis Island Dam (Lock & Dam No. 1), Ohio River, Pittsburgh Vic., Allegheny County, Pennsylvania, 1878–1885. U.S. Army Corps of Engineers, delineator, 1889. TC549 U53 Folio,plate no. 2,detail.

Relying in large part on European precedents, it was the first movable dam ever constructed by the U.S. Army Corps of Engineers, and the largest movable dam built in the nineteenth century. William Merrill (1837–1891) was the engineer-in-charge, and George Goethals (1858–1928; later known for the Panama Canal) also worked on the project.

The Corps was so pleased with its experiment at Davis Island that it extended the system down the length of the Ohio River, building twelve additional dams of the same design. Construction of the Emsworth Locks and Dams in 1921 submerged Davis Island Lock and Dam.

2-024. Plan of weir no.1, Davis Island Dam (Lock & Dam No. 1), Ohio River, Pittsburgh Vic., Allegheny County, Pennsylvania, 1878–1885. U.S. Army Corps of Engineers, delineator, 1889. TC549 U53 Folio,plate no. 12,detail.

2-025. Section through bear trap, Davis Island Dam (Lock & Dam No. 1), Ohio River, Pittsburgh Vic., Allegheny County, Pennsylvania, 1878–1885. U.S. Army Corps of Engineers, delineator, 1889. TC549 U53 Folio,plate no. 18,detail.

2-024

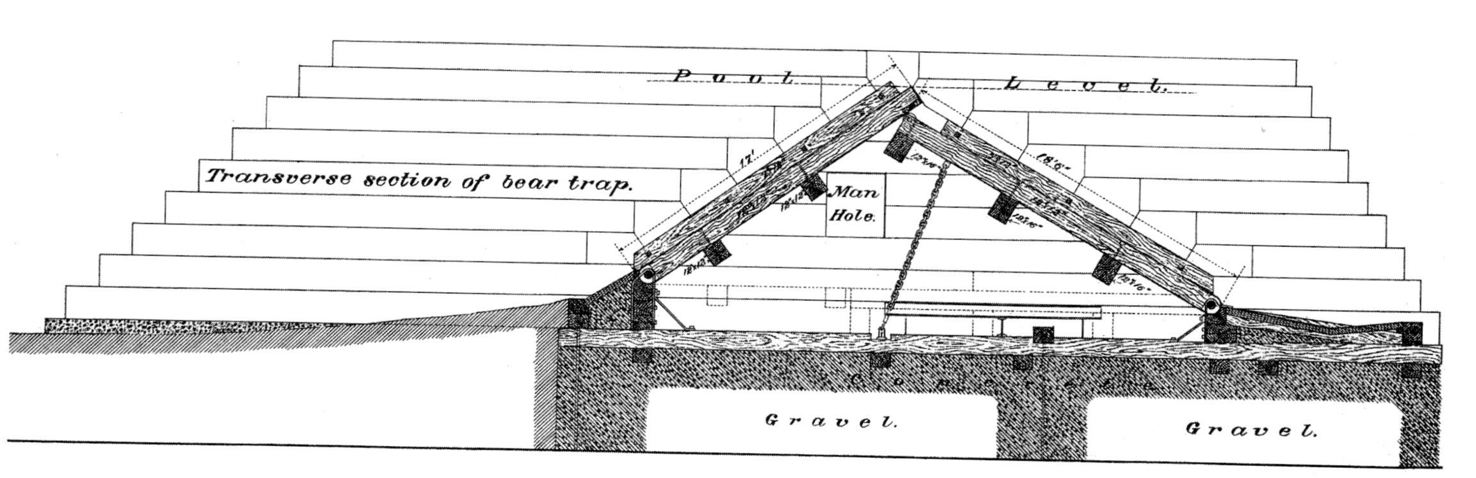

2-025

2-026

2-026. Lock & Dam No. 12, Ohio River Navigation System, Ohio River, Wheeling Vic., Ohio County, West Virginia, 1898–1908. William Edward Barrett, photographer, 1974. P&P,HAER,WVA,35-WHEEL.V,1-3.

2-027. View of wickets in decommissioned Lock & Dam No. 12, Ohio River Navigation System, Ohio River, Wheeling Vic., Ohio County, West Virginia, 1898–1908. William Edward Barrett, photographer, 1974. P&P,HAER,WVA,35-WHEEL.V,1, no. 1.

2-027

2-028

2-028. Coal fleets departing from Pittsburgh, Pennsylvania. U.S. Army Corps of Engineers, 1905. TC425 O4 1908.

2-029. Deckhands aboard a towboat on the Ohio River. Arthur S. Siegel, photographer, 1943. P&P,LC-USW3-031825-E.

2-029

2-030

2-031

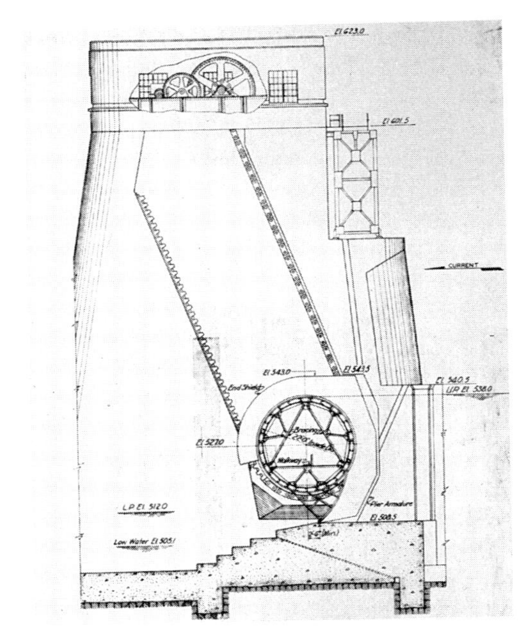

2-032

2-030. Site of Gallipolis Dam before construction, Ohio River. U.S. Army Corps of Engineers, 1935. P&P,HAER,WVA,27-GALIP.V,1, no. 3.

2-031. Gallipolis Lock & Dam, Ohio River Navigation System, Ohio River, Gallipolis, Gallia County, Ohio, 1935–1937. P&P,HAER,WVA,27-GALIP.V,1, no. 6.

Located just downstream from the junction of the Kanawha River with the Ohio, Gallipolis is one of the most impressive of the Corps' modern dams on the Ohio. It is a 1,225-foot-long movable dam, with eight steel roller gates which can be raised to allow floodwaters to pass through quickly.

2-032. Section through roller gate, Gallipolis Lock & Dam Ohio River Navigation System, Ohio River, Gallipolis, Gallia County, Ohio, 1935–1937. U.S. Army Corps of Engineers, delineator, 1935. P&P,HAER,WVA,27-GALIP.V,1A-25,detail.

2-033. Roller gates in raised position, Gallipolis Lock & Dam, Ohio River Navigation System, Ohio River, Gallipolis, Gallia County, Ohio, 1935–1937. U.S. Army Corps of Engineers, 1937. P&P,HAER,WVA,27-GALIP.V,1, no. 4.

2-034. Roller gates in lowered position, Gallipolis Lock & Dam, Ohio River Navigation System, Ohio River, Gallipolis, Gallia County, Ohio, 1935–1937. Arthur S. Siegel, photographer, 1943. P&P,LC-USW3-30361-E.

2-035. Smithland Lock & Dam, Ohio River Navigation System, Ohio River, Pope County, Illinois, 1971–1980 (replacing L&D 50, 51). U.S. Army Corps of Engineers, delineator, 1994. G&M,G1377.O3P53 U684 1994,chart no. 15.

2-033

2-034

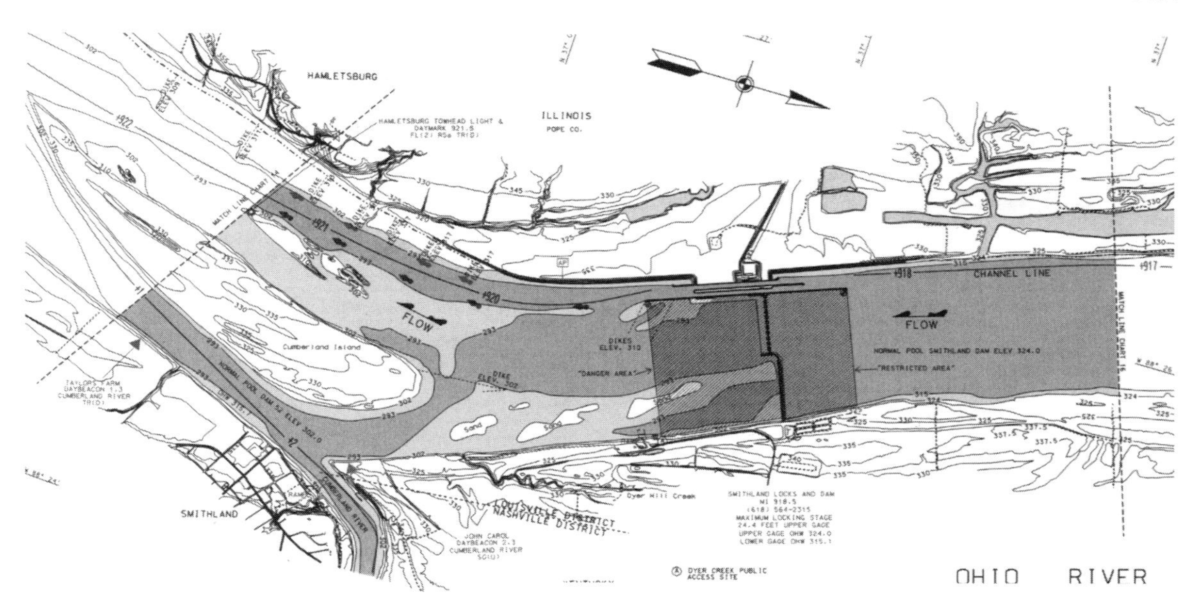

2-035

Lower Fox River Waterway Dams

The Fox River was a well-known route used by early missionaries and fur traders to reach the Mississippi River system from the Great Lakes. A barge canal links the Fox with the Wisconsin rivers at Portage, Columbia County, Wisconsin, forming a continuous waterway from Lake Michigan to the Mississippi.

The nine dams on the Lower Fox River from Lake Winnebago to Green Bay were built between 1836 and 1866, out of timber cribbing filled with stone and surfaced with a watertight wooden sheathing. Quarried stone dams replaced many of these in the 1870s, when the federal government obtained control of the waterway. From 1927 to 1940 they were rebuilt again, this time in concrete anchored to the riverbed. The modern dams have spillways to shed excess water and sluiceways with tainter gates, operated by a crab hoist that runs on rails along the crest.

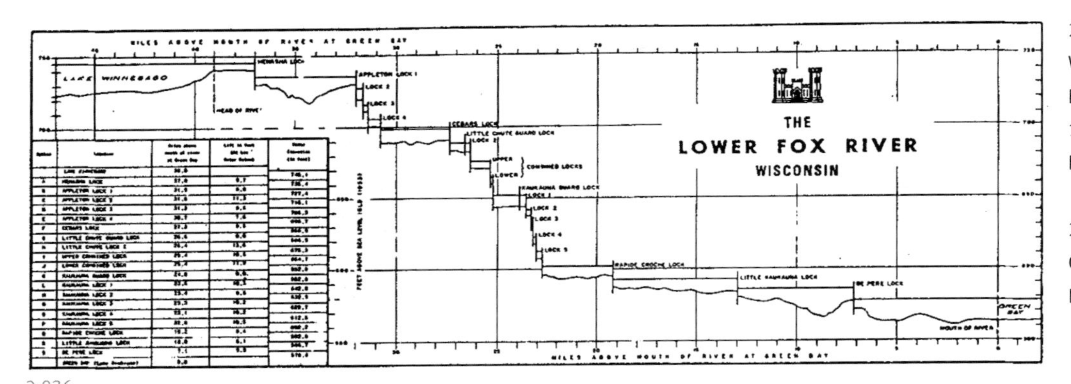

2-036

2-036. Elevation drop of Lower Fox River Waterway, from Lake Winnebago to Green Bay, Wisconsin, 1836–1866, rebuilt 1927–1941. 1995. P&P,HAER,WIS,044-KAUK,2-,p.3,detail.

2-037

2-037. Appleton Dam, Fox River, Appleton, Outagamie County, Wisconsin, 1850–1859 DETR, ca. 1898. P&P,LC-D4-4783.

2-038

2-038. Little Chute Lock & Dam, Fox River, Little Chute, Outagamie County, Wisconsin, 1933. Joseph Paskus, photographer, 1995. P&P,HAER,WIS,044-LITCH,2, no. 2.

2-039. Rapide Croche Lock & Dam, Fox River, Wrightstown, Brown County, Wisconsin, 1931. Joseph Paskus, photographer, 1995. P&P,HAER,WIS,5-WRITO,2, no. 2.

2-039

Saint Lawrence Seaway Dams

In its natural state, the Saint Lawrence River was navigable only as far as the Lachine Rapids at Montreal. The Lachine Canal locks were the first to be constructed on the river, which has since been transformed with an extensive system of dams and locks to become the Saint Lawrence Seaway (1953–1959), allowing ocean-going vessels to navigate from the Gulf of Saint Lawrence to Lake Ontario. The Saint Lawrence River is also an important source of hydroelectric power; one of the world's largest facilities is the Beauharnois Dam and Hydroelectric Plant near Montreal. Other dams in the seaway include the 3,200-foot-long Moses-Saunders Power Dam near Massena, New York, and Cornwall, Ontario, and its associated control structures, the Long Sault and Iroquois control dams. Agreements between the United States and Canada govern power distribution and navigation in the international section of the river.

2-040

2-041

2-040. "He seems to mean it," Canada threatens to go it alone on the Saint Lawrence Seaway. Leo Joseph Roche, artist, 1953. P&P,CD-Roche,no. 11(B size).

2-041. Construction of Iroquois Control Dam, Saint Lawrence Seaway, Dundas County, Ontario, and St. Lawrence County, New York, 1954–1958. P&P,LC-dig-ppmsca-17263.

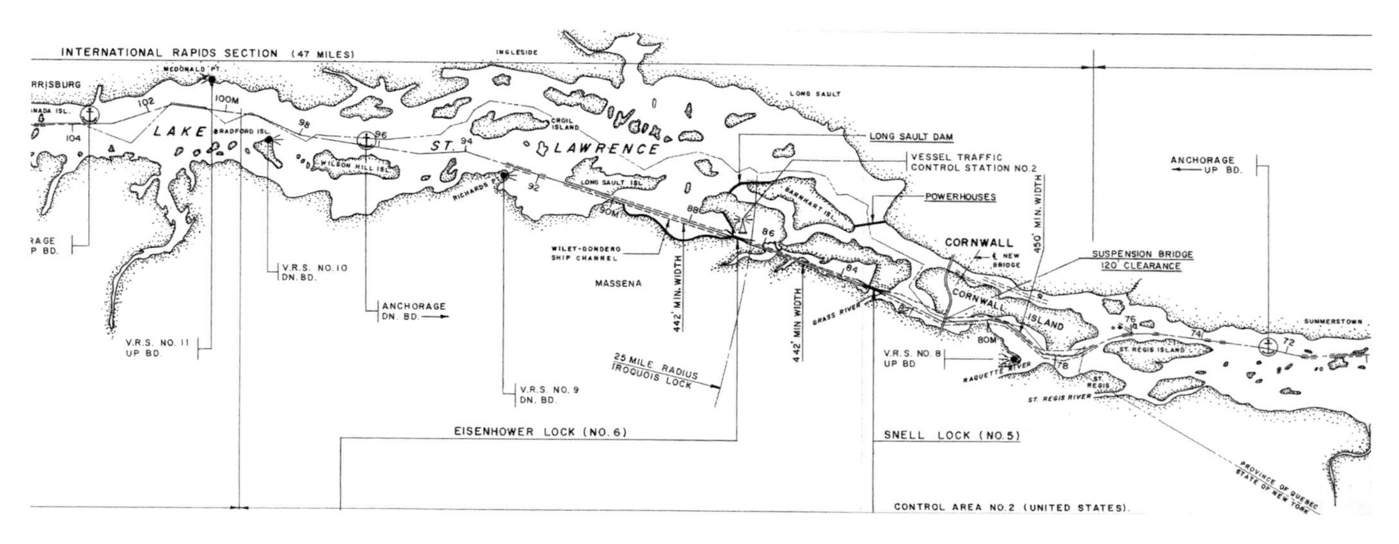

2-042

2-042. International Rapids section with Moses-Saunders Power Dam and Long Sault Dam, Saint Lawrence Seaway, Stormont County, Ontario, and St. Lawrence County, New York, 1954–1958. Saint Lawrence Seaway Authority, delineator, 1961. G&M,G3312 .S53 svar .S31,sheet no. 2,detail.

2-043. Construction of Long Sault Dam showing coffer dam, Saint Lawrence Seaway, Stormont County, Ontario, and St. Lawrence County, New York, 1954–1958. New York State Power Authority, 1957. P&P,LC-dig-ppmsca-17269.

2-043

2-044

2-044. Long Sault Dam, Saint Lawrence Seaway, Stormont County, Ontario, and St. Lawrence County, New York, 1954–1958. New York State Power Authority, 1964. P&P,LC-dig-ppmsca-17268.

2-045. Overview of dewatered river at site of Moses-Saunders Power Dam, Saint Lawrence Seaway, Stormont County, Ontario, and St. Lawrence County, New York, 1954–1958. New York State Power Authority, 1957. P&P,LC-dig-ppmsca-17280.

2-045

MILL DAMS

Westward settlement carried inland the many mill dams that were so characteristic of early settlement along the eastern seaboard. By the late eighteenth century, small mills and timber-crib dams or weirs that fed raceways dotted the western frontier.

2-046

2-046. Mill at Kanawha Falls, Kanawha River, Fayette County, West Virginia. Edward Beyer, delineator, 1857. P&P,LC-USZ62-51899.

2-047. Mills on the Black River, Jefferson County, New York, ca. 1820. Jacques G. Milbert, delineator, 1828. P&P,LOT 4397-A.

2-047

2-048

2-049

2-048. Dam at Otto, South Branch of Cattaraugus Creek, Cattaraugus County, New York. Unidentified photographer, ca. 1900. WC. P&P,LC-dig-ppmsca-17319.

2-049. Argo Dam, Huron River, Ann Arbor, Washtenaw County, Michigan, 1832. DETR, 1908. P&P,LC-D4-70766.

Built to power the flour mill of Swift & Company, the Argo Dam was one of the earliest mill dams in Ann Arbor. The milling industry prospered along the banks of the Huron until the early 1900s, when the Plains states began to dominate the agricultural market owing to new technology and cheaper transportation. From 1905 to 1927, Detroit Edison Power Company bought water rights along the Huron and replaced the older mill facilities with a series of modern dams and hydroelectric plants, including Barton Dam (1912) and a rebuilt Argo Dam (1913), as well as Geddes, Superior, Peninsula, Ford, and Belleville dams.

2-050

2-050. Chagrin Falls Dam, Chagrin River, Chagrin Falls, Cuyahoga County, Ohio, ca. 1850, breached 1994. WC. P&P,LC-dig-ppmsca-17320.

The falls of the Chagrin River east of Cleveland attracted settlement as early as the 1830s, powering grist and lumber mills. The Adams Paper Bag Company built this concrete mill dam.

2-051

2-051. Sorting grounds and sawmill, Michigan, ca. 1870. P&P,LC-dig-ppmsca-17296.

The lumbering regions of the Michigan peninsula are crossed by countless waterways, ponds, streams, and lakes. Virtually every settler in this region had some experience building dams to maintain water levels to float logs. As the lumbering industry developed, dam-building technology did as well, contributing to industrial development and rapid urban growth of Michigan.

Such nineteenth-century dams were timber-crib structures less than 10 feet high that fed mill races to power industries. The mills flourished in the mid to late nineteenth century, turning local agricultural products into consumer goods and exportable products. In the 1860s, steel turbines began to replace the earlier wooden water wheels, functioning at lower river levels and under icy conditions.

2-052

2-053

2-052. Mill pond, Grand Rapids, Kent County, Michigan. P&P,LC-dig-ppmsca-17304.

2-053. Paper mill dam, Inland Water Route, Petoskey, Emmet County, Michigan, ca. 1870. DETR, 1908. P&P,LC-D4-70723.

At the very top of Michigan's Lower Peninsula, the Inland Water Route is 38 miles of interconnecting waterways, beginning near Petoskey on Lake Michigan and ending near Cheboygan on Lake Huron. In the nineteenth century, logging was the biggest industry in the region, as lake freighters ferried lumber to ports around the Midwest. The Bear River in Petoskey was first dammed in 1855 for a grist mill, under the impetus of the Presbyterian Indian Mission. Paper mills followed, then a hydroelectric plant, until, in 1991, the dams were removed, returning the river to its natural state.

The gorges and steep chasms of upstate New York set the stage for the development of some of the highest-head hydroelectric plants in the country. Taking advantage of the tremendous drop in altitude in the Niagara River, Niagara Hydroelectric Plant No. 1 held lessons that were put to use in dam design across the country, but especially in the mountainous West, where high-head conditions were common.

Paradoxically, the Great Lakes region was also the site for the development of the lowest-head hydroelectric plants in the country (which took advantage of reliable stream flows) and long-distance transmission lines, to serve the industries and cities of Wisconsin and Michigan. These low-head hydroelectric plant designs are interesting because they don't require a reservoir to generate power, relying instead on the steady flow of water to run turbines. Also known as run-of-the-river dams, early examples like the Fox River dams in Appleton, Wisconsin (1882), did not generate much power compared to contemporaneous high-head plants like those at Niagara Falls. But by the time the Sault Sainte Marie hydroelectric facility was built in 1902, on the canals that drained Lake Superior into Lake Huron, it was the first low-head plant with directly connected turbines and generators and the largest low-head plant in United States (a distinction it still holds today), generating electricity for 50 percent of Michigan's Upper Peninsula. Low-head hydroelectric generation was also used at the American Falls Dam in Idaho, built in 1902 (see 7-017a–7-019) and in Michigan's Croton Dam of 1906 (2-062–2-064).

2-054. Niagara Falls, Niagara County, New York, 1882. Henry Wellge, delineator, 1881. G&M,G3804.N7A3 1882.W4 Rug 150.

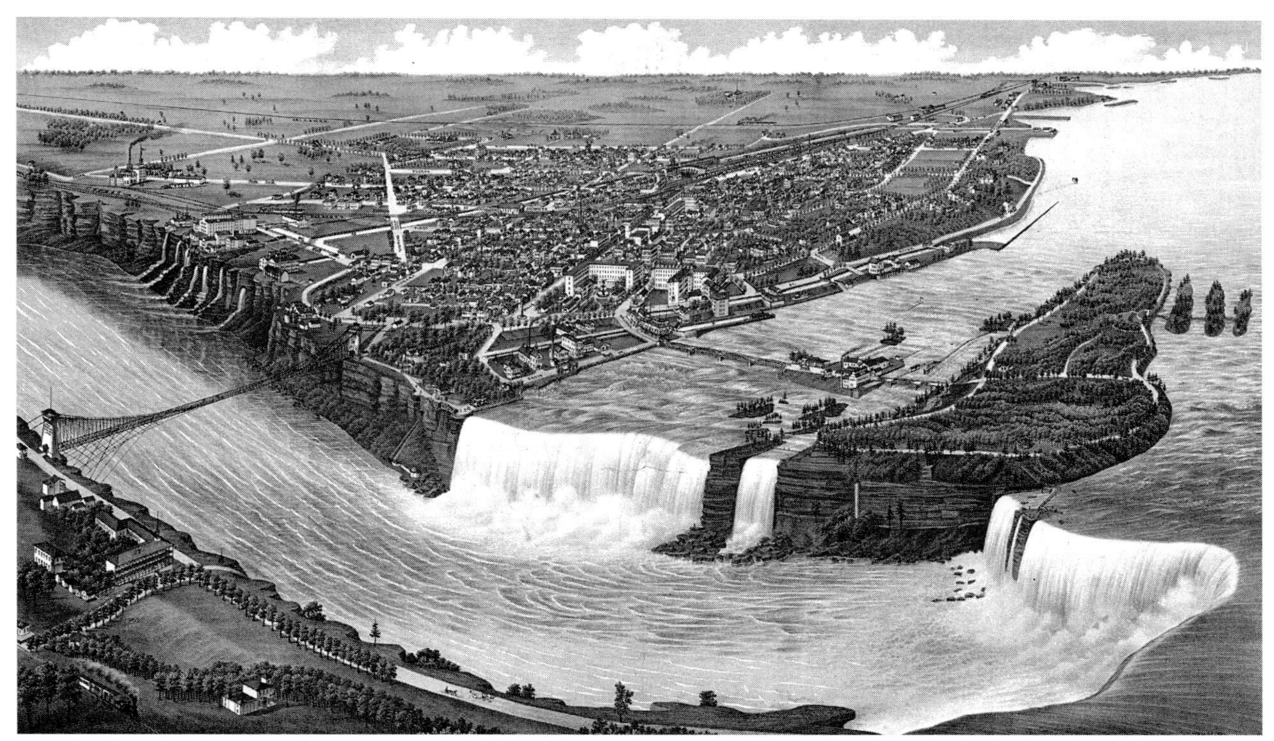

2-054

2-055

2-055. Mill district and power canal, Niagara Falls, Niagara County, New York, 1882. Henry Wellge, delineator, 1881. G&M,G3804.N7A3 1882.W4 Rug 150,detail.

The first hydropower canal was built at the Falls in 1860. It was purchased a decade later by Jacob Schoellkopf, who built one of the first hydroelectric generating stations in the world, providing street lighting in the town and power for several mills on the high bank (1881). The competing Niagara Falls Power Company, with substantial financing from New York financial elites, adopted a fantastic scheme drawn up years earlier to divert the entire Niagara River through power canals and tunnels to drive shaft turbines linked to factories on 1,500 acres of land. In 1894, the 6,700-foot-long power tunnel was completed.

2-056

2-056. Mill district showing overflows from power canals, Niagara River, Niagara Falls, Niagara County, New York. DETR, ca. 1900. P&P,LC-dig-det-4a31746.

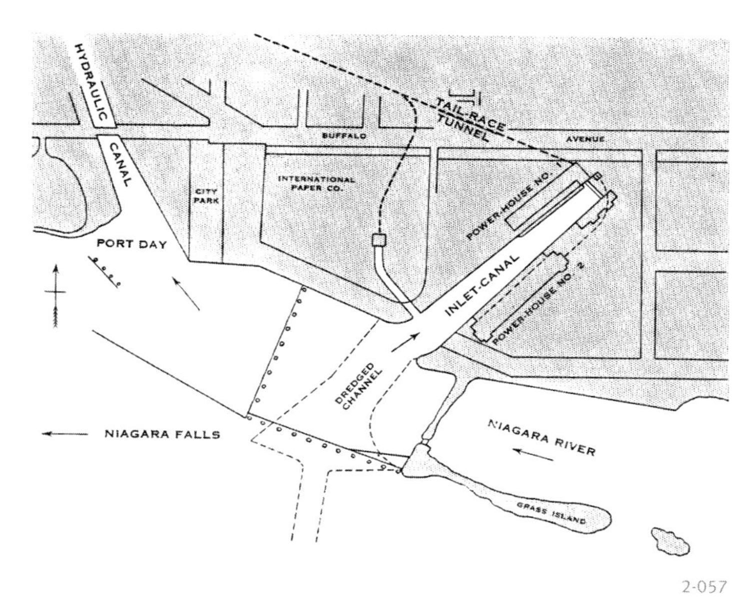

2-057

2-058

2-057. Plan of inlet canal, powerhouses and tailrace tunnel, Niagara Falls Power Company Powerhouse No. 1 (Edward D. Adams Plant), Niagara Falls, Niagara County, New York, 1894–1895, demolished 1961. P&P,HABS,NY,32-NIAF,3-6.

In 1895, the Niagara Falls Power Company completed its Powerhouse No. 1, putting 60-cycle, three-phase generators into operation. This was a high-head plant in which water descended 160 feet to power the turbines. Widely admired, it set a standard for high-head plants elsewhere. The completion of Powerhouse No. 2 (1901–1903) set a second record, for long-distance transmission of alternating current to the City of Buffalo (1896).

2-058. Niagara Falls Power Company Powerhouse No. 1, Niagara Falls, Niagara County, New York, 1894–1895, demolished 1961. Unidentified photographer, 1941. P&P,HABS,NY,32-NIAF,3, no. 12.

Powerhouses Nos. 1 and 2 were designed by the well-known New York architects McKim, Mead and White in a massive and dignified Romanesque style considered suitable for major civic buildings—a sign of the significance of the project.

2-059. Section through wheel pit, original design for Niagara Falls Power Company Powerhouse No. 1, Niagara Falls, Niagara County, New York, 1894–1895, demolished 1961. P&P,HABS,NY,32-NIAF,3-8.

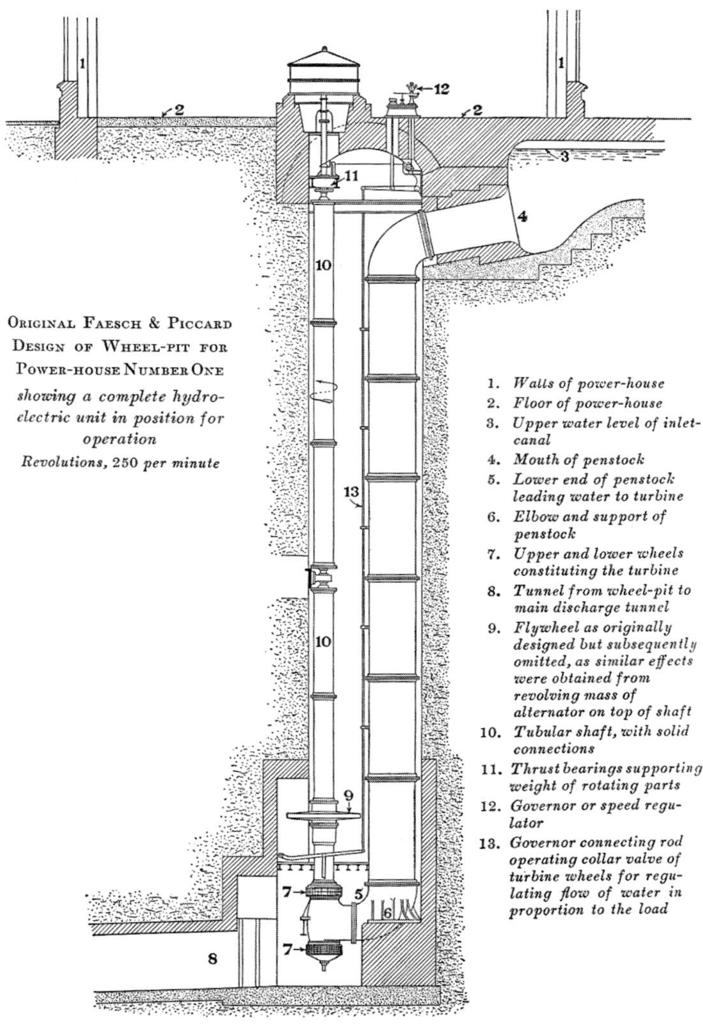

2-059

2-060. Trenton Falls Dam, West Canada Creek, Rome Vic., Oneida County, New York, 1899–1901. DETR, ca. 1906. P&P,LC-D4-16838.

At Trenton Falls, Canada Creek cut a three-mile limestone gorge, descending 250 feet in a series of cascades. In 1889, the Utica Electric Light and Power Company acquired water rights to the creek. Influenced by the Niagara Falls hydroelectric plant, Trenton Falls was the highest-head plant in the country at the time of its construction. An unusual feature is the auxiliary spillway cut into bedrock on the eastern side.

2-061. Sault Sainte Marie Hydroelectric Plant, Saint Mary's River between Lakes Superior and Huron, Sault Sainte Marie, Chippewa County, Michigan, 1900–1902. William Henry Jackson, photographer, 1902. P&P,LC-D4-14735.

Built by the Lake Superior Power Company to supply power for Union Carbide, the plant spans a power canal that parallels the Soo Locks. It was designed as a low-head plant, using horizontal shaft turbines to take advantage of the modest 20-foot drop between Lakes Superior and Huron. To reach the target output of 40,000 horsepower, eighty turbines were required, making the powerhouse a quarter of a mile long. It remains the largest low-head facility in the United States.

2-062. Spillway of Croton Dam and Hydroelectric Plant, Muskegon River, Croton, Newaygo County, Michigan, 1906–1908. Carla Anderson, photographer, 1994. P&P,HAER,MICH,62-CROTO.V,1, no. 2.

The Grand Rapids–Muskegon Power Company built this low-head dam to supply electricity to Grand Rapids, 35 miles away. A central spillway and sluiceway of concrete are flanked by earth embankment dams that illustrate the innovative hydraulic sluice construction technique. In this process, an earth and water slurry is pumped into troughs and deposited on the embankment in layers, allowing the water to gradually drain off and the fill to compact. The Croton Dam is a rare example of this technique—derived from placer mining technology—east of the Mississippi River.

2-060

2-061

2-062

2-063

2-063. Turbine installed in 1915, Croton Dam and Hydroelectric Plant, Muskegon River, Croton, Newaygo County, Michigan, 1906–1908. Carla Anderson, photographer, 1994. P&P,HAER,MICH,62-CROTO.V,1B, no. 8.

2-064. Manual wheel for raising tainter gates, Croton Dam and Hydroelectric Plant, Muskegon River, Croton, Newaygo County, Michigan, 1906–1908. Carla Anderson, photographer, 1994. P&P,HAER,MICH,62-CROTO.V,1C, no. 8.

2-064

2-065

2-066

2-065. Earth embankment, spillway, and powerhouse of Cooke Hydroelectric Plant, Au Sable River, Oscoda Vic., Iosco County, Michigan, 1909–1911. Clayton B. Fraser, photographer, 1995. P&P,HAER,MICH,35-OSCO.V,1, no. 1.

The Cooke Hydroelectric Plant was a landmark in the development of the Consumers Power Company—an interconnected power system through Michigan's Lower Peninsula. The steady flow of the Au Sable promised a reliable source of power, but the site was far from metropolitan and industrial markets, such as Flint, 125 miles away. The plant is significant for its high-voltage transmission lines of 140,000 volts, a world record at the time. The earthen embankment dams adjacent to the powerhouse reinforce the natural geography, guaranteeing a pool for supplying the turbines. The spillway and sluiceway are used only for overflow.

2-066. Horizontal turbine, Cooke Hydroelectric Plant, Au Sable River, Oscoda Vic., Iosco County, Michigan, 1909–1911. Clayton B. Fraser, photographer, 1995. P&P,HAER,MICH,35-OSCO.V,1C, no. 14.

2-067. Quemahoning Dam, Quemahoning Creek, Johnstown Vic., Somerset County, Pennsylvania, 1910–1913. Acme, 1937. P&P,LC-dig-ppmsca-17277.

This dam, built to supply water for the Cambria Steel Company's works in Johnstown, Pennsylvania, was the site of a clash between steel producers and the CIO (Congress of Industrial Organizations) under John L. Lewis, during the latter's efforts to unionize steel industries in the summer of 1937. As tensions rose, a supply line from the dam to the plant was dynamited, forcing the plant to temporarily close. This image shows Pennsylvania highway patrolmen guarding the dam from further tampering. To defuse the threat of continued violence, the governor of Pennsylvania placed Johnstown under martial law and ordered plant closure.

2-067

2-068

2-068. Dix Dam and Hydroelectric Plant, Dix River, Harrodsburg Vic., Mercer County, Kentucky, 1923–1925. Simmons Studio. P&P,LC-dig-ppmsca-17243.

The Dix River canyon had been a site for hydropower mills since the 1780s. The Kentucky Utilities Company built this hydroelectric dam to supply electricity to customers in four states. Promoted as the world's largest stone-filled dam, it also developed as a tourist attraction with a clubhouse, flowers in the hydroelectric plant, and landscaped grounds for picnicking and fishing.

2-069

2-069. Lake Lynn Dam and Hydroelectric Plant, Cheat River, Morgantown, Monongalia County, West Virginia, 1913–1914, completed 1925–1926. Jet Lowe, photographer, 1980. P&P,HAER,WVA,31-MORG.V,2, no. 6.

This dam and hydroelectric plant, which helped to bring electrification to West Virginia and southwestern Pennsylvania, represented the state of the art in hydroelectric power technology when it was begun in 1913. Originally known as the State Line Dam because of its location, it is a concrete gravity dam equipped with steel tainter gates along its crest to allow overflow in the event of a flood. It is still in use.

2-070

2-070. Housing for gate hoists, Lake Lynn Dam, Cheat River, Morgantown, Monongalia County, West Virginia, 1913–1926. Jet Lowe, photographer, 1980. P&P,HAER,WVA,31-MORG.V,2, no. 13.

2-071. Tainter gate, downstream side, Lake Lynn Dam, Cheat River, Morgantown, Monongalia County, West Virginia, 1913–1926. Jet Lowe, photographer, 1980. P&P,HAER,WVA,31-MORG.V,2, no. 16.

2-072. Generator hall, Lake Lynn Dam and Hydroelectric Plant, Cheat River, Morgantown. Monongalia County, West Virginia, 1913–1926. Jet Lowe, photographer, 1980. P&P,HAER,WVA,31-MORG.V,2, no. 35.

2-071

2-072

The Ohio River and its tributaries are notorious for floods. The 1913 flood that devastated the city of Dayton led to the establishment of the Miami River Conservancy District to provide flood protection on the Miami tributary of the Ohio River. This regional agency was exemplary for its careful attention to flood control planning and legislation. It pioneered the innovative use of earthen dams to retain floodwaters while remaining dry most of the year. Five of these dams (Englewood, Huffman, Taylorsville, Germantown, and Lockington dams) were built from 1918 to 1922, and President Franklin Roosevelt later asked their engineer, Arthur E. Morgan, to head up the newly established Tennessee Valley Authority.

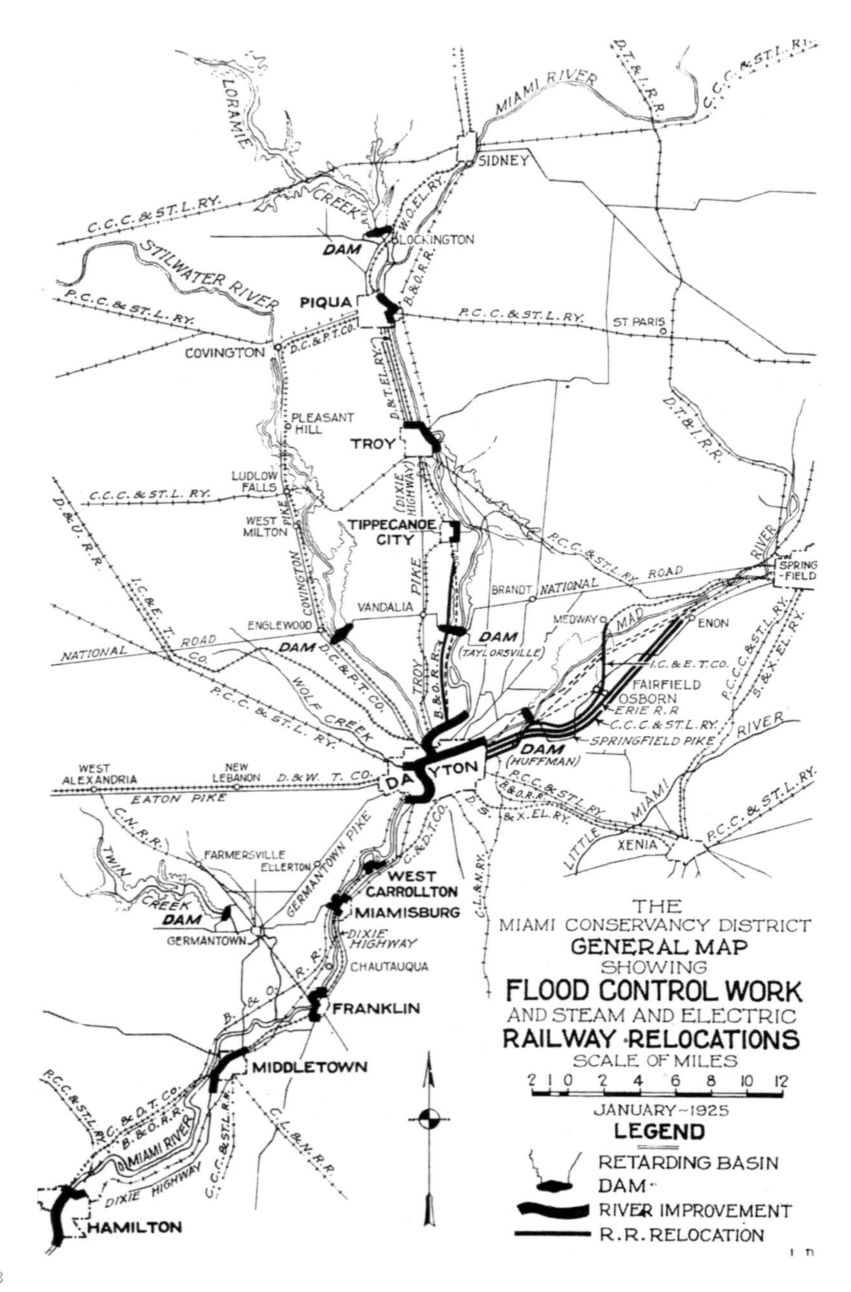

2-073. Overview of projects, Miami River Conservancy District, Miami River, Ohio. 1925. TC425 M45 M6,p. 230.

2-073

2-074. Miami River flood, Dayton, Montgomery County, Ohio, 1913. William F. Cappel, photographer, 1913. P&P,LC-dig-ppmsca-17310.

2-075. Depositing hydraulic fill, Miami River Conservancy District, Miami River, Montgomery County, Ohio, 1918–1922. TC425 M45 M6,p. 420.

2-076. Pool of hydraulic fill, Englewood Dam, Miami River Conservancy District, Miami River, Englewood, Montgomery County, Ohio. TC425 M45 M6,p. 434.

2-074

2-075

2-076

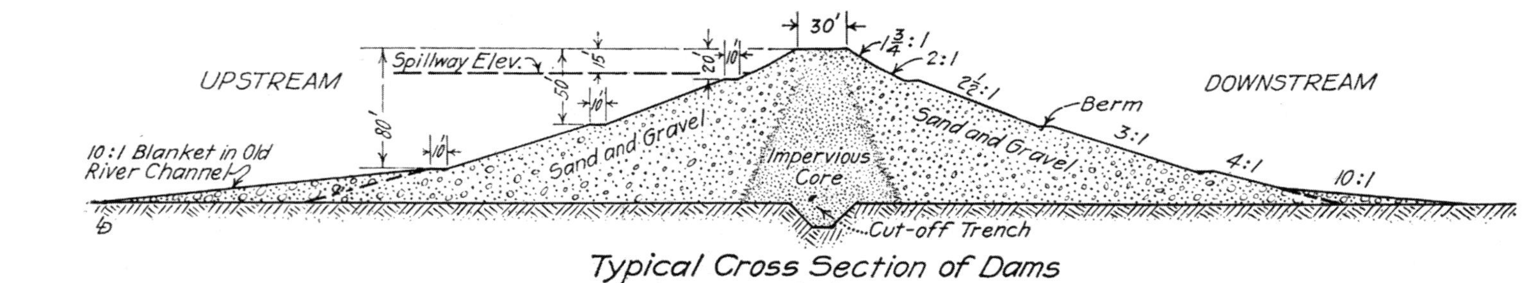

2-077

2-078

2-079

2-080

2-077. Typical cross-section of earthen dam, Miami River Conservancy District, Miami River, Montgomery County, Ohio, 1918–1922. 1925. TC425 M45 M6,p. 249.

2-078. Aerial view of Englewood Dam, Miami River Conservancy District, Miami River, Englewood, Montgomery County, Ohio, 1918–1922. TC425 M45 M6,p. 440.

2-079. Aerial view of Taylorsville Dam, Miami River Conservancy District, Miami River, Vandalia vic., Montgomery County, Ohio, 1918–1922. TC425 M45 M6,p. 466.

2-080. Aerial view of Huffman Dam, Miami River Conservancy District, Miami River, Fairborn vic., Greene County, Ohio, 1918–1922. TC425 M45 M6,p. 448.

TENNESSEE VALLEY AUTHORITY DAMS

Early hydroelectric developments in the Piedmont region of North Carolina and Georgia provide a context for the engineering and dam-building accomplishments of the Tennessee Valley Authority. These included the Marshall Hydroelectric Plant on the French Broad River (1908), the Santeetlah Hydroelectric Plant on the Cheoah River (1925–1928) and the 225-foot-tall Cheoah Hydroelectric Plant on the Little Tennessee River (1916–1919), which, at the time it was built, was the highest overfall dam in the world. Waterville Hydroelectric Plant on the Pigeon River (1927–1930) included a tunnel blasted through six miles of rock, with a drop of 863 feet, making it the highest-head hydroelectric plant east of the Rockies for a decade. Its remote location and challenging terrain attracted national attention.

The Tennessee Valley Authority (TVA) was a federal program established in 1933 that embraced an enormous scope: from dam design for flood control, power generation, and navigation, to resettlement, public works projects, model town planning, model farming practices, rural electrification, craft therapy, freeway design, and chemical and agricultural engineering. Over the first six years of its existence, its triumvirate of three directors led it in many directions. By the beginning of World War II, the main function of the TVA was power generation for the war effort, to support fertilizer, chemical, aluminum, and manufacturing industries, leading to nuclear research

2-081

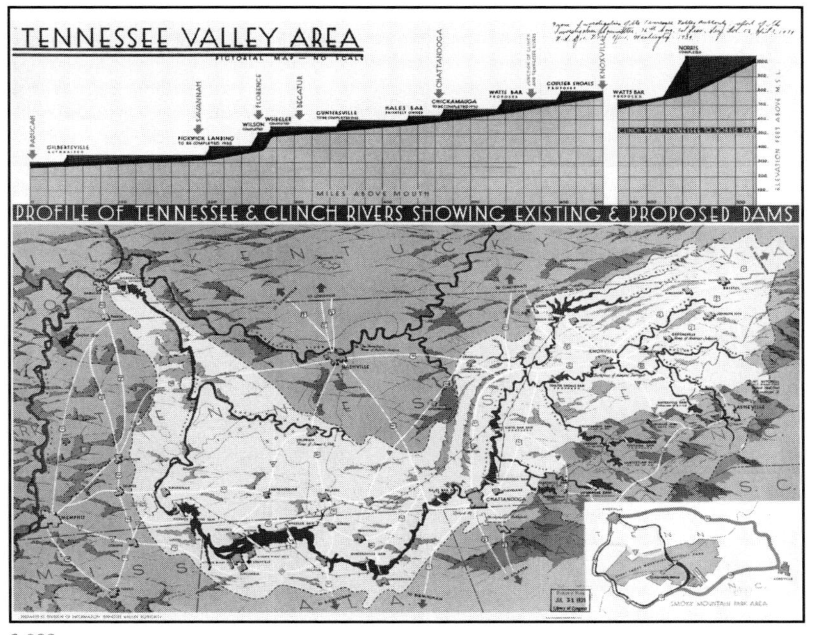

2-082

2-081. Wheeler Lock and Dam, Tennessee Valley Authority, Tennessee River, Florence Vic., Lawrence County, Alabama, 1933– 1936. Tennessee Valley Authority, ca. 1936. P&P,LC-dig-fsa-8e00609.

2-082. Map of Tennessee Valley Authority, Tennessee Valley Authority, delineator, 1939. G&M,US Tenn Valley (Region), Pictorial 1939,detail.

2-083

facilities such as the one at Oak Ridge, fossil fuel plants, nuclear power plants, and industrial development. The TVA built fourteen dams on the Tennessee River between 1933 and 1973, including Kentucky, Pickwick Landing, Wilson, Wheeler, Guntersville, Chickamauga, Watts Bar, Fort Loudon, and Cherokee. The agency built a further twelve dams on tributary rivers, including Apalachia and Hiwassee dams on the Apalachia River, Fontana on the Little Tennessee River, Douglas on the French Broad River, and Norris on the Clinch River.

2-083. Wilson (Muscle Shoals) Lock and Dam, Tennessee Valley Authority, Tennessee River, Florence, Lauderdale County, Alabama, 1917–1925. George Lee Bracey, photographer, 1927. P&P,LC-dig-pan-6a15190.

While privately financed dams had existed in the Appalachian foothills since the nineteenth century, and some of these stayed in private hands, many were taken over and incorporated into the Authority. The federal government (through the U.S. Army Corps of Engineers) also built a number of dams on rivers that link up to the Tennessee system, such as the Cumberland in Tennessee and the Tombigbee–Black Warrior–Alabama river system in Alabama.

The TVA began with Wilson Dam, built during World War I to provide power for a nitrate munitions plant. Wilson Dam was the subject of much controversy in the years following its construction, as the federal government debated its fate. Henry Ford famously offered to purchase it for a fraction of its construction cost, to use it to power an automobile assembly plant. He envisioned a number of smaller car component plants upstream that would be manned by employees who would also farm on the side, as agriculturalist-laborers. Car parts would be shipped downstream on the Tennessee River, as if it were a giant conveyor belt. Under the vigorous initiative of Senator George Norris of Nebraska, the government decided to keep Wilson Dam and build a number of additional hydroelectric dams on the Tennessee River. By entering into electricity generation, it argued, government could set a "yardstick" to establish a fair cost for electricity production and ensure a wider distribution network that would include even the most rural areas. Private power companies were outraged.

2-084

2-084. "Heartbreak"—Power trust weeping in front of Wilson Dam, soon to be operated by the U.S. government. Rollin Kirby, artist, 1933. P&P,CD 1-Kirby,no. 65 (B size).

Norris Dam was the first project built by the TVA, on a tributary of the Tennessee River several miles north of Knoxville. The TVA director in charge of construction was Arthur E. Morgan, an engineer who had gained national renown for his system of using dry earthen dams for flood control in Ohio's Miami River valley. A Quaker and a passionate advocate for cooperative ideals, he instituted a number of socially progressive policies at the core of the TVA's mission, including merit-based hiring, limited working hours, and the establishment of a permanent community for dam workers in place of the usual work camp. The result at Norris Dam and Town was an exceptionally integrated vision of technological accomplishment, community development, and resource conservation.

Under pressure to get Norris Dam under way, Morgan used an engineering scheme

2-085. Before and after photographs of Norris Dam, Tennessee Valley Authority, Clinch River, Knoxville Vic., Anderson County, Tennessee, 1933–1936. W. M. Cline, photographer, 1938. P&P,LC-dig-ppmsca-17365.

2-085

drawn up for the site by the U.S. Army Corps of Engineers several years earlier. A number of added features heightened the visual impact of the dam and its accommodations for visitors—the public appearance of the TVA's flagship project never being far from Morgan's mind. A Hungarian émigré architect, Roland Wank, was responsible for coordinating the visual appearance of the dam, which set a standard for design in all the TVA's subsequent projects.

2-086

2-087

2-088

2-086. Dam and powerhouse, Norris Dam, Tennessee Valley Authority, Clinch River, Knoxville Vic., Anderson County, Tennessee, 1933–1936. Tennessee Valley Authority, ca. 1936. P&P,LC-dig-fsa-8e00652.

2-087. Installing a generator, Norris Dam, Tennessee Valley Authority, Clinch River, Knoxville Vic., Anderson County, Tennessee, 1933–1936. Tennessee Valley Authority, ca. 1935. P&P,LC-dig-ppmsca-17335.

2-088. Powerhouse, Norris Dam, Tennessee Valley Authority, Clinch River, Knoxville Vic., Anderson County, Tennessee, 1933–1936. Tennessee Valley Authority, 1936. P&P,LC-dig-fsa-8e00552.

2-089

2-090

2-089. Night view, Norris Dam, Tennessee Valley Authority, Clinch River, Knoxville Vic., Anderson County, Tennessee, 1933–1936. Tennessee Valley Authority, 1936. P&P,LC-dig-fsa-8e00551.

2-090. Visitors' building, Norris Dam, Tennessee Valley Authority, Clinch River, Knoxville Vic., Anderson County, Tennessee, 1933–1936. Tennessee Valley Authority, 1936. P&P,LC-dig-fsa-8e00648.

2-091. Overlook terrace at visitors' building, Norris Dam, Tennessee Valley Authority, Clinch River, Knoxville Vic., Anderson County, Tennessee, 1933–1936. Tennessee Valley Authority, 1936. P&P,LC-dig-fsa-8e00651.

2-092. Town of Norris, Knoxville Vic., Anderson County, Tennessee, 1933–1936. Tennessee Valley Authority, 1936. P&P,dig-fsa-8e00687.

2-091

2-092

Located on the main stem of the Tennessee River, Wheeler Dam incorporated navigation locks as well as a hydroelectric plant and, as was the case with all the TVA's projects, a visitors' facility. The following images, taken by TVA staff photographers, show the dual emphasis on labor and modernity that characterizes most TVA publicity photos.

2-093

2-093. Workers installing speed ring in a generator, Wheeler Lock and Dam, Tennessee Valley Authority, Tennessee River, Florence Vic., Lawrence County, Alabama, 1933–1936. Tennessee Valley Authority, ca. 1935. P&P,LC-USZ62-92782.

2-094. Visitors' lobby, Wheeler Lock and Dam, Tennessee Valley Authority, Tennessee River, Florence Vic., Lawrence County, Alabama, 1933–1936. Tennessee Valley Authority, ca. 1936. P&P,LC-dig-fsa-8e00558.

2-095. Pickwick Landing Lock and Dam, Tennessee Valley Authority, Tennessee River, Hardin County, Tennessee, 1934–1938. Tennessee Valley Authority, ca. 1953. P&P,LC-USZ62-65884.

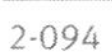

2-094

2-095

Another main-stem dam, 1.5 miles long, Pickwick was built in stages, starting with the lock and earth dams abutting the riverbanks, followed by the powerhouse and, last, the spillway. Its stillwater lake extended 53 miles upstream, affecting 23,000 acres of land and requiring the relocation of over 500 families and 400 graves. Such relocations were a recurring feature of TVA projects, as the river and tributary valleys had been densely inhabited for centuries. Lobby displays extolled the benefits brought by the TVA to farmers, industry, and navigation in the region.

2-096

2-096. Reception wing and powerhouse, Pickwick Landing Lock and Dam, Tennessee Valley Authority, Tennessee River, Hardin County, Tennessee, 1934–1938. Tennessee Valley Authority. P&P,LC-dig-fsa-8e00560.

2-097

2-097. Visitors' reception room, Pickwick Landing Lock and Dam, Tennessee Valley Authority, Tennessee River, Hardin County, Tennessee, 1934–1938. Tennessee Valley Authority. P&P,LC-dig-fsa-8e00597.

2-098

2-099

2-100

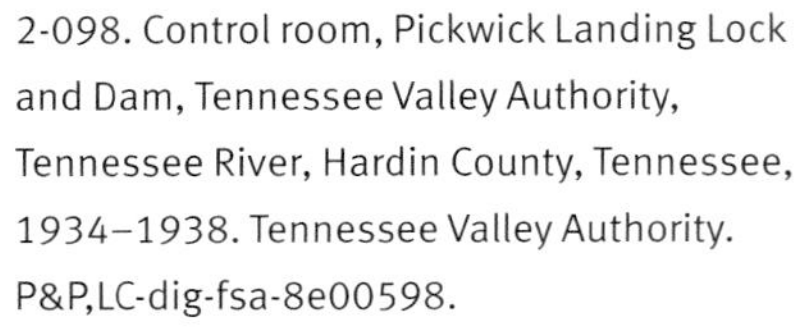

2-098. Control room, Pickwick Landing Lock and Dam, Tennessee Valley Authority, Tennessee River, Hardin County, Tennessee, 1934–1938. Tennessee Valley Authority. P&P,LC-dig-fsa-8e00598.

2-099. Inside the powerhouse, Pickwick Landing Lock and Dam, Tennessee Valley Authority, Tennessee River, Hardin County, Tennessee, 1934–1938. Tennessee Valley Authority. P&P,LC-dig-fsa-8e00613.

2-100. Concrete buttresses on spillway section, Pickwick Landing Lock and Dam, Tennessee Valley Authority, Tennessee River, Hardin County, Tennessee, 1934–1938. Tennessee Valley Authority. P&P,LC-dig-fsa-8e00605.

2-101. Guntersville Lock and Dam, Tennessee Valley Authority, Tennessee River, Guntersville Vic., Marshall County, Alabama, 1935–1939. Tennessee Valley Authority, 1940. P&P,LC-dig-fsa-8e00600.

2-102. Powerhouse, Guntersville Lock and Dam, Tennessee Valley Authority, Tennessee River, Guntersville Vic., Marshall County, Alabama, 1935–1939. Tennessee Valley Authority, 1940. P&P,LC-dig-ppmsca-17350.

2-103. Inside powerhouse looking at visitors' gallery, Guntersville Lock and Dam, Tennessee Valley Authority, Tennessee River, Guntersville Vic., Marshall County, Alabama, 1935–1939. Tennessee Valley Authority, 1940. P&P,LC-dig-fsa-8e00601.

Great care was taken with architectural composition in the TVA's dams. TVA architects strived for a modern, stripped-down look: concealing roof trusses in the power house, streamlining the housing of gantry cranes, experimenting with board-formed concrete and generous daylighting in the inhabited spaces, and developing color palettes for the tiled interiors.

2-101

2-102

2-103

2-104

2-105

2-104. Chickamauga Lock and Dam, Tennessee Valley Authority, Tennessee River, Chattanooga, Hamilton County, Tennessee, 1936–1940. Tennessee Valley Authority, ca. 1940. P&P,LC-dig-fsa-8e00547.

All TVA dams were lit for visual effect. This dam, just upstream from the city of Chattanooga, was easy to visit and shows a corresponding care taken with the tourist areas. Specially designed lighting fixtures inside and outside the building promoted the use of electricity, as did the frequent use of aluminum, glass, bakelite, and other modern materials.

2-105. Entry to powerhouse and visitors' reception, Chickamauga Lock and Dam, Tennessee Valley Authority, Tennessee River, Chattanooga, Hamilton County, Tennessee, 1936–1940. Tennessee Valley Authority, ca. 1940. P&P, LC-dig-ppmsca-17349.

2-106. Visitors' reception, Chickamauga Lock and Dam, Tennessee Valley Authority, Tennessee River, Chattanooga, Hamilton County, Tennessee, 1936–1940. Tennessee Valley Authority, ca. 1940. P&P,LC-USW33-15617-ZC.

2-107. Inside powerhouse toward visitors' overlook, Chickamauga Lock and Dam, Tennessee Valley Authority, Tennessee River, Chattanooga, Hamilton County, Tennessee, 1936–1940. Tennessee Valley Authority, ca. 1940. P&P,LC-dig-fsa-8e00549.

2-108. Visitors' reception building at lock, Chickamauga Lock and Dam, Tennessee Valley Authority, Tennessee River, Chattanooga, Hamilton County, Tennessee, 1936–1940. Tennessee Valley Authority, ca. 1942. P&P,LC-dig-fsa-8e00592.

2-106

2-107

2-108

2-109

2-110

2-109. Hiwassee Dam, Tennessee Valley Authority, Apalachia River, Murphy Vic., Cherokee County, North Carolina, 1936–1940. Tennessee Valley Authority, ca. 1940. P&P,LC-dig-fsa-8e00541.

A reservoir dam built on a tributary of the Tennessee, Hiwassee shares some design features of Norris, particularly its overflow spillway (2-082). In 1956, the TVA installed reversible pump-turbines in Hiwassee, the first of many such integrated pump-turbines installed later in plants across the country. During periods of low electricity usage, these turbines pump water up to the reservoir so that it can be available at times of peak demand.

2-110. Visitors' lookout, Hiwassee Dam, Tennessee Valley Authority, Apalachia River, Murphy Vic., Cherokee County, North Carolina, 1936–1940. Tennessee Valley Authority, ca. 1940. P&P,LC-dig-fsa-8e00691.

The visitors approach was over the roof of the generator hall, which they could view from an overlook once inside.

2-111

2-111. Control building and visitors' reception on bluff above powerhouse, Watts Bar Lock and Dam, Tennessee Valley Authority, Tennessee River, Decatur Vic., Meigs County, Tennessee, 1939–1942. Tennessee Valley Authority, ca. 1941. P&P,LC-dig-fsa-8e00668.

Watts Bar powerhouse was integrated into the main body of the dam, while the control building and visitors' areas were far above the cliff. A bridge was built across the dam at a later date.

2-112. Watts Bar Lock and Dam, Tennessee Valley Authority, Tennessee River, Decatur Vic., Meigs County, Tennessee, 1939–1942. Tennessee Valley Authority, ca. 1941. P&P,LC-dig-ppmsca-17345.

2-112

2-113

2-114

2-113. Apalachia Dam, Tennessee Valley Authority, Hiwassee River, Murphy Vic., Cherokee County, North Carolina, 1941–1943. Tennessee Valley Authority, ca. 1954. P&P,LC-dig-ppmsca-17356.

2-114. Tops of draft tube liners before installation of the speed rings, Cherokee Dam, Tennessee Valley Authority, Holston River, Rutledge Vic., Jefferson County, Tennessee, 1940–1941. Tennessee Valley Authority, 1941. P&P,LC-dig-ppmsca-17358.

2-115

2-115. Cherokee Dam, Tennessee Valley Authority, Holston River, Rutledge Vic., Jefferson County, Tennessee, 1940–1941. Tennessee Valley Authority, 1942. P&P,LC-dig-fsa-8e00584.

2-116. Workers installing spiral case, Cherokee Dam, Tennessee Valley Authority, Holston River, Rutledge Vic., Jefferson County, Tennessee, 1940–1941. Tennessee Valley Authority, 1941. P&P,LC-dig-ppmsca-17357.

2-116

TVA press releases used photographs of workers, some seemingly staged, to express the industrial power of the United States as a democratic country, in contrast to Germany, Japan, or even the Soviet Union. The architectural critic Lewis Mumford extolled the TVA dams as "democratic pyramids." Not everyone was treated equally, however. The TVA maintained southern Jim Crow laws requiring separate facilities for African American employees that were generally smaller and less well equipped than those for whites. Cherokee Indians also worked on TVA projects, especially at Fontana, near the Cherokee Reservation. TVA dam building escalated as the country geared up for war, and it played a key role in the development of the Oak Ridge nuclear weapons lab.

2-117

2-117. Negro drillers at Fort Loudon Dam, Tennessee Valley Authority, Tennessee River, Loudon Vic., Loudon County, Tennessee, 1940– 1943. Jack Delano, photographer, 1942. P&P,LC-USW3-006549-D.

2-118. Bucketman, Douglas Dam, Tennessee Valley Authority, French Broad River, Knoxville Vic., Jefferson County, Tennessee, 1942–1943. Alfred T. Palmer, photographer, 1942. P&P,LC-USE6-D-007339.

2-119. Carpenter, Douglas Dam, Tennessee Valley Authority, French Broad River, Knoxville Vic., Jefferson County, Tennessee, 1942–1943. Alfred T. Palmer, photographer, 1942. P&P,LC-USE6-D-007301.

2-118

2-119

2-120

2-120. Drawing of Fontana Dam, Tennessee Valley Authority, Little Tennessee River, Graham County, North Carolina, 1942– 1944. TVA Architect's Office, delineator, ca. 1941. P&P,LC-dig-fsa-8e00585.

Fontana Dam, built during the peak years of World War II and at the height of rationing, employed concrete rather than steel as much as possible. TVA architects nonetheless managed to include a funicular railway that allowed visitors a breathtaking descent from the crest of the dam to the powerhouse at its base.

Norris Dam's progressive town planning (2-092) was not repeated in later dam sites, after the departure of Arthur Morgan from the TVA. Under the leadership of Director David Lilienthal, the TVA focused on building consumer markets for electricity and serving developing wartime industries, particularly in aluminum, weapons manufacture, and atomic energy. Fontana Dam's large town site had modular prefabricated trailer homes that could be used in other locations once the dam was complete (2-121–2-124).

2-121

2-124

2-122

2-123

2-121. Half of an experimental trailer-house en route from the factory in Michigan, Fontana Dam, Tennessee Valley Authority, Little Tennessee River, Graham County, North Carolina, 1942–1944. Tennessee Valley Authority, 1942. P&P,LC-dig-fsa-8e09028.

2-122. Second section joined to first, Fontana Dam, Tennessee Valley Authority, Little Tennessee River, Graham County, North Carolina, 1942–1944. Tennessee Valley Authority, 1942. P&P,LC-dig-fsa-8e09031.

2-123. Completed trailer-house, Fontana Dam, Tennessee Valley Authority, Little Tennessee River, Graham County, North Carolina, 1942–1944. Tennessee Valley Authority, 1942. P&P,LC-dig-fsa-8e09027.

2-124. Trailer-houses re-installed at White Pine, North Carolina, for Douglas Dam, Tennessee Valley Authority, French Broad River, Knoxville Vic., Jefferson County, Tennessee, 1942–1943. Tennessee Valley Authority, 1942. P&P,LC-dig-fsa-8e00669.

DAMS IN REMOTE LOCATIONS

The reinforced-concrete, multiple-arch dam is rare in the United States, and examples that do exist are mostly in the West (see Section 8). A structurally efficient design that requires little concrete, it was a good choice for the Victoria Dam and Hydroelectric Plant, built in a remote location on the Upper Peninsula of Michigan. Originally built by the Copper Range Company as a hydroelectric facility for its mining operations, it is now connected to the regional power grid.

2-125

2-125. Victoria Dam and Hydroelectric Plant, Ontonagon River, Ontonagon County, Michigan, 1929–1931. Eric Munch, photographer, 1991. P&P,HAER,MICH,66-ROCK.V,1, no. 1.

2-126. Crest, Victoria Dam and Hydro Plant, Ontonagon River, Ontonagon County, Michigan, 1929–1931. Eric Munch, photographer, 1991. P&P,HAER,MICH,66-ROCK.V,1, no. 5.

2-126

2-127

2-128

2-127. Buttressed arches, Victoria Dam and Hydro Plant, Ontonagon River, Ontonagon County, Michigan, 1929–1931. Eric Munch, photographer, 1991. P&P,HAER,MICH, 66-ROCK.V,1, no. 8.

2-128. Redridge Steel Dam, Salmon Trout River, Houghton County, Michigan, 1901. Jet Lowe, photographer, 1978. P&P,HAER,MICH,31-BEHIL, 1, no. 1.

This steel dam, even rarer than the multiple arch dam, was only the second of its kind in the country. A replacement for a timber-crib structure from 1894, it was commissioned by the Atlantic and Baltic mining companies to supply water for their stamping mills in Michigan's Upper Peninsula. Because of the remote location and lack of suitable stone, they decided to build a steel gravity dam. It was designed by J. F. Jackson, engineer for the Wisconsin Bridge and Iron Company, which had built the first steel dam three years earlier in Arizona. The structure, which is 1,006 feet long, incorporates steel plates that are ⅜ inch thick and 8 feet wide, shaped with a slight convexity on the reservoir side and supported by a framework of steel beams and struts.

A DAM DISASTER

Although seasonal floods are a part of the Ohio River landscape, the most notorious dam disaster in this region—and, in fact, in the country—was not the result of spring flooding. The Johnstown Dam was built in 1839 to supply a canal and was abandoned for this use in 1857. Eighteen years later, it was purchased by a hunting and fishing club, who raised the dam height a few feet and screened off its spillway to increase the lake. As a result, overflow water eroded the embankment and the entire dam collapsed, utterly destroying the city of Johnstown, Pennsylvania, which lay in the path of the floodwaters, and killing over two thousand people. Although the original design for the masonry gravity dam was sound, the disaster underscored the importance of regulatory oversight for the construction, operation, and modification of dams.

2-129

2-129. "The bursted dam," Johnstown Dam, South Fork of Little Conemaugh River, Johnstown, Cambria County, Pennsylvania, 1839–1853, modified 1879–1881, failed 1889. George Barker, photographer, 1889. P&P,LC-USZ62-94747.

2-130. "The Johnstown Calamity, a slightly damaged house," 1889. George Barker, photographer, 1889. P&P,LC-USZ62-46831.

2-131. "Main Street after the flood," Johnstown, Cambria County, Pennsylvania, 1889. Langile and Darling, photographers, 1889. P&P,LC-USZ62-24368.

2-130

2-131

MISSISSIPPI

THE MISSISSIPPI RIVER RISES IN THE LAKES of Minnesota and flows south, meeting its major tributaries, the Missouri and the Ohio rivers, halfway along its journey to the Gulf of Mexico. Its drainage basin is the third largest in the world, exceeded only by those of the Amazon and Congo rivers. As the central river artery of the United States, it is one of the world's busiest commercial waterways. The word "Mississippi" comes from a French rendering of an Ojibwe word for "big river."

The flow of the river is greatest in the spring, when heavy rainfall and melting snow on tributaries cause the main stream to rise and frequently overflow its banks and levees, inundating vast areas of the plain. For thousands of years, flooding was a reliable feature of the Mississippi. Natural levees, built up from sediment deposited in times of flood, border the river for much of its length. During floods, these banks break in crevasses, inundating valley lands on each side of the river.

3-001. Dike plowed around farmland for flood protection, Des Moines, Polk County, Iowa, 1950. Don Ultang, photographer, 1950. P&P,LC-USZ62-110681.

Agriculture began thousands of years ago in the Mississippi River basin. Native American cultures prospered in this region, constructing towns and villages on

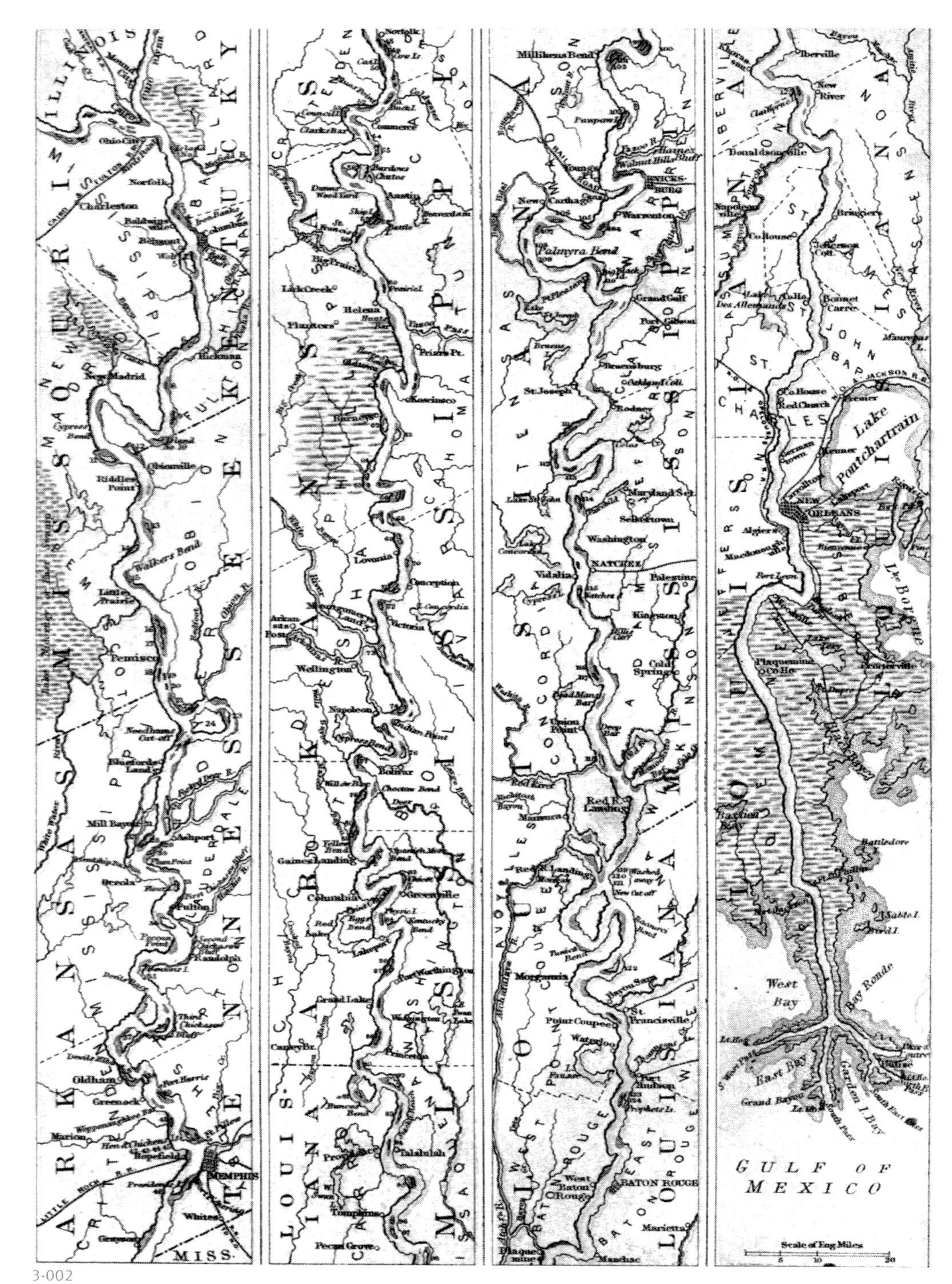

3-002

3-002. Chart of the Mississippi River from the Ohio River to the Gulf of Mexico. Jacob Wells, delineator, 1863. G&M,G4042.M5 1863.W4 CW 47.7.

earthwork mounds that rose above the river's floodplain. At their peak, in the Mississippian period (AD 700–1550), the population of mound-builders numbered in the millions and stretched across the river's tributaries from the Appalachian Mountains to the eastern edge of the Great Plains. Chroniclers of the de Soto expedition of the mid-sixteenth century repeatedly commented upon the density of the population and the abundance of food in this region.

3-003. Great Mound at Marietta, Washington County, Ohio. Charles Sullivan, lithographer, 1848. P&P,LC-USZ62-74133.

3-003

A major artery for the Native Americans and the fur-trading French, in the nineteenth century the river was the principal outlet for the newly settled areas of mid-America; exports were floated downstream, and imports were towed upstream on rafts and keelboats. By mid-century, the valley was sufficiently populated that floods in 1849 and 1850 caused widespread property damage, and levee-building began to become more organized, with the establishment of levee districts to coordinate repairs. In 1879, Congress established the Mississippi River Commission to facilitate navigation on the river and stabilize its banks at times of flood. In keeping with Congress's mandate to promote interstate trade, the focus of the U.S. Army Corps of Engineers in this period was keeping the river navigable. That involved accurate charting of the river's channels, construction of wing dams to speed up its flow, snag clearing and channel dredging, and construction of a number of reservoir dams in the river's headwaters in Minnesota.

Yet, by the end of the nineteenth century, pressures for flood protection in the rich farmlands of the river's lower reaches led the Mississippi River Commission to shift its focus from maintaining navigation routes to levee-building, the so-called "levees-only" policy.[1] Flooding, of course, continued along the river, with major occurrences in 1882, 1912, and 1913. In the flood of 1927, the river broke over its banks in hundreds of places and flooded tens of thousands of square miles, devastating crops and industries and displacing half a million people. The following year, Congress passed the Flood Control Act of 1928 and authorized the construction of dams on the Upper Mississippi and floodways on the lower reaches of the river. This became the massive Mississippi River and Tributaries Project.

Mississippi River improvements have enormously increased freight traffic on the river—principally grains, coal and coke, petroleum products, sand and gravel, chemicals, and building materials. The Corps of Engineers uses the river's junction with the Ohio to divide it into two administrative districts: the Upper and Lower Mississippi. The section above Saint Louis is developed with locks, dams, and canals to maintain navigable depths. The last dam on the Mississippi is just below its confluence with the Missouri. Below this, the river flows freely, although it is retained in the riverbed with levees.

UPPER MISSISSIPPI RIVER BASIN

Kaskaskia, Illinois, Rock, Wisconsin, Chippewa–Red Cedar–Flambeau, Saint Croix, Minnesota, Des Moines, and Salt Rivers

The northernmost region of the Upper Mississippi River is heavily wooded like the Great Lakes region and interwoven with waterways, which have served as logging routes and fostered sawmills and related industries. The rivers that feed the Upper Mississippi flow through the agricultural lands of Illinois, Iowa, and southern Wisconsin. In this part of the Mississippi River valley, several projects have aimed to connect the Great Lakes with the Mississippi, including the Fox River waterway and the Sanitary and Ship Canal (connecting Lake Michigan to the Wisconsin and Illinois rivers, respectively). The Mississippi itself was channeled in the early part of the twentieth century, using the lessons the U.S. Army Corps of Engineers had learned on the Ohio River.

3-004

3-004. Scene on the Upper Mississippi River. F. Sala & Company, lithographer, 1860. P&P,LC-USZ62-94754.

LOWER MISSISSIPPI RIVER BASIN

Ouachita-Black, Saint Francis, Yazoo, Tensas, and Atchafalaya Rivers

Below its junction with the Ohio River, the Mississippi is nearly a mile wide, meandering in great loops across a broad alluvial plain marked with oxbow lakes and marshes that are remnants of the river's former channels. Natural levees, built up from sediment deposited in times of flood, border the river for much of its length, so that in places the surface of the Mississippi is higher than its surrounding plain, a phenomenon visible in the Saint Francis, Black, Yazoo, and Tensas river basins. Breaks in the levees—called crevasses—frequently flood the fertile bottomlands of these low-lying areas.

3-005. On the busy levee of the Mississippi, Saint Louis, Missouri. Underwood & Underwood, 1903. P&P,LC-dig-stereo-1s01724.

3-005

After receiving the Arkansas and Red rivers, the Mississippi enters its delta, flowing into the Gulf of Mexico through a number of distributaries, the most important being the Atchafalaya River and Bayou Lafourche. The main stream continues southeast through the delta, past Baton Rouge and New Orleans, as it enters the Gulf. Concerns that the Mississippi River might abandon this course and divert through the faster-flowing Atchafalaya River led to the construction of a series of dams, locks, and canals—known as the Old River Control Structure—to keep the river in its present course.

WISCONSIN LOGGING DAMS

Wisconsin's logging industry was an enormous economic force in the upper Mississippi River region. Seemingly endless forests, a steady inflow of settlers wanting to clear land, abundant rivers to transport timber downstream, and most important, the state's strategic

3-006. Freshets in the West—great jam of logs at Chippewa Falls boom, Wisconsin, 1869. N. A. Preston, photographer, 1869. P&P,LC-dig-cph-3c05558.

3-006

location feeding into the Mississippi River basin, combined by the mid-nineteenth century to make Wisconsin one of the major lumber-producing states in the country. Early logging centered on the Wisconsin and Wolf rivers, but by mid-century the Black and Chippewa rivers in the northwestern part of the state were the preeminent logging regions. Dozens of smaller logging companies combined into a shipping conglomerate led by Frederick Weyerhaeuser to send timber down the Mississippi to the city of Saint Louis and beyond.

Loggers built "splashes," or driving dams, to hold back small streams so that water could be released into the channel during a log drive. In 1876, Weyerhaeser formed the Chippewa River Improvement and Log Driving Company, which built 148 dams in the valley. There were many ways to build these dams, depending on size and the materials available; some common types included stone- or earth-filled crib dams, wooden rafter dams (which used the force of water to hold the dam in place), and post dams.

3-007. Logging in the alluvial lands of the Mississippi River. Brinkerhoff and Barnett, ca. 1901. P&P,LC-dig-cph-3c04328.

3-007

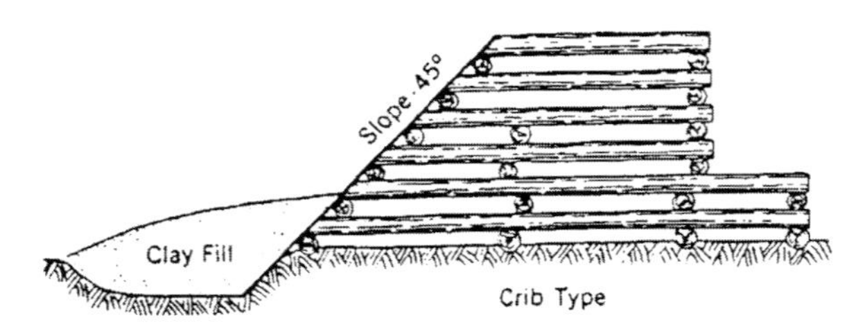

3-008

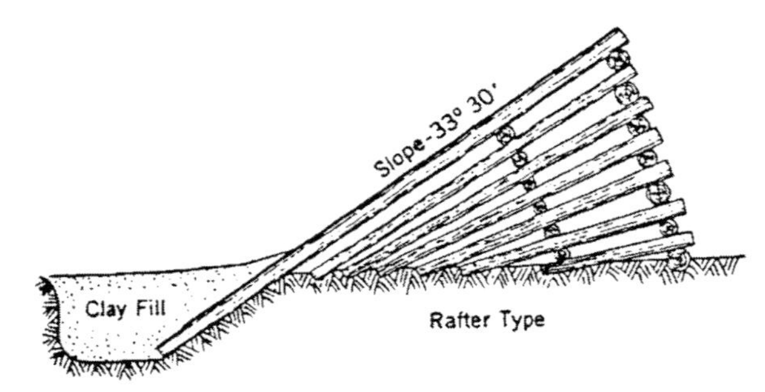

3-009

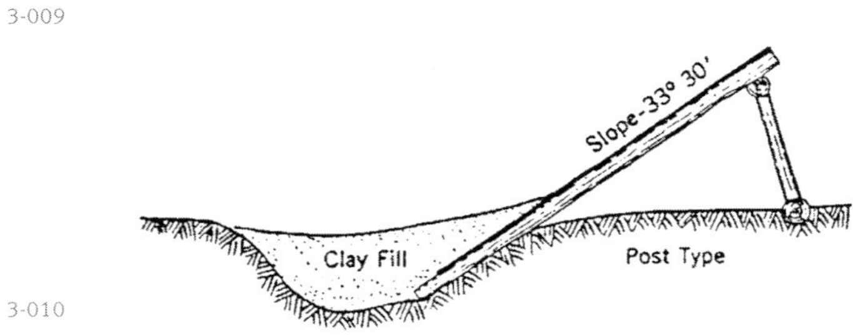

3-010

3-008. Typical crib driving dam, Chippewa and Flambeau rivers, Wisconsin, 1860–1900. P&P,HAER,WIS,50-PAFA.V, 1-p. 35.

3-009. Typical rafter driving dam, Chippewa and Flambeau rivers, Wisconsin, 1860–1900. P&P,HAER,WIS,50-PAFA.V, 1-p. 36.

3-010. Typical post driving dam, Chippewa and Flambeau rivers, Wisconsin, 1860–1900. P&P,HAER,WIS,50-PAFA.V, 1-p. 37.

3-011

3-011. Round Lake Logging Dam, South Fork of Flambeau River, Park Falls Vic., Price County, Wisconsin, 1883–1886. Raymond H. Merritt, photographer, 1980. P&P,HAER,WIS,50-PAFA.V,1, no. 10.

This vernacular, unengineered structure was built for "booming"—that is, the collection and sorting of timber in dammed-up ponds before it was driven downstream to the mills.

MILL DAMS AND HYDROELECTRIC DAMS

The steadily flowing rivers that fed into the Upper Mississippi not only served for transporting logs and people. They also provided a reliable source of hydropower for sawmills and, later, paper mills, flour mills, machine shops, foundries, and factories. These industries led to the rapid growth of towns and cities along the rivers.

3-012. Lumber mill dams, Rock River, Jefferson, Jefferson County, Wisconsin. Louis Kurz, delineator, ca. 1857. P&P,LC-USZ62-15358.

3-013. Pecatonica Dam, Pecatonica River, Freeport, Stephenson County, Illinois, ca. 1870. WC. P&P,LC-dig-ppmsca-17355.

3-012

3-013

3-014

3-014. Saint Cloud (Watob) Dam, Mississippi River, Saint Cloud, Stearns County, Minnesota, 1876, modified 1908. Hugh Spencer, photographer, 1908. P&P,LC-dig-ppmsca-17354.

Since its settlement in the 1850s, Saint Cloud's location near the upper limit of Mississippi River traffic made it the Hudson's Bay Company's choice of terminus for unloading its furs and outfitting its traders. As the town grew and developed industrial manufactures, the increased demand for steam power led to the city's first dam, which the Saint Cloud Water Power and Mill Company built in 1876. The dam supplied five mills with power and sold electricity to the city. In 1908, the dam was enlarged, and a steam plant was added in 1914. It is still used to generate electricity today.

3-015

3-015. Dells Paper and Pulp Company Dam, Chippewa River, Eau Claire, Eau Claire County, Wisconsin, ca. 1890. WC., ca. 1890–1940. P&P,LC-dig-cph-3c16796.

Although the Chippewa River watershed in Wisconsin was renowned as a logging region, many other industries developed here, starting with lumber-related activities such as sawmills, paper mills, and furniture and piano factories, and growing to include textile and food industries and heavy manufacturing. The Dells Paper and Pulp Company built this dam to power its plant.

3-016

3-016. The Falls of Saint Anthony, Minnesota. P&P,LC-USZ62-17052.

3-017. Lower Dam at Saint Anthony Falls, Mississippi River, Minneapolis, Hennepin County, Minnesota, 1895–1897, demolished 1959. DETR, ca. 1908. P&P,LC-D4-70644.

Mid-nineteenth-century Minneapolis had numerous mills along the Mississippi, fed by small dams built between the riverbanks and Hennepin and Nicollet islands. Water powered lumber and flour mills, machine shops, foundries, and factories. In 1881, Pillsbury Company constructed a huge mill complex on the river; in 1897, the company completed the Lower Dam and a hydroelectric plant just below Saint Anthony Falls. A gravity overflow structure, the Lower Dam was 1,085 feet long and 15 feet high, built of locally quarried limestone and faced with granite from Saint Cloud. A spillway at one end supplied water to the hydroelectric plant. The Twin City Rapid Transit Company purchased power from the plant for its newly electrified streetcar lines. The dam was destroyed in the 1950s to prepare for the construction of the current Lower Lock and Dam.

3-017

3-018. Aerial view of dams at Saint Anthony Falls and milling district, Minneapolis, Hennepin County, Minnesota. George Miles Ryan, photographer, ca. 1950. P&P,LC-dig-ppsmca-17359.

3-018

3-019

3-020

3-019. Cedar Falls Dam, Red Cedar River, Menomonie Vic., Dunn County, Wisconsin, 1910–1915. WC. P&P,LC-dig-ppmsca-17353.

During the nineteenth century, the Red Cedar River served as a logging run for a Menomonie-based firm that was the largest lumber producer in the United States. The river was dammed in Menomonie as early as 1848, and a similar timber-crib structure upstream at Cedar Falls powered a sawmill. The original Cedar Falls dam was replaced with a concrete structure in 1910, and new generators were added from 1912 to 1915. Since then, the plant has operated largely unchanged. A number of such plants were built in the 1900s on the Saint Croix, Apple, Chippewa, and Red Cedar rivers, supplying electricity to local industries and to Minneapolis. The original generators in these plants were among the first alternating current machines manufactured.

3-020. Wisconsin Rapids Dam and Hydroelectric Plant, Wisconsin River, Wisconsin Rapids, Wood County, Wisconsin, 1902–1904. WC. P&P,LC-dig-ppmsca-17318.

This postcard image marked the construction of a new dam in 1904 at a point just downstream from the Wisconsin River Crossing, a wide area in the flood plain. In the 1890s, growing industrial demand for electrical power on the Wisconsin River led a small group of investors to consolidate many of the hydroelectric dams along the river into the Consolidated Water Power Company. In 1904, this company completed construction of a hydroelectric dam and a paper and pulp mill near the towns of Grand Rapids and Centralia. The mill included the world's first electrically powered paper machines.

Built to carry Chicago's sewage away from the city's water supply in Lake Michigan, the Chicago Drainage Canal was the largest sanitation engineering project of the nineteenth century. After noting that the ridge separating the Great Lakes watershed from that of the Mississippi River was only 8 feet high and 12 miles west of the lakeshore, engineers made a plan to cut a canal through the ridge, carrying sewage away from the lake and down to the Des Plaines River, a tributary of the Illinois and, ultimately, the Mississippi.

It was the largest earth-moving project in the world, employing new methods and machinery that became known as the "Chicago School of earth moving" (and that were subsequently employed in the construction of the Panama Canal). The canal was a continuous waterway without locks, designed to carry an enormous amount of water—a fact that led to the dubious claim by the Sanitary District that sewage would be "sanitized" by the time it reached Joliet. The quantity of sewage carried by the new

3-021

3-021. The *Marine Angel*, largest vessel to travel the Mississippi River and the Illinois Waterway, rounds a sharp bend in the Chicago River, Chicago, Cook County, Illinois. AP, 1953. P&P,LC-USZ62-119799.

canal reversed the flow of the Chicago River. A few decades later, the Supreme Court ordered a reduction in the amount of Lake Michigan water that Chicago could use to flush the canal, forcing the city to construct sewage treatment plants.

From Chicago, the canal followed the Des Plaines riverbed, paralleling the older Illinois & Michigan Canal for 28 miles to Lockport, Illinois. Seven years after the canal's completion, a lock was built to negotiate the 36-foot descent to Joliet, allowing navigation all the way to the Mississippi. The canal's controlling works, the great lock and a hydroelectric plant, are in Lockport. The Illinois Waterway, built in the 1930s, consisted of five locks and dams on the Illinois River to deepen the navigational channel from the Mississippi River to the Chicago Drainage Canal, renamed the Sanitary and Ship Canal.

3-022

3-022. Dam and controlling works at Lockport, Chicago Drainage Canal, from Chicago to Lockport, Illinois, 1892–1900. DETR, ca. 1908. P&P,LC-D4-70186.

3-023. Power plant and great lock, Chicago Drainage Canal, Lockport, Will County, Illinois, 1907. DETR, ca. 1908. P&P,LC-D4-15607L.

3-024. Power plant and great lock, Chicago Drainage Canal, Lockport, Will County, Illinois, 1907. DETR, ca. 1908. P&P,LC-D4-15607 R.

3-023

3-024

U.S. ARMY CORPS OF ENGINEERS PROJECTS ON THE MISSISSIPPI RIVER

The twin goals of improving navigation and flood control underpin federal projects on the Mississippi. Early federal work on the Mississippi aimed toward improving the river's navigability, while serious flood control work began only in the mid-twentieth century, in the wake of the Great Flood of 1927.

The first U.S. Army Corps of Engineers projects on the Mississippi were surveys of the rapids at Des Moines and Rock Island (1829), where low water and a rocky riverbed made the Mississippi virtually impassable. From 1837 to 1866, the Corps tried to excavate a 4-foot channel through these stretches, finally abandoning the effort in favor of a canal that bypassed the Des Moines Rapids at Keokuk, Iowa (1877). Congress continued its involvement in promoting commerce on the river by establishing the Mississippi River Commission in 1879. This agency, comprised of engineers from the Army Corps, the Coast and Geodetic Survey, and civilian life, was charged with improving the river's navigation channel and stabilizing its banks to minimize flood damage.

To maintain navigable depths on the main body of the river, the Army Corps engaged in three activities: levee repair, the construction of wing dams, and dredging. Levee repair involved reinforcing the river's natural banks as they caved in or broke open in the form of crevasses during floods. By adding artificial levees to close the river's secondary channels, the Army Corps effectively created a continuous earthen dam that runs the

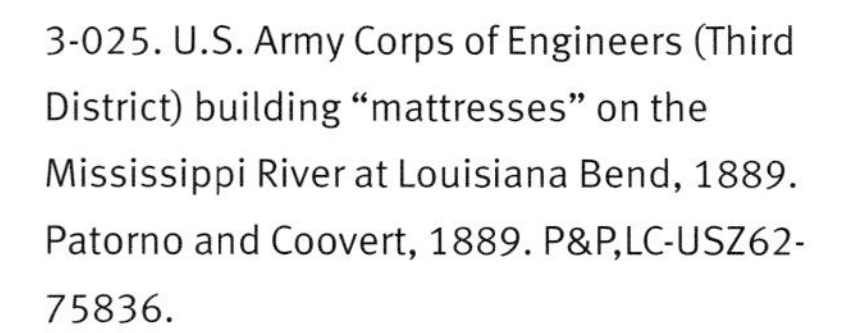

3-025. U.S. Army Corps of Engineers (Third District) building "mattresses" on the Mississippi River at Louisiana Bend, 1889. Patorno and Coovert, 1889. P&P,LC-USZ62-75836.

3-025

entire length of the Mississippi. The Army Corps built wing dams to direct the river's flow and force it to run faster and deeper. These were woven mattresses of willow and maple saplings that were towed into position by barges and gradually loaded with rock as they sank into the river. At the shallowest, siltiest sections of the river—its confluences and delta regions—the Army Corps had to dredge, employing hydraulic suction and hopper dredges from the 1870s onward.

3-026

3-027

3-026. Dredging to supply dirt for levee, Mississippi River, Geismar Vic., Ascension Parish, Louisiana, 1927. Unidentified photographer, 1927. P&P,LC-USZ62-75845.

3-027. Building a levee, Mississippi River levees. J. O. Davidson, delineator, 1884. P&P,LC-dig-cph-3c08307.

Mapping, levee maintenance, and dredging continued through the turn of the century. Army Corps projects from this period include the Mississippi River Headwaters Reservoir dams in north-central Minnesota, built to impound spring run-off for later use. In 1907, Congress authorized a 6-foot navigation channel on the Mississippi. This project was still under way when it was superseded twenty years later by the Nine-Foot Channel Project.

3-028. Laborers going to work on Levee Mile 14, Mississippi River below Helena, Arkansas. Unidentified photographer, 1927. P&P,LC-USZ62-129988.

3-029. Map of Mississippi River showing caving banks, Hickman, Fulton County, Kentucky. 1892. G&M,G1375.U544 1892 Folio.

3-028

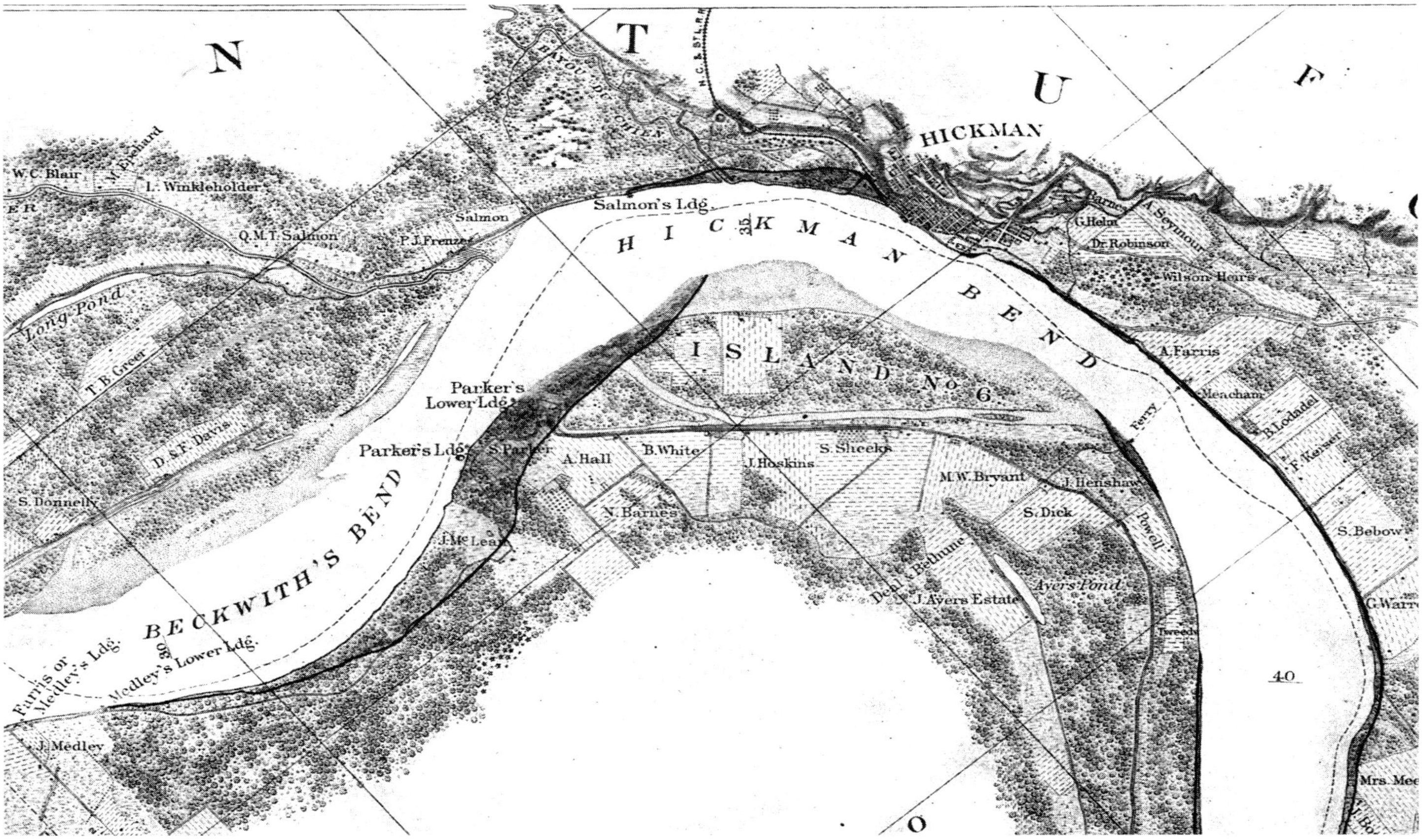

3-029

MISSISSIPPI RIVER HEADWATERS RESERVOIR DAMS

These dams in north-central Minnesota were built at the close of the nineteenth century to supplement the flow of the Upper Mississippi River during periods of low water. They continue to be the source of a dispute between the Ojibwe nation and the U.S. government over damages. The original timber construction of these dams, dating from 1881 to 1893, was replaced between 1903 and 1912 with reinforced concrete dams.

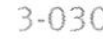

3-030. Headwaters of the Mississippi River, Lake Itasca, Itasca County, Minnesota. Arthur Rothstein, photographer, 1936. P&P,LC-USF34-005332-D.

3-031. Lake Winnibigoshish Reservoir Dam, Deer River Vic., Itasca County, Minnesota, 1881–1883. U.S. Army Corps of Engineers, 1884. P&P,HAER,MINN,31-DERIV.V,1, no. 7.

3-031

3-032. Site plan of Lake Winnibigoshish Reservoir Dam, Deer River Vic., Itasca County, Minnesota, 1881–1883. U.S. Army Corps of Engineers, delineator, 1883. P&P,HAER,MINN,31-DERIV.V,1-10.

3-033. Leech Lake Reservoir Dam, Federal Dam Vic., Cass County, Minnesota, 1882–1884. U.S. Army Corps of Engineers, 1884. P&P,HAER,MINN,11-FEDAM.V,1, no. 7.

3-034. Sandy Lake Reservoir Dam & Lock, McGregor Vic., Aitkin County, Minnesota, 1893. U.S. Army Corps of Engineers, 1895. P&P,HAER,MINN,1-MCGRE.V,1, no. 7.

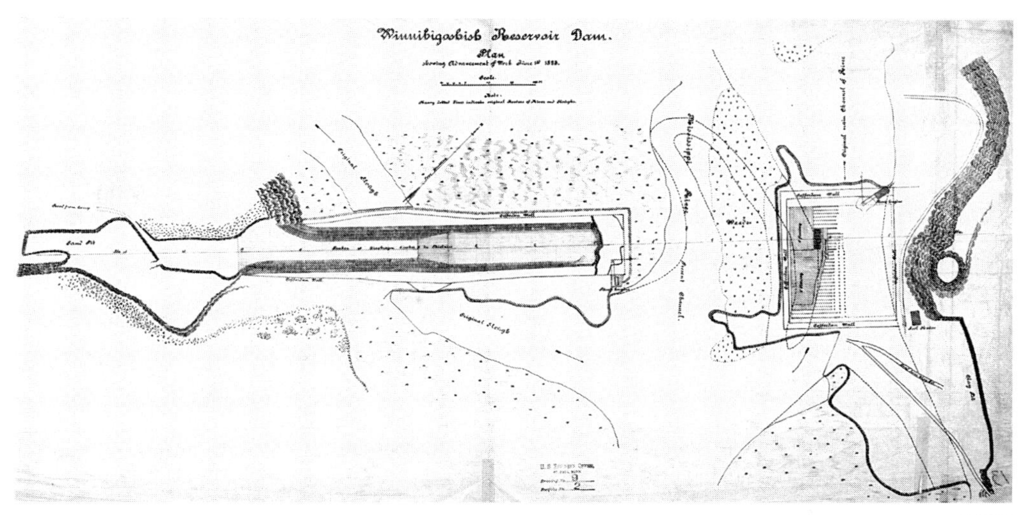

3-032

3-033

3-034

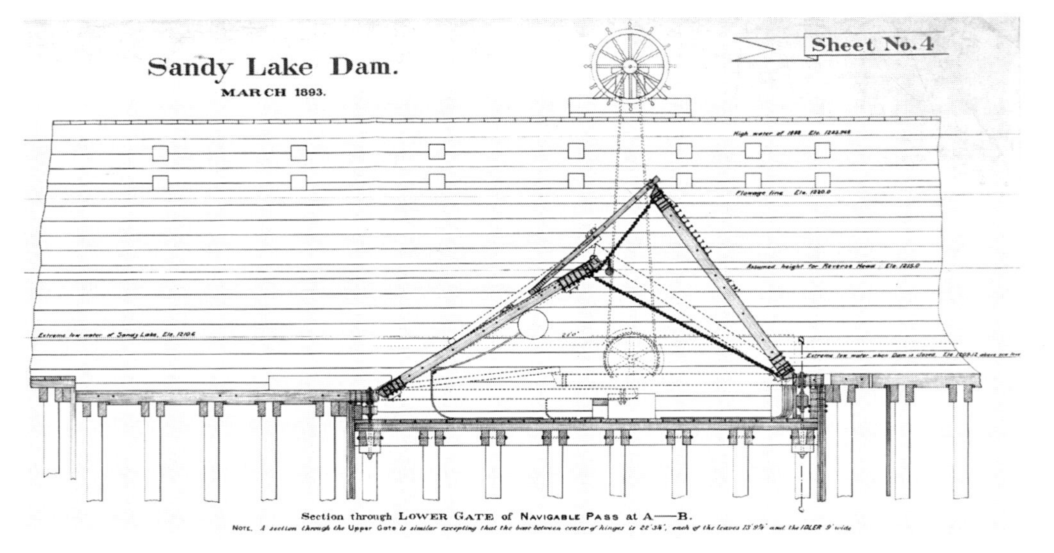

3-035

3-036

3-037

3-035. Section of gate at navigable pass, Sandy Lake Reservoir Dam & Lock, McGregor Vic., Aitkin County, Minnesota, 1893. U.S. Army Corps of Engineers, delineator, 1893. P&P,HAER,MINN,1-MCGRE.V,1-17,detail.

3-036. Reconstruction under way, Leech Lake Reservoir Dam, Federal Dam Vic., Cass County, Minnesota, 1882–1884, reconstruction 1900–1903. U.S. Army Corps of Engineers, 1901. P&P,HAER,MINN,11-FEDAM.V,1, no. 8.

3-037. Reconstructed Leech Lake Reservoir Dam, Federal Dam Vic., Cass County, Minnesota, 1882–1884, reconstruction 1900–1903. U.S. Army Corps of Engineers, 1937. P&P,HAER,MINN,11-FEDAM.V,1, no. 9.

3-038. Pine River Reservoir Dam, Crosslake Vic., Crow Wing County, Minnesota, 1884–1886, reconstruction 1905–1907. Burt Levy, photographer, 1993. P&P,HAER,MINN,18-CROLK.V,1, no. 3.

3-039. Section through gate, Pine River Reservoir Dam, Crosslake Vic., Crow Wing County, Minnesota, 1884–1886, reconstruction 1905–1907. U.S. Army Corps of Engineers, delineator, 1905. P&P,HAER,MINN,10-CROLK.V,1-8,detail.

3-040. Fish ladder, Pine River Reservoir Dam, Crosslake Vic., Crow Wing County, Minnesota, 1884–1886, reconstruction 1905–1907. Burt Levy, photographer, 1993. P&P,HAER,MINN,18-CROLK.V,1, no. 4.

3-041. Section through fish ladder, Pine River Reservoir Dam, Crosslake Vic., Crow Wing County, Minnesota, 1884–1886, reconstruction 1905–1907. U.S. Army Corps of Engineers, delineator, 1911. P&P,HAER,MINN,10-CROLK.V,1-9,detail.

3-038

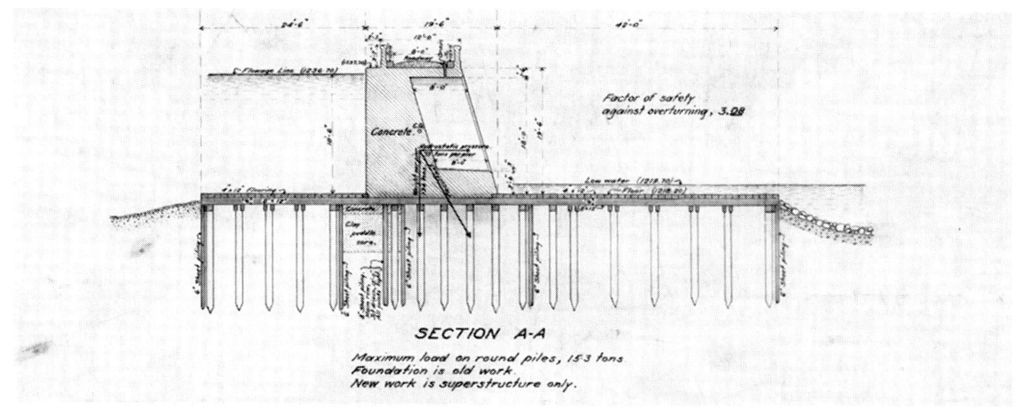

3-039

3-040

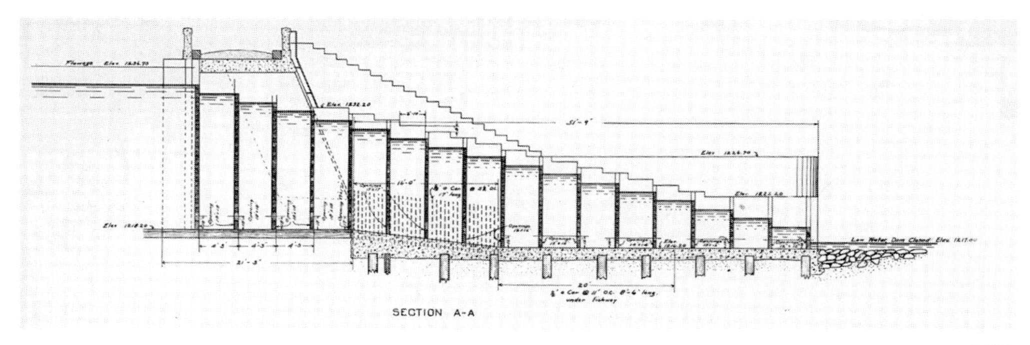

3-041

The U.S. Army Corps of Engineers' Nine-Foot Channel Project between Minneapolis and Saint Louis aimed to modernize and rationalize navigation on the Mississippi. This project added more than twenty multipurpose dams in addition to the three dams that had been built early in the century: the massive Keokuk Dam over the Des Moines Rapids in Iowa (1914), and two dams in the Minneapolis–Saint Paul region (1899–1917).

Modeled after the Corps' successful canalization of the Ohio River, the Nine-Foot Channel on the Mississippi was an even larger and more demanding undertaking. The river presented many unusual conditions, and the Corps closely studied its hydrology, using large-scale models to mimic the river's behavior. The engineers experimented with dam designs, developing and refining the movable dam, and improving hydraulic systems for locks.

Although the project's primary goal was to improve commerce on the river, because it was constructed during the Great Depression it also became one of the country's largest public works projects, occurring at over twenty sites up and down the river simultaneously. On its completion, the Nine-Foot Channel Project turned the upper reaches of the Mississippi River into an intracontinental canal. As a result, commercial freight traffic on the Mississippi rose from 30 million tons in 1940 to almost 400 million in 1984.

3-042

3-042. Twin Cities Lock and Dam No. 1, Mississippi River, Saint Paul, Ramsey County, Minnesota, 1899–1917, modified 1932. U.S. Army Corps of Engineers, 1936. P&P,HAER,MINN,62-SAIPA,33, no. 17.

This unique dam is an Ambursen-type concrete overflow dam with eight sluiceways and an inflatable flashboard system to raise water levels serving the adjoining hydroelectric plant owned and operated by Ford Motor Company. Rivalry between the cities of Minneapolis and Saint Paul caused Congress to authorize two locks and dams for this section of the Mississippi River. One of these was later dismantled because of changing public attitudes toward natural resources.

3-043. Dam collapse during construction, Twin Cities Lock and Dam No. 1, Mississippi River, Saint Paul, Ramsey County, Minnesota, 1899–1917. U.S. Army Corps of Engineers, 1916. P&P,HAER,MINN,62-SAIPA,33, no. 24.

3-044. Aerial view of Keokuk Lock and Dam No. 19, Mississippi River, Keokuk, Lee County, Iowa, canal 1867–1877, dry dock 1883–1889, dam and hydroelectric plant 1910–1914, lock rebuilt 1952–1957. U.S. Army Corps of Engineers, 1982. P&P,HAER,IOWA,56-KEOK,3, no. 54.

3-043

3-044

In 1866, the U.S. Army Corps of Engineers began its first major project on the Upper Mississippi: the construction of a canal and three locks to bypass the 24-foot fall of the Des Moines Rapids. In 1883–1889, the Corps added a dry dock. At the turn of the century, the Keokuk and Hamilton Water Power Company asked Congress for permission to build a hydroelectric dam at the site, with a large navigation lock that would be turned over to the Corps. When completed, the mile-long Keokuk Dam and Power Plant was an extraordinary accomplishment in providing large-scale, low-head hydroelectric power. It set a number of national and world records—the longest concrete dam, the largest hydropower plant (at 125 megawatts), one of two sites in North America generating 25-cycle power, the largest privately funded construction project. In short, it was big and newsworthy. Keokuk's designer, Hugh L. Cooper (1865–1937), went on to become the consulting engineer for Wilson Dam on the Tennessee River (see 2-088–2-084) and Dneprostroi Dam in Ukraine. Keokuk is a run-of-river plant, with no water storage. Its first major customer was Union Electric Company of Saint Louis, 150 miles away. Today the plant provides electricity to over 2.5 million people in Missouri, Illinois, and Iowa.

3-045. Keokuk Lock and Dam No. 19 under construction, Mississippi River, Keokuk, Lee County, Iowa, 1910–1914. Anschutz, 1911. P&P,LC-DIG-ggbain-50127.

3-045

3-046. Placing concrete, Keokuk Lock and Dam No. 19, Mississippi River, Keokuk, Lee County, Iowa, 1910–1914. Anschutz, 1911. P&P,LC-DIG-ggbain-50129.

3-047. Completed powerhouse, Keokuk Lock and Dam No. 19, Mississippi River, Keokuk, Lee County, Iowa, 1910–1914. WC, ca. 1914. P&P,LC-dig-ppmsca-17352.

3-048. Generator hall, Keokuk Lock and Dam No. 19, Mississippi River, Keokuk, Lee County, Iowa, 1910–1914. Peter A. Rathbun, photographer, 1987. P&P,HAER,IOWA,56-KEOK,3, no. 11.

3-046

3-047

3-048

3-049

3-050

3-051

3-049. Powerhouse at night, Keokuk Lock and Dam No. 19, Mississippi River, Keokuk, Lee County, Iowa, 1910–1914. 1913. P&P,LC-USZ62-59720.

3-050. Lock and Dam No. 2, Mississippi River, between Hastings, Dakota County, Minnesota, and Pierce County, Wisconsin, 1928–1930. U.S. Army Corps of Engineers, 1937. P&P,LC-dig-fsa-8e01388.

3-051. Lock and Dam No. 4, Mississippi River, between Alma, Wisconsin, and Wabasha County, Minnesota, 1930–1935. U.S. Army Corps of Engineers, 1935. P&P, LC-dig-fsa-8e01389.

3-052

3-052. Lock and Dam No. 5, Mississippi River, Winona Vic., between Winona County, Minnesota, and Buffalo County, Wisconsin, 1930–1935. U.S. Army Corps of Engineers, 1939. P&P,LC-dig-fsa-8e01387.

3-053. Lock and Dam No. 5a, Mississippi River, Winona Vic., between Winona County, Minnesota, and Buffalo County, Wisconsin, 1932–1936. AP, 1936. P&P,LC-dig-ppmsca-17291.

3-053

3-054

3-055

3-054. Lock and Dam No. 6, Mississippi River, Trempealeau Vic., between Trempealeau County, Wisconsin, and Winona County, Minnesota, 1933–1938. U.S. Army Corps of Engineers, 1936. P&P,HAER,WIS,61-TREM.V,1, no. 67.

3-055. Lock and Dam No. 8, Mississippi River, between Genoa, Vernon County, Wisconsin, and Houston County, Minnesota, 1933–1938. U.S. Army Corps of Engineers, 1937. P&P,HAER,WIS,62-GEN.V,1, no. 73.

3-056

3-057

3-056. Lock and Dam No. 15 (Rock Island Rapids) under construction, Mississippi River, between Davenport, Iowa, and Rock Island County, Illinois, 1931–1934. U.S. Army Corps of Engineers, 1934. P&P,LC-dig-ppmsca-17351.

3-057. Lock and Dam No. 26 under construction, showing cofferdam, Mississippi River, between Alton, Madison County, Illinois, and Saint Charles County, Missouri, 1934–1938. H. Aylette Meade, photographer, 1936. P&P,HAER,ILL,60-ALT,3, no. 56.

3-058

3-058. Site plan showing channels and foundation piers, Lock and Dam No. 10, Mississippi River, Gutenburg Vic., between Clayton County, Iowa, and Grant County, Wisconsin, 1934–1937. U.S. Army Corps of Engineers, delineator, 1934. P&P,HAER,IOWA,22-GUTBU,1-81.

3-059. Steel sheet piling in spillway, Lock and Dam No. 16, Mississippi River, Muscatine Vic., between Muscatine County, Iowa, and Rock Island County, Illinois, 1933–1937. U.S. Army Corps of Engineers, 1935. P&P,HAER,IOWA,70-MUSCA.V,1, no. 9.

3-059

3-060

3-060. Placing concrete with hand-held buggies, Lock and Dam No. 26, Mississippi River, between Alton, Madison County, Illinois, and Saint Charles County, Missouri, 1934–1938. U.S. Army Corps of Engineers, 1935. P&P,HAER,ILL,60-ALT,3, no. 49.

3-061. C-Lift concrete formwork, Lock and Dam No. 27, Mississippi River, between Granite City, Madison County, Illinois, and Saint Louis County, Missouri, 1947–1964. U.S. Army Corps of Engineers, 1948. P&P,HAER,ILL,60-GRACI,2, no. 56.

3-061

Construction of the Dams

The Corps began its work on the Nine-Foot Channel at Dam No. 15 (3-056), located at Rock Island Rapids in Iowa, the most serious obstacle to navigation remaining in the 1930s. In designing the dams, the Corps had to consider some special characteristics of the Mississippi. It was too shallow to use submersible gates of the kind the Corps had used on the Ohio. Its ferocious floods suggested that any gates had to be able to raise entirely out of the water. Gates also had to be strong enough to resist ice floes and jams, and allow passage for fish, silt, and debris. By incorporating two types of gates in its Mississippi dams, the Corps hoped to adapt to the river's many requirements.

The roller gate is a hollow cylindrical gate that can be mechanically raised or lowered on tracks. A Swedish invention, it operates well under icy conditions—its large spans (over 80 feet) between the supporting piers allow for the passage of ice floes, and its curved form offers little resistance to water (or debris) passing underneath. This was the Corps' choice for Lock and Dam No. 15, located on a narrow rapids subject to ice and debris jams. With eleven 100-foot-long roller gates, it remains the largest roller dam in the United States.

The tainter gate is an American development of an earlier radial gate design. Hinged on each side to piers, it pivots up or down, presenting a convex face on its upstream side.

3-062. Plan and downstream elevation of Lock and Dam No. 9, Mississippi River, Lynxville Vic., between Crawford County, Wisconsin, and Allamakee County, Iowa, 1936–1938. U.S. Army Corps of Engineers, delineator, 1936. P&P,HAER,WIS,12-LYNX.V,1-75,detail.

3-063. Section A-A through roller gate, Lock and Dam No. 9, Mississippi River, Lynxville Vic., between Crawford County, Wisconsin, and Allamakee County, Iowa, 1936–1938. U.S. Army Corps of Engineers, delineator, 1936. P&P,HAER,WIS,12-LYNX.V,1-75,detail.

3-064. Section B-B through tainter gate, Lock and Dam No. 9, Mississippi River, Lynxville Vic., between Crawford County, Wisconsin, and Allamakee County, Iowa, 1936–1938. U.S. Army Corps of Engineers, delineator, 1936. P&P,HAER,WIS,12-LYNX.V,1-75,detail.

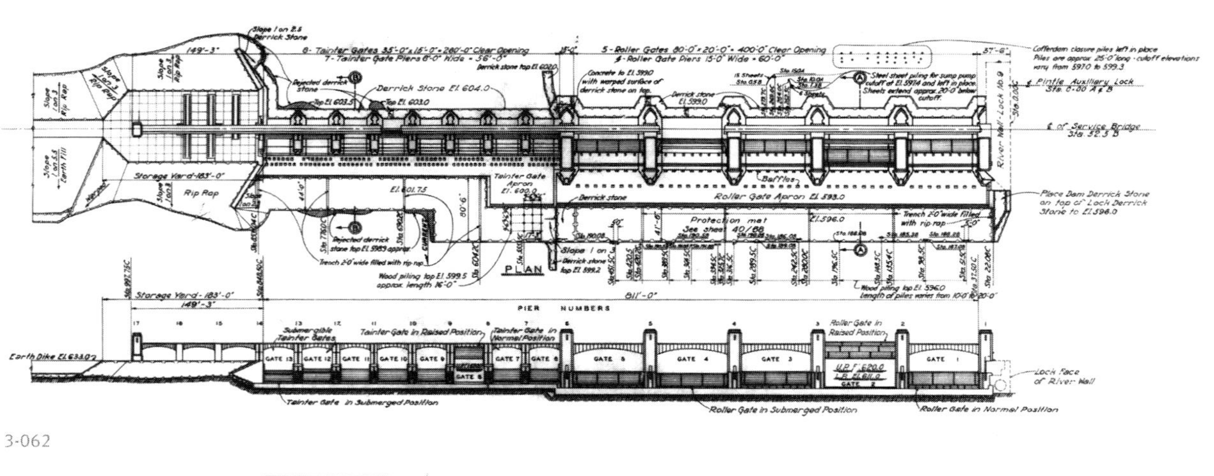

3-062

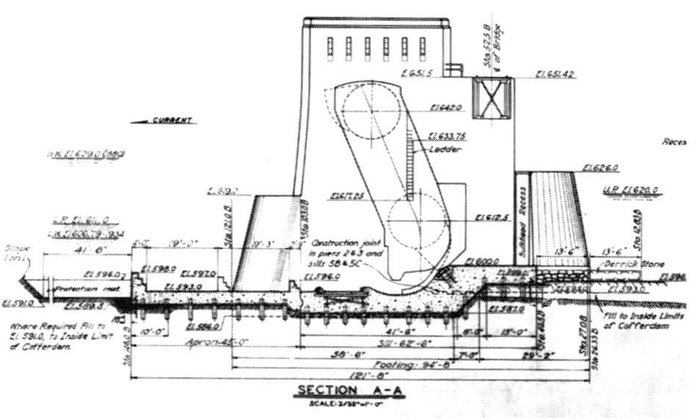

3-063

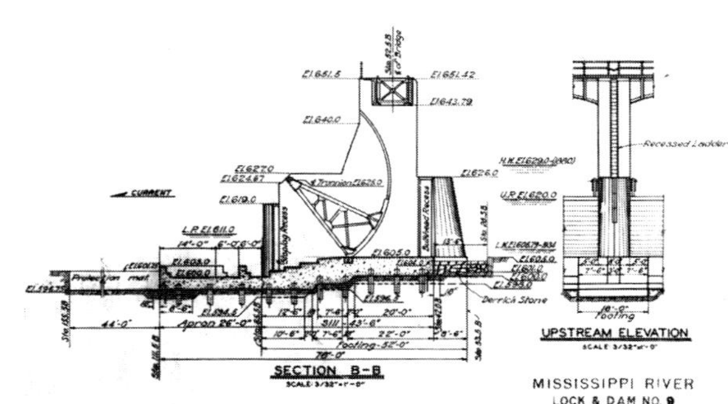

3-064

When it is hoisted, rushing water helps to raise it, while it closes under its own weight. Although the Corps had tried it out in 1889, in the Nine-Foot Channel Project the Corps significantly modified and refined it, eventually choosing it as the preferred gate in most of the dams. The Corps engineers structurally stiffened these gates, experimented with their degree of curvature, and developed submersible versions, creating tainter gates of an unprecedented size and strength.

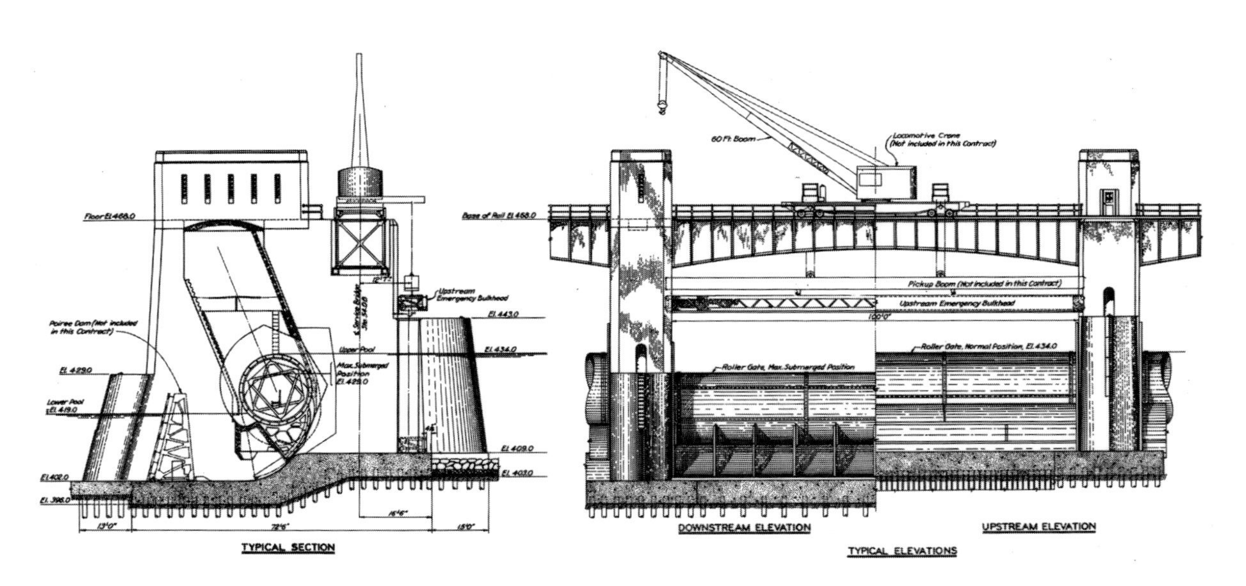

3-065

3-065. Section and elevation of roller gate, Lock and Dam No. 25, Mississippi River, between Cap-au-Gris, Lincoln County, Missouri, and Calhoun County, Illinois, 1935–1939. U.S. Army Corps of Engineers, delineator, 1937. P&P,HAER,MO,57-CAG,1-72.

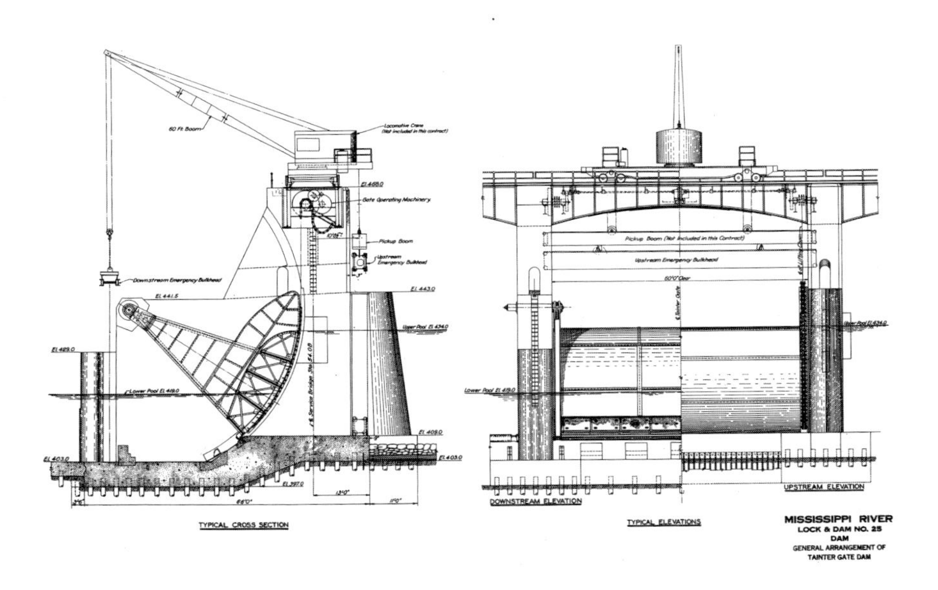

3-066

3-066. Section and elevation of tainter gate, Lock and Dam No. 25, Mississippi River, between Cap-au-Gris, Lincoln County, Missouri, and Calhoun County, Illinois, 1935–1939. U.S. Army Corps of Engineers, delineator, 1937. P&P,HAER,MO,57-CAG,1-73.

3-067

3-068

3-067. Roller gates under construction, Lock and Dam No. 25, Mississippi River, between Cap-au-Gris, Lincoln County, Missouri, and Calhoun County, Illinois, 1935–1939. U.S. Army Corps of Engineers, 1938. P&P,HAER,MO,57-CAG,1, no. 59.

3-068. Tainter gates under construction, Lock and Dam No. 25, Mississippi River, between Cap-au-Gris, Lincoln County, Missouri, and Calhoun County, Illinois, 1935–1939. U.S. Army Corps of Engineers, 1937. P&P,HAER,MO,57-CAG,1, no. 54.

3-069

3-069. Head house with roller gate lifting machinery, Lock and Dam No. 9, Mississippi River, Lynxville Vic., between Crawford County, Wisconsin, and Allamakee County, Iowa, 1936–1938. Clayton B. Fraser, photographer, 1987. P&P,HAER,WIS,12-LYNX.V,1, no. 12.

3-070. Roller gate, showing operating house atop pier, Lock and Dam No. 25, Mississippi River, between Cap-au-Gris, Lincoln County, Missouri, and Calhoun County, Illinois, 1935–1939. John P. Herr, photographer, 1988. P&P,HAER,MO,57-CAG,1, no. 8.

3-070

3-071

3-073

3-072

3-071. Inside headhouse, looking down at roller gate chain, Lock and Dam No. 10, Mississippi River, Gutenburg Vic., between Clayton County, Iowa, and Grant County, Wisconsin, 1934–1937. Clayton B. Fraser, photographer, 1987. P&P,HAER,IOWA,22-GUTBU,1, no. 16.

3-072. Early version of tainter gate, Lock and Dam No. 8, Mississippi River, between Genoa, Vernon County, Wisconsin, and Houston County, Minnesota, 1933–1938. John P. Herr, photographer, 1988. P&P,HAER,WIS,62-GEN.V,1, no. 15.

3-073. Detail view of tainter gauges, Lock and Dam No. 25, Mississippi River, between Cap-au-Gris, Lincoln County, Missouri, and Calhoun County, Illinois, 1935–1939. John P. Herr, photographer, 1988. P&P,HAER,MO, 57-CAG,1, no. 7.

While flooding was a perennial feature of life in the Mississippi River valley, the flood of 1927 was unique in its scale and the public outcry that followed. The river breached its levees in hundreds of places, flooding 27,000 square miles up to depths of 30 feet. Crops were destroyed, industries and transportation were paralyzed, hundreds of lives were lost, and more than 600,000 people were displaced. The following year, Congress passed the Flood Control Act of 1928. This represented a departure from the levee-only policy that had characterized work on the river up to this time. It led to the construction of twenty three flood control and navigation dams on the Upper Mississippi and a number of floodways on the lower reaches of the river, as well as levee construction, channel stabilization, and tributary basin works such as dams, pumping plants, and auxiliary channels.

By the 1960s, the Army Corps had completed the major elements of its plans to prevent river flooding, and it turned its attention to hurricane protection for the coastal delta. Dogged by conflicting demands for environmental conservation and hurricane protection, implementation of the plan was slow and piecemeal.[2] The shortcomings of the system became tragically evident in the disastrous Hurricane Katrina of 2005. In this storm, the coastal defenses of New Orleans were breached in numerous places, flooding 80 percent of the city and killing over a thousand people in what became the costliest natural disaster in the country's history.

3-074

3-074. Flood on the Katy No.1, Mississippi River, 1904. Gannaway, 1904. P&P,LC-dig-cph-3a21072.

3-075

3-076

3-075. Arkansas City in flood, Desha County, Arkansas, 1927. U.S. Army Air Corps, 1927. P&P,LCUSZ62-129240.

3-076. "Flivvers rode at anchor in this imitation Venice," Greenville, Washington County, Mississippi, 1927. Unidentified photographer, 1927. P&P,LC-USZ62-129263.

Main-Stem Levee System

The main-stem levee system, begun after the Great Flood of 1927, follows the main stem of the Mississippi River for 1,000 miles and extends up the Arkansas and Red rivers and down the Atchafalaya basin an additional 600 miles. Built by the federal government, the levees are maintained locally, except in major floods.

3-077. Placing earth to control seepage through the levee, between Baton Rouge and New Orleans, Mississippi River, 1927. Unidentified photographer, 1927. P&P,LC-USZ62-75856.

3-078. Levee protected from wave wash, Mississippi River, Baton Rouge, Louisiana, 1927. Ewing Incorporated, 1927. P&P,LC-USZ62-129496.

3-077

3-078

3-079

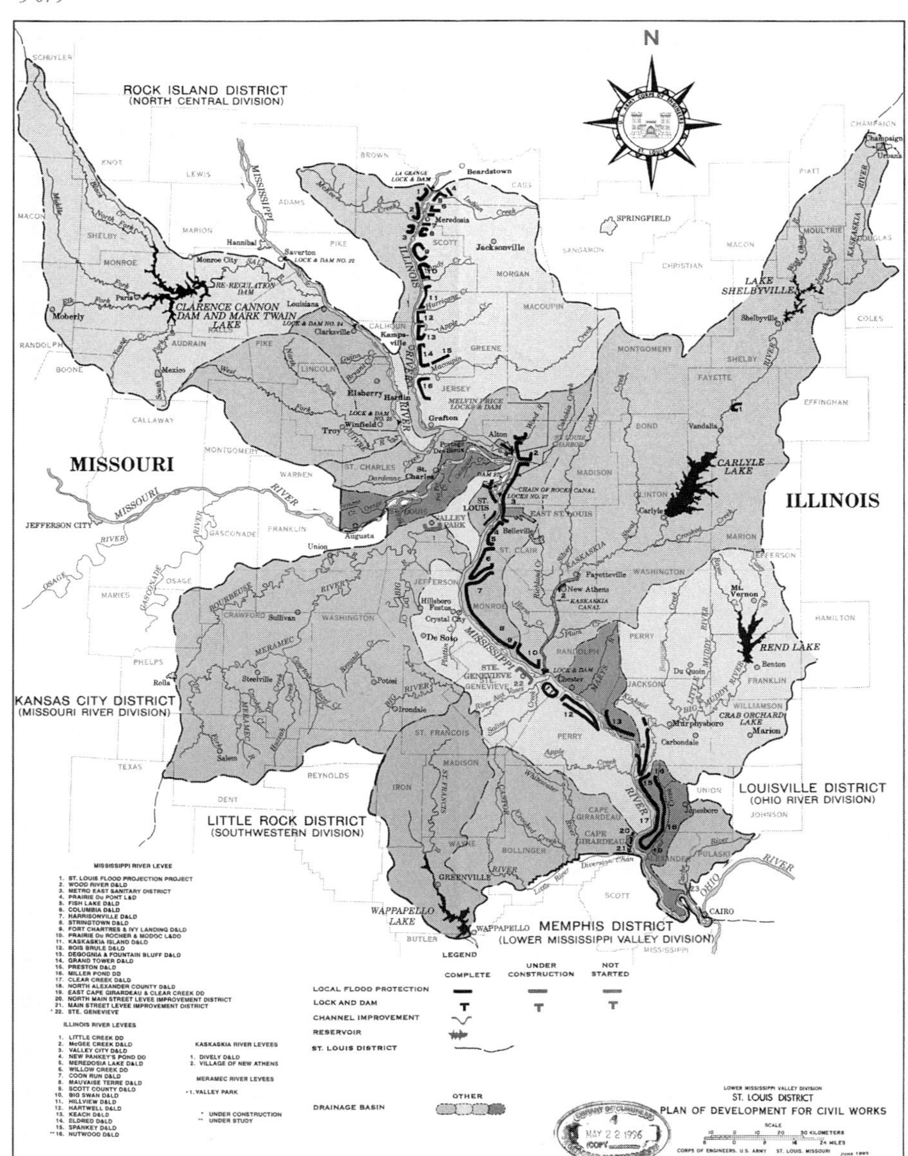

3-080

3-079. Sandbagging the levee, Mississippi River, Cairo, Alexander County, Illinois, 1937. Unidentified photographer, 1937. P&P,LC-dig-cph-3a16705.

3-080. Plan of development for civil works, Lower Mississippi River valley. U.S. Army Corps of Engineers, delineator, 1995. G&M,G4042.M5N22 1995.U5.

Floodways

The U.S. Army Corps of Engineers constructed a number of floodways to divert the flow of the Mississippi River at critical points on its route, such as its confluence with the Ohio River (Cairo–New Madrid Floodway) and the Red River (Atchafalaya and Morganza Floodways), and above the city of New Orleans (Bonnet Carré Spillway).

3-081

3-081. Levee dynamited to save Cairo, Illinois, 1936. Office of War Information/AP, 1936. P&P,LC-dig-fsa-8e03735.

3-082

3-082. Bonnet Carré Spillway, Mississippi River, Saint Charles Parish, Louisiana, 1929–1931. Acme, 1936. P&P,LC-dig-ppmsca-17235.

The Bonnet Carré Spillway protects New Orleans from Mississippi floods by diverting water into Lake Pontchartrain and the Gulf of Mexico. First used during the flood of 1937, this floodway has been opened several times since then to lower river levels at New Orleans.

3-083

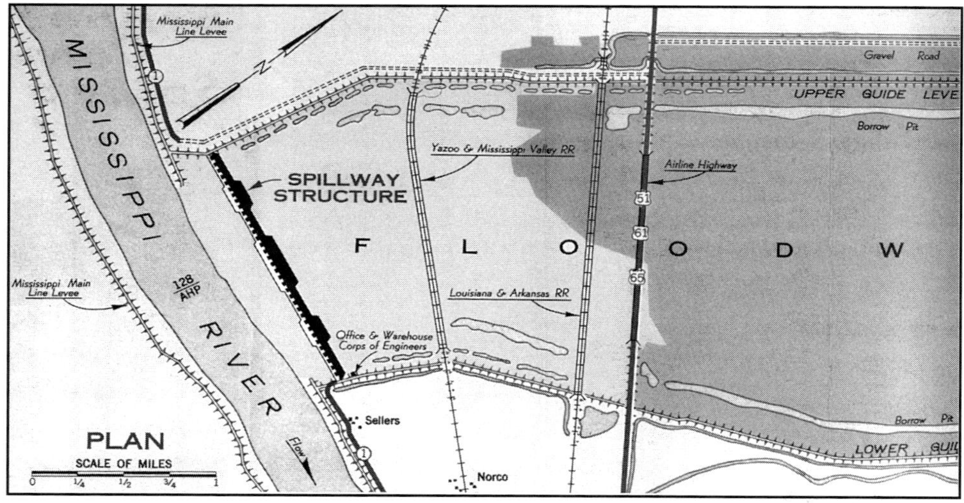

3-084

3-083. Opening the Bonnet Carré Spillway for a 20-foot flood, Mississippi River, Saint Charles Parish, Louisiana, 1929–1936. Acme, 1950. P&P,LC-dig-ppmsca-17236.

3-084. Plan, Bonnet Carré Spillway, Mississippi River, Saint Charles Parish, Louisiana, 1929–1931, 1936, 1950. U.S. Army Corps of Engineers, delineator, 1950. G&M,G1361.S1C4 2003,p. 280.

3-085. Dry season, Bonnet Carré Spillway, Mississippi River, Saint Charles Parish, Louisiana, 1929–1931. U.S. Army Corps of Engineers. G&M,G4012.B675 1987.U5, detail.

3-086. Opening for flood of 1983, Bonnet Carré Spillway, Mississippi River, Saint Charles Parish, Louisiana, 1929–1931. U.S. Army Corps of Engineers, 1983. G&M, G4012.B675 1987.U5,detail.

3-085

3-086

Another significant result of the flood of 1927 was a shift in the amount of water drained by the Atchafalaya. This river had captured the flow of the Red River in 1940 and was becoming one of the five largest rivers in the country. By 1945, a third of the Mississippi's flow ran through the Atchafalaya River basin. Because the Atchafalaya has a shorter and steeper route to the Gulf than does the Mississippi, it became clear that the Mississippi was going to divert its course entirely, threatening the economic viability of New Orleans and Baton Rouge as well as the many industries that line the Mississippi's riverbanks. In 1950, the Corps of Engineers began to undertake the process of halting this diversion, guaranteeing that two-thirds of the Mississippi's flow would continue to follow the old riverbed—essentially, as John McPhee has said in *The Control of Nature* (1989), preserving the flow of the year 1950 in perpetuity. The Old River Control Structure allows a portion of the Mississippi's flow to run into the Atchafalaya River, while a lock and dam permit navigation between the two waterways. Three other floodways were built in the Atchafalaya River basin as well.

3-087

3-087. Old River Control Structure, Mississippi River, Pointe Coupee Parish, Louisiana, 1955–1963. U.S. Army Corps of Engineers, 1990. G&M,G1361.S1C4 2003,p. 249.

ANOTHER U.S. ARMY CORPS OF ENGINEERS DAM

The Ouachita River of Arkansas flows southeast through a cotton-producing region to join the Red River after a course of 600 miles. Below its confluence with the Tensas River in Louisiana, it is called the Black River. Because the river served as a navigation route from the late eighteenth century, the U.S. Army Corps of Engineers built four locks and dams on the Ouachita between Camden, Arkansas, and Jonesville, Louisiana. Dam No. 8 is a good remaining example of the movable wicket–type construction that the Corps employed on the Ohio River. In addition to the lock, the dam has several sections, each fitted with chanoine wickets that can be raised or lowered to adjust water levels. These sections are a navigable pass, an overflow weir, and a drift pass.

3-088. Construction of Ouachita River Lock and Dam No. 8, Ouachita River, Calion, Union County, Arkansas, 1907–1916. U.S. Army Corps of Engineers, 1911. P&P,HAER,ARK,7-CAL.V,1, no. 23.

3-088

3-089

3-090

3-089. Construction of aprons through drift and weir passes, Ouachita River Lock and Dam No. 8, Ouachita River, Calion, Union County, Arkansas, 1907–1916. U.S. Army Corps of Engineers, 1911. P&P,HAER,ARK,7-CAL.V,1, no. 26.

3-090. Drift pass section, Ouachita River Lock and Dam No. 8, Ouachita River, Calion, Union County, Arkansas, 1907–1916. Howard Naylor, photographer, 1983. P&P,HAER,ARK,7-CAL.V,1, no. 4.

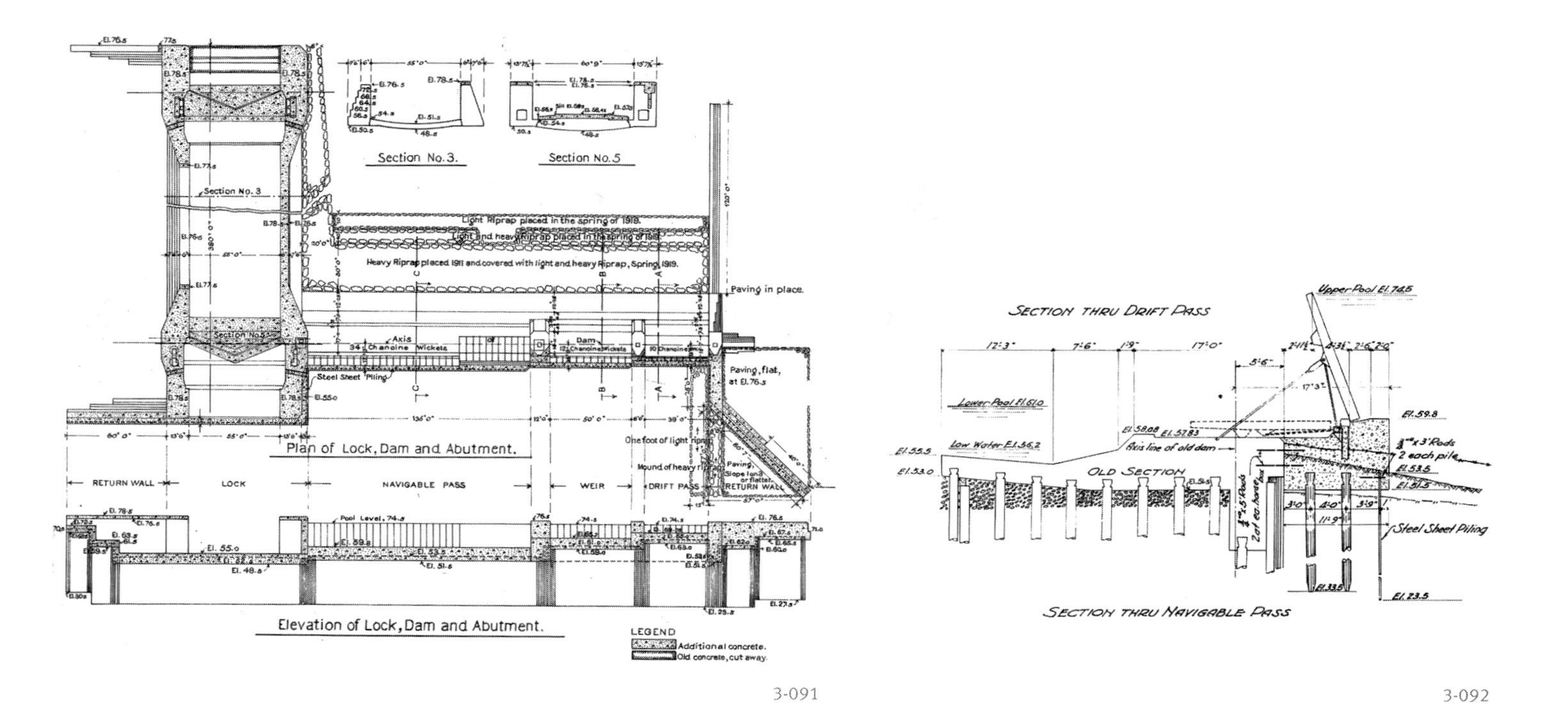

3-091

3-092

3-091. Plan and elevation of Ouachita River Lock and Dam No. 8, Ouachita River, Calion, Union County, Arkansas, 1907– 1916. U.S. Army Corps of Engineers, delineator, 1921. P&P,HAER,ARK,7-CAL,V,1-33,detail.

3-092. Section through drift pass section showing chanoine wickets, Ouachita River Lock and Dam No. 8, Ouachita River, Calion, Union County, Arkansas, 1907–1916. U.S. Army Corps of Engineers, delineator, 1921. P&P,HAER,ARK,7-CAL,V,1-40,detail.

3-093. Detail of chanoine wicket, Ouachita River Lock and Dam No. 8, Ouachita River, Calion, Union County, Arkansas, 1907–1916. U.S. Army Corps of Engineers, delineator, 1921. P&P,HAER,ARK,7-CAL.V,1-42,detail.

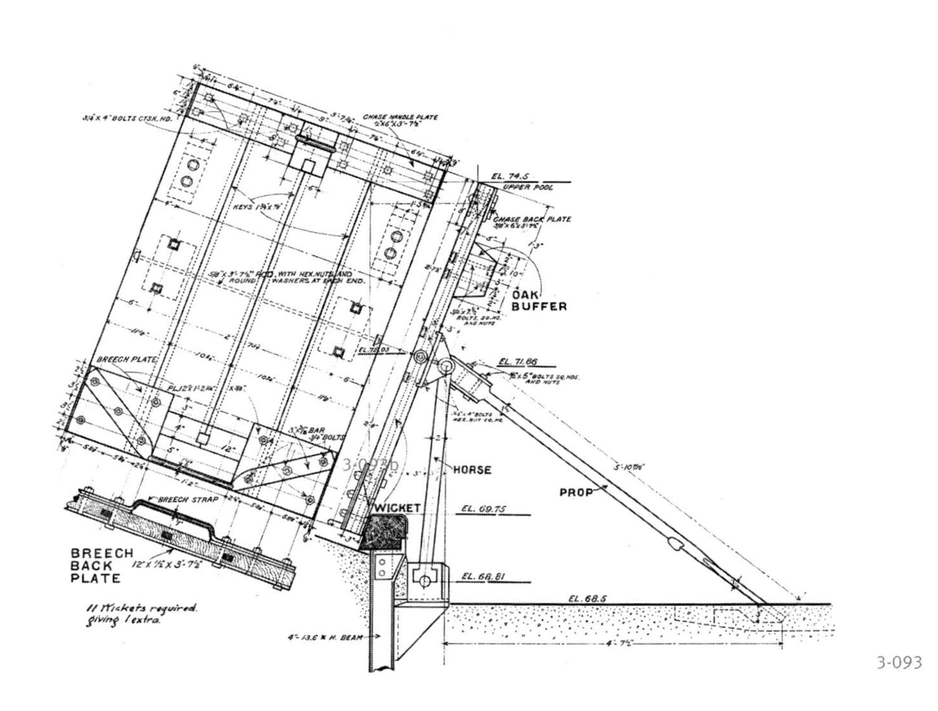

3-093

SECTION 3 NOTES

1. Albert E. Cowdrey, *Land's End: A History of The New Orleans District, U.S. Army Corps of Engineers, and Its Life Long Battle with the Lower Mississippi and Other Rivers Wending Their Way to the Sea* (New Orleans, LA: U.S. Army Corps of Engineers, 1977), 32.

2. Damon Manders, "Weathering the Storm: New Orleans District and Hurricane Protection Projects," in *The Bayou Builders: A History of the U.S. Army Corps of Engineers, New Orleans district, 1976–2000*, draft, Alabama A&M University Research Institute, 2002. www.hq.usace.army.mil/history/Hurricane_files/ NOUpdate4.pdf

MISSOURI

MISSOURI RIVER BASIN

Jefferson, Madison, Gallatin, Saskatchewan, Osage, Kansas, Smoky Hill, Republican, Platte, Cheyenne, Belle Fourche, Yellowstone, Bighorn, Tongue, and Powder Rivers

As settlers moved westward, one of the most notable characteristics of the landscape they encountered was its aridity. Because the natural precipitation on the Great Plains was insufficient to sustain agriculture, new settlers frequently found themselves in court arguing for access to rivers and streams. Each state had different laws regarding water appropriation. In Montana, where landowners simply had to post a notice to assert water rights, over-appropriation of water led to frequent litigation. Colorado's solution to the problem was a "first in time, first in right" system of water allocation: the person who first claimed and used a water source had a permanent right to the amount of water he or she used. This Doctrine of Prior Appropriation (of 1876) further distinguishes between senior rights (those with first claim to the water) and junior rights (those entitled to unused surplus) and establishes criteria to define the "beneficial" use of water, with first priority going to household uses, followed by agriculture, and last,

4-001. Opening day, Dam 96, Upper Souris National Wildlife Refuge Dams, Upper Souris River, Foxholm Vic., Ward County, North Dakota, 1935–1936. Unidentified photographer, 1936. P&P,HAER,ND,51-FOX,V1-C, no. 12.

manufacturing. These rights are private property that can be sold, rented, and inherited. Water rights played a key role in western settlement, and governmental regulation and development of water were hugely influential in creating the American West.

The West is full of paradoxes—one the one hand, there is the ideal of the lonely rancher, cowboy, or trapper, and on the other, the enormous influence of the federal government and industrial capitalism. The upper reaches of the Missouri watershed were centers for mining. Hydraulic mining, in which water displaces massive quantities of earth, involved flume and channel building, settling basins, soil liquification, and stabilization techniques—all of which were brought directly into dam making. Mining concerns also provided markets for hydroelectricity, since the smelters and production plants of large operations like Anaconda required enormous amounts of power, often in remote locations where the importation of coal was prohibitively expensive. Irrigation is another theme that characterizes this landscape, as surface water and eventually ground water irrigation promised prosperity and growth. Water battles were to color much of the western landscape: in the Great Plains, the Texas high plains, the Rocky Mountains, the Great Basin, and California.

4-002

4-002. Mandan man on bluff overlooking the Missouri River. Edward S. Curtis, photographer, 1908. P&P,LC-USZ62-46989.

At 2,565 miles, the Missouri is the longest river in the United States. When it merges with the Mississippi at Saint Louis, the Missouri carries more water. Its headwaters—the Jefferson, Madison, and Gallatin rivers—originate in the Rocky Mountains and meet at Three Forks, Montana. At Great Falls, the river enters a 10-mile stretch of cataracts that prevented navigation to the upper river, so Fort Benton, Montana, became the head of navigation for nineteenth-century riverboats. Below Fort Benton, the Missouri follows a meandering course east across the prairie before entering the Mississippi River near Saint Louis. Until very recently, the Missouri was a sprawling river nicknamed "Big Muddy" for its heavy load of silt. Periodically, it would flood, drenching farms and towns along its banks, displacing thousands, and depositing fresh topsoil along its banks.

Today large multipurpose dams regulate the Missouri's flow by capturing floodwaters for irrigation, flood control, and power generation. Since the dams have no locks, Sioux City is the head of navigation for a 9-foot channel maintained over the 760-mile stretch downstream to the Mississippi. The damming of the river has greatly reduced wildlife and wetlands.

4-003. Mill dam at Walhalla, Pembina River, a tributary of the Red River of the North, North Dakota. WC. P&P,LC-dig-ppmsca-17366.

4-003

THE SOURIS, RED, AND RAINY RIVERS

The Souris, Red, and Rainy rivers flow northward into Canada, feeding Lake Winnipeg and ultimately the Hudson Bay system. These rivers lie within the bed of the huge prehistoric glacial Lake Agassiz, the remnants of which form the great lakes of Manitoba—Lakes Winnipeg, Winnipegosis, and Manitoba—and the Lake of the Woods between Minnesota and Ontario. French place names throughout this region remind us that these rivers were traveled by the *coureurs du bois*, early fur traders who followed the rivers and lakes of the continent westward. These waters are also on the great avian migration routes north and south; in fact, the 1930s-era Civilian Conservation Corps dams on the Souris River and its tributary, the Des Lacs, created one of the first constructed wetlands for wildlife conservation in the United States. The Red River, named after the reddish-brown silt it carries, was an important transportation link between Lake Winnipeg and the Mississippi River system. Its fertile valley produces cereals, potatoes, and sugar beets and supports cattle raising.

4-004

4-004. The Red River flooding Winnipeg, Manitoba. Wide World, 1950. P&P,LC-dig-ppmsca-17367.

Early efforts in water supply on the eastern slopes of the Rocky Mountains were directed toward irrigating dry lands for agriculture. These were not simple projects; in many cases, they involved diverting rivers from one watershed to another. Such agricultural waterworks rapidly became intertwined with municipal water supply. Examples include the 24-mile-long City Ditch, built to deliver South Platte River water to Denver (1867), the water diversion projects constructed by the Union Colony at Greeley (1870), and the Larimer and Weld Canal, also known as the Eaton Ditch (1876). With Colorado's accession to

4-005

4-005. Fort Morgan Canal System, South Platte River, Fort Morgan, Morgan County, Colorado, 1883–1884. Louis C. McClure, photographer, ca. 1910. P&P,LC-dig-ppmsca-17314.

Fort Morgan's canal system was developed by Abner S. Baker, an entrepreneur and developer who acquired lands at the site of the former Fort Morgan, with the aim of establishing a new town. The irrigation system was essential to promoting settlement. Diversion dams on the South Platte River fed the canals.

4-006

4-006. Fort Morgan Canal System, South Platte River, Fort Morgan, Morgan County, Colorado, 1883–1884. Louis C. McClure, photographer, ca. 1910. P&P,LC-dig-ppmsca-17315.

statehood, canal building accelerated in projects such as the Rio Grande Canal (1881–1884), the North Poudre Canal (1882), and the High Line Canal, which diverted water from the South Platte River to Denver and other communities (1883). Dams built for these projects range from small diversion dams to massive masonry or earth reservoir dams on the eastern slopes of the Rocky Mountains. This category also includes smaller reservoir dams built to supply the needs of railroad companies or isolated ranches.

4-007

4-007. High Line Canal Diversion Dam, South Platte River to Second Creek, Douglas, Arapahoe, and Adams Counties, Colorado, 1879–1883. Clayton B. Fraser, photographer, 1994. P&P,HAER,COLO 16-DENV,64-1.

The High Line Canal dates from Colorado's most prolific period of canal building. It was built by the "English Company" with the support of the Union Pacific Railroad to supply water to railroad-owned farmland in eastern Colorado. Eighty-four miles long, the High Line Canal was the largest canal ever built in Colorado, longer even than the Bureau of Reclamation's costly cross-mountain diversion projects of fifty years later. The project's headworks originally included a timber diversion dam with sandstone masonry piers and a sluice gate that fed a 540-foot granite tunnel, wooden flumes, and earthen canals. The irregular flows of the South Platte and engineering flaws dogged the success of the canal. After several years, High Line Canal investors built the earthfill Antero Reservoir Dam on the upper reaches of the South Platte River.

4-008

4-008. Flume, High Line Canal, South Platte River to Second Creek, Douglas, Arapahoe, and Adams Counties, Colorado, 1879–1883. Clayton B. Fraser, photographer, 1994. P&P,HAER,COLO 16-DENV,64-3.

4-009. Castlewood Dam, Cherry Creek, Denver Vic., Douglas County, Colorado, 1890, failed 1933. AP, 1933. P&P,LC-dig-ppmsca-17237.

Investors hoping to facilitate agricultural development in the county financed this dam south of Denver. Built on a poor foundation, it collapsed in 1933, killing five people and flooding downtown Denver.

4-010. Bluebird (Billings and Arbuckle No. 2) Dam, Ouzel Creek, Boulder County, Colorado 1902–1904, rebuilt 1923. Arnold Thallheimer, photographer, 1985. P&P, HAER,COLO,7-ALSPA.V,1-1.

This privately built irrigation dam for a ranch is an early-twentieth-century example of an arched concrete dam.

4-009

4-010

4-011

4-011. Aerial view of Three Bears Lake showing Burlington Northern tracks, Glacier National Park, Glacier County, Montana, 1902. George McFarland, photographer, 1991. P&P,HAER,MONT,18-EAGLPA.V,1, no. 3.

4-012

4-012. Three Bears Lake Dam, Glacier National Park, Glacier County, Montana, 1902. George McFarland, photographer, 1991. P&P,HAER,MONT,18-EAGLPA.V,1A, no. 3.

This earthen dam with wood-board facing and wooden pipe system supplied water for steam locomotives of the Great Northern Railroad.

HYDROELECTRIC DAMS BUILT FOR INDUSTRY

The rich mineral resources of the West proved irresistible incentives to the full-scale industrial exploitation of western landscapes and water resources. The industrial development of this region was rapid, and water power was essential not only to power mills, process ores, and supply the domestic needs of settlements, but also to generate hydroelectricity. In Montana, copper was one of the big-ticket items, followed by aluminum and other minerals. Many innovative dam and diversion projects, particularly in the mountainous western reaches of the state, were financed by industrialists used to planning big, taking risks, and trying to stay ahead of the competition.

4-013. Rocky Ford Dam, Big Blue River, Rocky Ford Power Company, Riley County, Manhattan, Kansas, 1866. Unidentified photographer, ca. 1900. WC. P&P,SSF-Dams-Kansas-Rocky Ford.

The steady flow of the Big Blue River led to its sobriquet "the Merrimack of Kansas." This 10-foot-high and 342-foot-long timber-crib dam was built for a flour mill. It was replaced with a 500-foot-long reinforced concrete dam and hydroelectric plant in 1908–1910.

4-013

4-014

4-014. Kearney Canal Hydroelectric Dam, Platte River, Kearney, Buffalo County, Nebraska, 1882–1888. K. O. Holmes, publisher. P&P,LC-dig-ppmsca-17324.

This 24-mile-long canal supplied waterpower and irrigation to the short-lived industrial boom of Kearney, Nebraska. Hydroelectric power was added in 1888, leading to the establishment of a cotton mill and electrification of the city, followed by the national depression five years later and an industrial bust by 1901.The Central Colorado Power Company built Kossler and Barker reservoirs, connected by a 12-mile pipeline, to supply a hydroelectric power plant in Boulder Canyon 17 miles downstream.

Hydroelectric Dams of Great Falls, Montana

The town of Great Falls, Montana, developed around a 10-mile stretch of cataracts where the Missouri cuts through a shift in geologic formation, changing from a meandering river to a fast-moving torrent in a deep gorge. Over this distance, the river drops 500 feet in a series of rapids and five waterfalls—the largest being the Great Falls of the Missouri encountered by Lewis and Clark in 1804. In the 1890s, investors in the region's copper, silver, gold, and coal mining industries began to harness these waterfalls for hydropower. The dam furthest upstream is Black Eagle Falls, followed by Rainbow, Cochrane, Ryan, and Moreny; all are significant for their association with the industrial development of Montana.

4-015

4-015. View of Great Falls, Montana, 1891. American Publishing Company, 1891. G&M,G4254.G7A3 1891. A6.

4-016

4-016. Black Eagle Falls Dam and Hydroelectric Plant, Missouri River, Great Falls, Cascade County, Montana, ca. 1890, rebuilt 1926–1927. American Publishing Company, 1891. G&M,G4254. G7A3 1891. A6,detail.

4-017

4-017. Water wheels in the Boston and Great Falls Electric Light and Power Company's Station, Great Falls, Cascade County, Montana. Unidentified photographer, 1895. P&P,LC-dig-ppmsca-17294.

4-018. Powerhouse, Ryan Hydroelectric Plant, Missouri River, Great Falls Vic., Cascade County, Montana, 1913–1936. Kristi Hager, photographer, 1996. P&P,HAER,MONT,7-GREFA.V,3A, no. 1.

4-018

Hydroelectric Dams Serving Helena and Butte, Montana

By the 1890s, Butte's copper mines were Montana's largest industry, and they had an ever-growing need for water, power, and hydroelectricity. The terrain in Silver Bow County was not ideal for dam building, and both innovation and failure marked the projects. The Montana Power Company's 500-foot-long timber-crib dam on the Big Hole River partially collapsed during the year of its completion (1898). The first Hauser dam (1905–1907), built for the Amalgamated Copper Company of Butte, was a steel dam designed by J. F. Jackson and built by the Wisconsin Bridge and Iron Company, which also built the Redridge Steel Dam in Wisconsin's copper-mining country (see 2-128). In 1908, Hauser Dam failed at its foundations. It was rebuilt in 1909–1912 as a concrete gravity structure. Public concerns about dam safety led to the introduction of a bill in Montana's legislature that would make dam owners liable for economic losses resulting from a dam's collapse. It failed to pass.

Dams were only one part of the complex systems of tunnels and mine shafts, flumes, ditches, pumping stations, diversion projects, hydroelectric plants, and mills of all kinds that required water supply in the mining regions of Montana. As mineshafts bored deeper into the bedrock of the Silver Bow basin, aquifers draining into the mines became polluted

4-019. Nunn Hydroelectric Plant, Madison River, Ennis Vic., Madison County, Montana, 1900–1901, modified 1908–1909. Pete Norbeck, photographer, 1990. P&P,HAER,MONT,29-ENNIS.V,1, no. 2.

Lucius Nunn, an early hydropower developer in the western states, had experience developing hydropower for gold mines in Colorado before he began projects in the Butte region, including the first Madison Dam (1900–1901), which supplied this hydroelectric plant through a wooden diversion dam and a wooden box flume; the Madison-Missouri Hydroelectric Plant (1902); and the second Madison dam (1904).

4-019

from tailings, the residues of industrial ore processing, and residential sewage. Water retrieved from the mines was unusable for drinking or irrigation. Consequently, a pumping station on the Big Hole River was constructed to supply drinking water to Butte (1898). This involved an 840-foot pump on the river to feed reservoirs on both sides of the Continental Divide. It was delivered through redwood flumes to a 15-square-mile area in Butte and its suburbs.

Samuel Hauser, a former governor of Montana territory and a banker for the mining industry, built the first large-scale hydroelectric facility in Montana, Canyon Ferry Hydroelectric Plant (1892–1898), on the Missouri River. Serving Helena and later Butte, it supplied power to both of the rival copper kings, William Clark in Butte and Marcus Daly in Anaconda. Hauser subsequently constructed two more dams on the Missouri, Hauser (1905–1907) and Holter (1908–1910) dams. The generating equipment in these projects—camelback turbines and an oil-pressure governance system to regulate turbine-generator speed—was state of the art at that time. The original Canyon Ferry dam was replaced by a Bureau of Reclamation dam of the same name in 1953–1954.

4-020. Drumlummon mine, Marysville, Lewis and Clark County, Montana, ca. 1890. Unidentified photographer, ca. 1890. HAW/DPL,X-62605.

This image—taken just north of Helena, Montana—shows that trestles, flumes, and earth moving were commonplace elements of the mining landscape in the region.

4-020

4-021. Holter Hydroelectric Plant, Missouri River, Wolf Creek Vic., Lewis and Clark County, Montana, 1908–1910, 1916–1918. Kristi Hager, photographer, 2000. P&P,HAER,MT-94-I, no. 2.

4-022. Cross-section, Hebgen Dam, Madison River, West Yellowstone Vic., Gallatin County, Montana, 1914–1915. P&P,LC-dig-ppmsca-17253.

The Montana Power Company built this rock-fill dam with a concrete core in the upper reaches of the Madison River, on the borders of Yellowstone Park. In 1959, it was damaged during a severe earthquake, but it did not collapse. However, 80 million tons of earth slid into the riverbed downstream, blocking the flow of the Madison River. To provide a reliable outlet for the rising waters and to prevent further failure, the U.S. Army Corps of Engineers quickly built a new spillway in the natural dam. The reservoir is now known as Earthquake Lake.

4-023. Bagnell Dam, Osage River, Bagnell Vic., Miller County, Missouri, 1924–1926, 1929–1931. AP, 1931. P&P,LC-dig-ppmsca-17233.

Built by the Missouri Hydroelectric Power Company to supply power to southeastern Missouri, this dam impounds Lake of the Ozarks on the Osage River, a tributary of the Missouri.

4-021

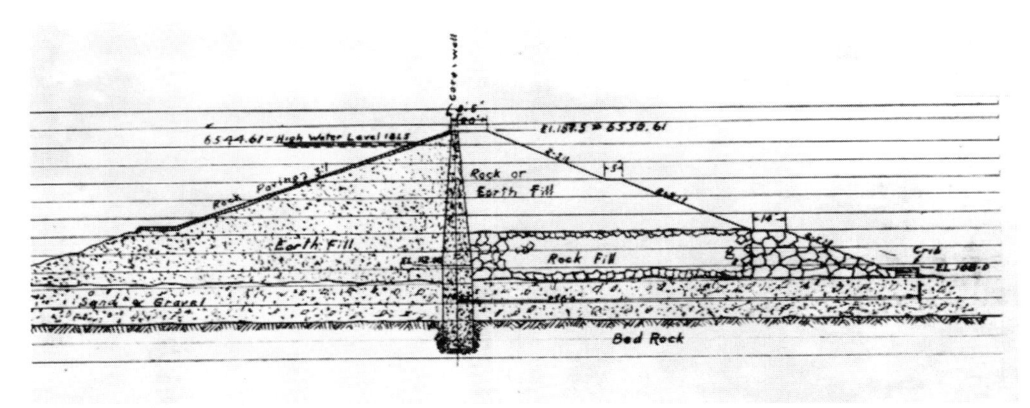

4-022

4-023

As settlement took root in the Trans-Mississippi West, there was increasing public dissatisfaction with the federal government's lack of involvement in irrigation projects. With its mandate for the regulation of interstate commerce, Congress had already subsidized canals, harbors, roads, and railroads. By 1900, western senators and congressmen had begun to use the political system more effectively to meet the interests of their constituents, introducing pro-irrigation positions to the platforms of both major parties and working as a block to ensure their wishes were respected by eastern representatives. With the passage of the Reclamation Act in 1901, the federal government began to play a key role in developing and financing irrigation projects in the West.

Of the first five projects of the new United States Reclamation Service, two were in the Missouri River region. The Milk River Project in Montana was designed to divert a portion of

4-024

4-024. President Theodore Roosevelt making one of his characteristic ten-minute speeches from the back of a train, North Platte, Lincoln County, Nebraska. H. C. White Company, 1905. P&P,LC-USZ61-2074.

4-025. Construction of Pathfinder Dam, North Platte River, Natrona County, Wyoming, 1905–1909, modified 1915, 1922, 1931, 1950, 1961, 1997. Bureau of Reclamation, 1908. HD1695 W4 R68 2006,p. 121,detail.

This 214-foot-high arched gravity dam is located at the confluence of the Sweetwater and North Platte rivers. Its remote location posed significant problems for its construction: materials had to be brought in from Casper, 45 miles away over difficult terrain. The dam is built of locally quarried cyclopean masonry, and wood rather than coal-burning steam engines powered the construction machinery. Sub-zero temperatures and spring floods further dogged the project.

4-026. Buffalo Bill (Shoshone) Dam, Shoshone River, Cody Vic., Park County, Wyoming, 1905–1910. HD1695 W4 R68 2006,p. 93.

Originally called Shoshone Dam, this was one of the thinnest concrete arch dams built by the Reclamation Service; at the time of its construction, it was the highest concrete arch dam in the world. Forty years later, it was renamed for William Cody, who had owned the property before transferring it to the Reclamation Service. An early project of the Service, it supported irrigation and provided hydroelectric power.

4-025

4-026

4-027

4-028

4-027. Upstream face, Buffalo Bill Dam, Shoshone River, Cody Vic., Park County, Wyoming, 1905– 1910. HAER, 1983. P&P,HAER, WYO,14-CODY,1, no. 4.

4-028. Headgate to east tunnel, Buffalo Bill Dam, Shoshone River, Cody Vic., Park County, Wyoming, 1905–1910. HAER, 1983. P&P,HAER,WYO,14-CODY,1, no. 8.

the Saint Mary River into the Milk River as it flows into Canada, before returning south to the United States to irrigate 120,000 acres of land. The North Platte Project captures the headwaters of the North Platte River in the spectacular masonry-arch Pathfinder Dam in Wyoming (1905–1909) to serve fields 111 miles away in Nebraska. Other early Reclamation projects in this region include the Shoshone Project in northwestern Wyoming, with its concrete gravity arch Buffalo Bill Dam (1905–1910), and the Belle Fourche Project, with its large earthfill Belle Fourche (Orman) Dam (1906–1911), near the Black Hills of South Dakota; as well as the Lower Yellowstone Reclamation Project (later integrated into the Pick-Sloan Missouri Basin Program). Because the Reclamation Service financed its programs through the sale of federal lands associated with each irrigation district, a number of projects were built on former Indian lands, such as the Huntley Project in southern Montana, developed on lands ceded by the Crow Indians in 1904; and the Riverton Project, built on a reclaimed portion of the Wind River Indian Reservation in 1905.

In 1868, the Shoshone Indians were assigned reservations in central Wyoming, and over the subsequent forty-year period these lands were reduced for a number of reasons. In 1905, Congress determined the Wind River Indian Reservation held "excess" lands of more than 1.4 million acres, so Congress transferred ownership of these to the Reclamation Service for development as an irrigation district. The key structure in the Riverton Project was the 37-foot-high concrete gravity Wind River Diversion Dam (1924–1925),

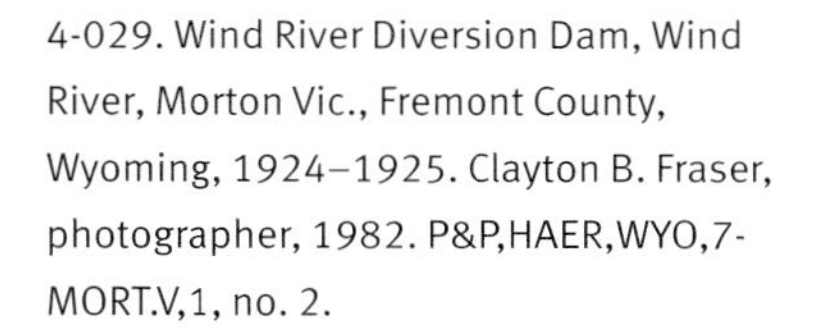

4-029. Wind River Diversion Dam, Wind River, Morton Vic., Fremont County, Wyoming, 1924–1925. Clayton B. Fraser, photographer, 1982. P&P,HAER,WYO,7-MORT.V,1, no. 2.

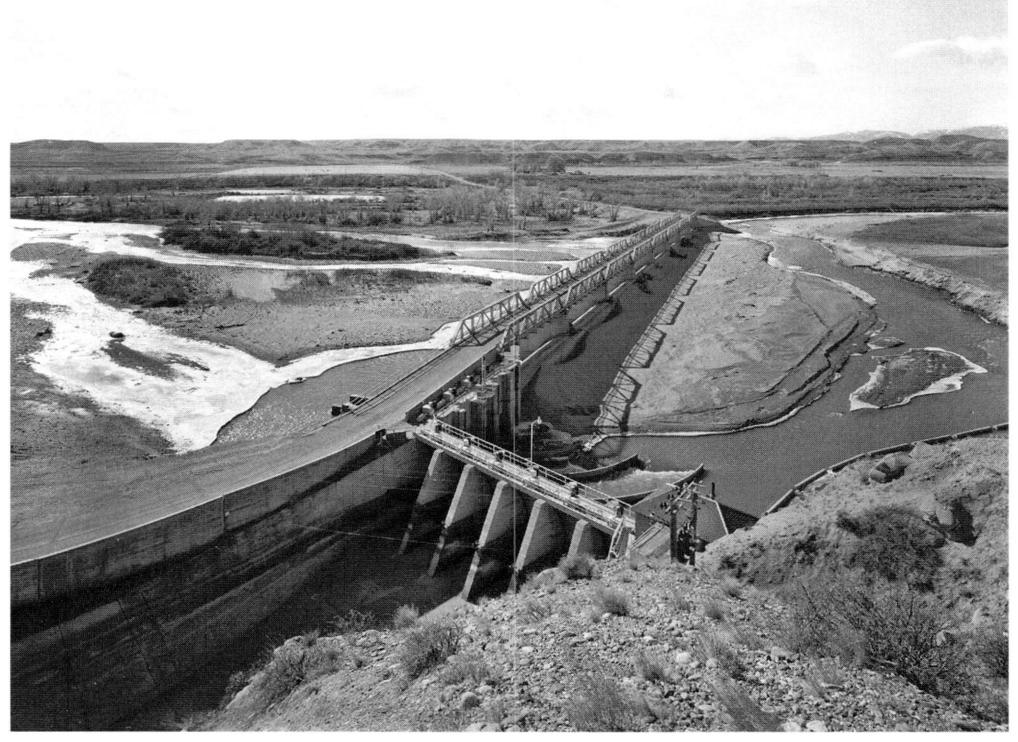

4-029

fitted with a logway and sluiceway and traversed by a steel highway bridge. Built to supply all the water in the district, the dam feeds the Wyoming Canal (1920–1951) and Pilot Canal (1926–1947) to irrigate lands held by Shoshone and Arapahoe Indians and more recent settlers.

The Shoshone have continued to protest the over-appropriation of the Wind River for irrigation, arguing that some of the water should remain in the river. The State of Wyoming, following Colorado's Prior Appropriation Doctrine, had granted early settlers the first water rights, finding that traditional Native uses of water for drinking, cooking, fishing, and cleaning were not "uses" comparable to agriculture. Subsequent litigation awarded Native tribes the right to exclude non-Native irrigators downstream in dry years, and the dispute between traditional and agricultural uses for the water of the Wind River continues today.

4-030

4-030. Wind River Diversion Dam, Wind River, Morton Vic., Fremont County, Wyoming, 1924–1925. Clayton B. Fraser, photographer, 1982. P&P,HAER,WYO,7-MORT.V,1, no. 1.

NEW DEAL DAMS

Drought and the Great Depression spurred the federal government to support irrigation and resettlement projects in the arid western states. The aim was to reduce erosion on the plains and support farming and ranching in sparsely settled regions of the country. And with the establishment of the Public Works Administration (PWA) in 1933—a New Deal agency charged with building public works to provide employment and revive American industry—dam building received a major impetus. Contracting with private firms for the construction of public works, the PWA epitomized Franklin Roosevelt's notion of "priming the pump" to encourage economic growth. Between 1933 and 1939, under the leadership of Harold Ickes, the PWA funded more than 34,000 projects—mostly roads and school buildings, but also airports, hydroelectric dams, bridges, and ships. Some of the best-known PWA-funded dams include the Grand Coulee Dam on the Columbia River (see 7-078–7-088), Tygart Dam in West Virginia, Keystone Dam in Oklahoma, Santee-Cooper dams in South Carolina (see 1-088–1-089), and Buchanan Dam in Texas (see 5-049a–5-050b). The larger projects were usually the result of intensive lobbying by the states to the federal government. One of the biggest was Fort Peck Dam

4-031. Installation of steel core wall, Fort Peck Dam, Missouri River, Valley County, Montana, 1933–1940. Acme, 1935. P&P,NYWTS-Subj.-Dams-Fort Peck-Montana.

4-031

on the Missouri River in Montana.

Four miles long, half a mile wide, and 250 feet high, Fort Peck Dam is one of the largest earthen dams in the world. At the time of its construction, it was five times bigger than the largest dam up to then, Gatun Dam on the Panama Canal. Although hydraulic mining technologies were a familiar sight in the West from the Rocky Mountains of Montana and Colorado to the Sierra Nevada of California, it wasn't until 1907 that the technology was proposed as sound engineering practice for dam construction. The process involves depositing layers of waterborne clay and silt in such a way that water can drain off and the puddled clay core of the dam dries into a solid, impermeable mass. The economy of the technique is that the construction materials are close at hand to the dam site, and they are transported by water rather than by men and machines.

A number of features were innovative in Fort Peck Dam, including the steel sheet pilings driven into its core across its entire length. Four dredges excavated the riverbed upstream and sent water-borne slurry to the dam site, depositing it in alluvial beaches as the water drained back into the riverbed. During construction, the Missouri River was diverted through tunnels. A spillway of sixteen gates allows a hundred-year flood to move through the dam without spilling over its crest, a potentially catastrophic occurrence on an earthen dam.

As the first main-stem dam built on the Missouri, Fort Peck Dam was funded by the Public Works Administration and built by the U.S. Army Corps of Engineers. It brought

4-032

4-032. Sutherland Reservoir Dam, South Platte River, Sutherland Vic., Lincoln County, Nebraska, 1933–1935. John Vachon, photographer, 1938. P&P,LC-USF34-008849-D.

employment and training to tens of thousands of people and resulted in the construction of the Fort Peck town site in 1934. Unfortunately, while Montana state law gave hiring preference to men with families, the Corps of Engineers had only built three hundred family residences in the town. As a result, eighteen shantytowns sprang up near the dam, with names like New Deal and Delano Heights. The spillway of the dam was featured on the cover of *Life* magazine's inaugural issue in 1936.

4-033

4-033. Kingsley Dam, North Platte River, Ogallala Vic., Keith County, Nebraska, 1936–1941. Acme, 1938. P&P,LC-dig-ppmsca-17264.

Central Nebraska Public Power and Irrigation District built this 2-mile-long earthen dam with PWA funding. Backers of the Kingsley Dam and those of the Sutherland Project competed, at times bitterly, to secure federal funding for their neighboring irrigation ventures. Ultimately, both were built. Like Fort Peck, Kingsley Dam had a core of sheet steel pilings and was constructed using hydraulic earth fill. The tower on the right is a morning glory spillway under construction.

4-034

4-034. Kingsley Dam, North Platte River, Ogallala Vic., Keith County, Nebraska, 1936–1941. Edwin C. Hunton, photographer, 1947. P&P,LC-dig-ppmsca-17333.

4-035

4-035. Twin Lakes Dam, Big Goose Creek, Sheridan County, Wyoming, 1936–1937. Richard Collier, photographer, 1989. P&P,HAER,WYO,052-SHER.V,2A, no. 1.

The PWA built this earthen dam faced with rip-rap as a municipal reservoir for the city of Sheridan.

Proposed Relocation of Valve Control House

GENERAL PLAN OF DAM AND APPURTENANT STRUCTURES

Note: Alternate Location of Outlet conduits used in construction

4-036

4-036. Plan, Twin Lakes Dam, Big Goose Creek, Sheridan County, Wyoming, 1936–1937. Daniel J. McQuaid, delineator, 1936. P&P,HAER,WYO,017-SHER.V,2-1,detail.

Note proposed change in Valve Control House Location

Selected Graded Materials

More pervious materials

VERTICAL SECTION A—A

EARTH TYPE OF DAM

Reinforced concrete chamber with Flash boards for submerging valves during winter months

4-037

4-037. Section, Twin Lakes Dam, Big Goose Creek, Sheridan County, Wyoming, 1936–1937. Daniel J. McQuaid, delineator, 1936. P&P,HAER,WYO,017-SHER.V,2-1,detail.

WPA Stock Water Dams

Under Roosevelt's New Deal, a number of government agencies worked together to develop national infrastructure and provide employment during the Great Depression. The Public Works Administration (1933–1939), for example, paid for the construction of public works (such as roads, bridges, airports, utilities, dams, parks, and public buildings) by contracting private firms to carry out the work. The Works Progress Administration, or WPA (1935–1943), was an employment program directed primarily toward the construction of public facilities and infrastructure. The WPA provided labor for hundreds of stock water dams and irrigation projects throughout the Missouri River basin — in central North Dakota on the Missouri River, in the states abutting the Black Hills of South Dakota, on the Platte and Missouri River basins in Nebraska and Iowa, and as far south as the Marais des Cygnes watershed in eastern Kansas.

4-038

4-039

4-038. Farmer checking the depth of moisture in the soil, Sheridan County, Kansas. Russell Lee, photographer, 1939. P&P,LC-USF33-012355-M4.

4-039. WPA workers repair and enlarge an old dam, Harvey, Wells County, North Dakota, 1936. Acme, 1936. P&P,LC-dig-ppmsca-17272.

4-040. Stock water dam under construction, Pennington County, South Dakota. Arthur Rothstein, photographer, 1936. P&P,LC-USF34-004391-D.

4-040

4-041

4-042

4-043

4-041. Laying stone on the face of a stock water dam, Pine Ridge Land Utilization Project, Sioux County, Nebraska. Arthur Rothstein, photographer, 1936. P&P,LC-USF34-004371.

4-042. Drought committee inspects dam, Rapid City, Pennington County, South Dakota. Arthur Rothstein, photographer, 1936. P&P,LC-USF34-005246-D.

4-043. President Franklin Roosevelt greeted by workers on a stock water dam, Mandan Vic., North Dakota. Arthur Rothstein, photographer, 1936. P&P,LC-USF34-005296-E.

Civilian Conservation Corps Dams

Another New Deal dam-building program was the Civilian Conservation Corps (CCC), a program that brought young unemployed men from urban areas to work on conservation projects across the country. These dams on the Souris River in North Dakota created habitat for migratory waterfowl, forming part of a national wildlife refuge system established during this period. Their construction is typical of federally-built conservation dams of the 1930s.

4-044. CCC workers collecting field stone used in the construction of spillways and control structures, Upper Souris National Wildlife Refuge Dams, Souris River Basin, Ward County, North Dakota, 1935–1936. Unidentified photographer, 1936. P&P,HAER,ND,51-FOX.V,1, no. 5.

4-045. Crest of Dam 341, J. Clark Salyer National Wildlife Refuge, Lower Souris River, Bottineau County, North Dakota, 1935–1936. Frederick L. Quivik, photographer, 1990. P&P,HAER,ND,5-KRA.V,1-D, no. 1.

4-046

4-046. Outlet works, Dam 341, J. Clark Salyer National Wildlife Refuge, Lower Souris River, Bottineau County, North Dakota, 1935–1936. Frederick L. Quivik, photographer, 1990. P&P,HAER,ND,5-KRA.V,1-D, no. 6.

4-047. Plan of typical gate control structure, J. Clark Salyer National Wildlife Refuge Dams, Lower Souris River, Bottineau County, North Dakota, 1935–1936. Bureau of Agricultural Engineering, delineator. P&P,HAER,ND,5-KRA.V,1-6,detail.

4-048. Section through typical gate control structure, J. Clark Salyer National Wildlife Refuge, Lower Souris River, Bottineau County, North Dakota, 1935–1936. Bureau of Agricultural Engineering, delineator. P&P,HAER,ND,5-KRA.V,1-6,detail.

PLAN OF GATE STRUCTURE

4-047

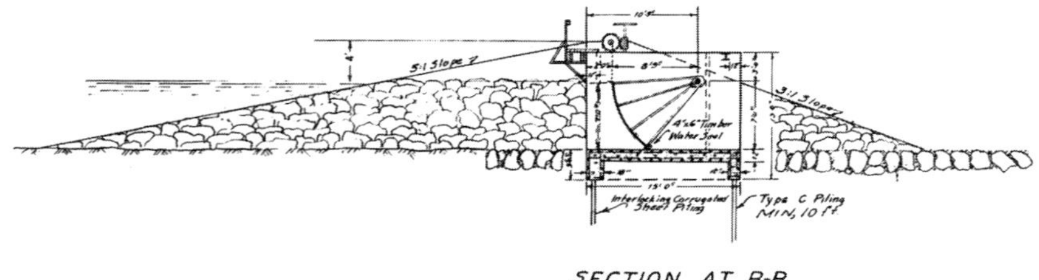

4-048

4-049. Rock work at outlet gates, Dam 357, J. Clark Salyer National Wildlife Refuge, Lower Souris River, Bottineau County, North Dakota, 1935–1936. Unidentified photographer, 1936. P&P,HAER,ND,5-KRA.V,1-E, no. 15.

4-050. Completed spillway wall, Dam 357, J. Clark Salyer National Wildlife Refuge, Lower Souris River, Bottineau County, North Dakota, 1935–1936. Unidentified photographer, 1936. P&P,HAER,ND,5-KRA.V,1-E, no. 12.

4-051. View of Dam 83 from the lookout tower, Upper Souris National Wildlife Refuge Dams, Upper Souris River, Ward County, North Dakota, 1935–1936. Unidentified photographer, 1936. P&P,HAER,ND,51-FOX.V,1-A, no. 13.

4-052. Section, Dam 83, Upper Souris National Wildlife Refuge Dams, Upper Souris River, Ward County, North Dakota, 1935–1936. Chief Engineer, U.S. Fish & Wildlife Service Regional Office, delineator, 1937. P&P,HAER,ND,51-FOX.V,1-A-16,detail.

4-049

4-050

4-051

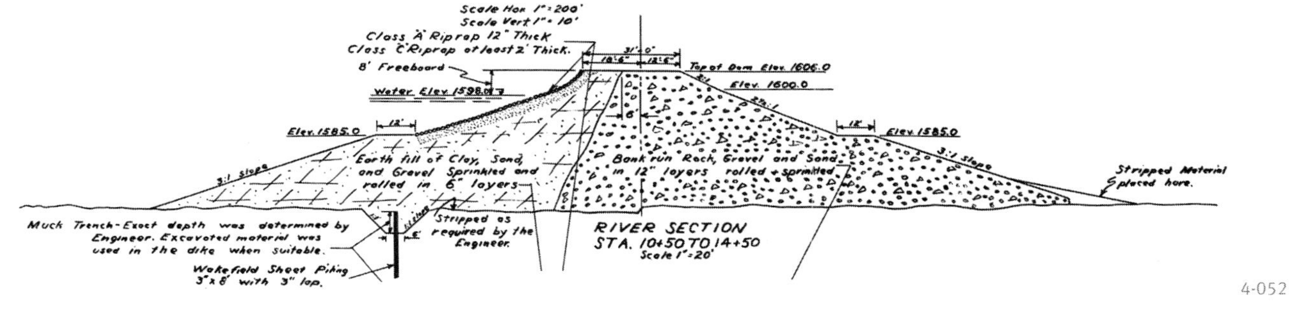

4-052

4-053. Dam 87, Upper Souris National Wildlife Refuge Dams, Upper Souris River, Foxholm Vic., Ward County, North Dakota, 1935–1936. Unidentified photographer, 1936. P&P,HAER,ND,51-FOX.V,1-B, no. 12.

4-054. Opening day, Dam 96, Upper Souris National Wildlife Refuge Dams, Upper Souris River, Foxholm Vic., Ward County, North Dakota, 1935–1936. Unidentified photographer, 1936. P&P,HAER,ND,51-FOX,V1-C, no. 12.

4-053

4-054

MISSOURI RIVER BASIN (PICK-SLOAN) PROJECT DAMS

Disastrous floods on the Missouri in 1943 led the two major federal agencies involved in dam building to propose competing plans for the control of the river. The U.S. Army Corps of Engineers' proposals, presented by Brigadier General Lewis Pick, emphasized flood control and navigation; the Bureau of Reclamation's plan, proposed by William Sloan, focused on irrigation, power generation, conservation, and recreation. Since both plans had congressional supporters, President Franklin Roosevelt required a negotiated compromise that resulted in the Pick-Sloan Project of 1944. The plan called for building almost one hundred reservoirs on the Missouri and its tributaries, with hundreds of miles of levees and floodwalls throughout the basin. It anticipated thousands of barges transporting millions of tons of grain out of the Midwest to ports in New Orleans. It called for irrigation channels watering 30 million acres of farmland. Finally, it generated huge amounts of electricity to supply cities and industries throughout the basin. This was truly a multipurpose project, and one that had epic repercussions. Six main-stem dams were completed on the Missouri, upstream and downstream from Fort Peck Dam (Canyon Ferry, Garrison, Oahe, Big Bend, Fort Randall, and Gavins Point), and eighty other dams have been built on the river's tributaries, severely damaging fish and wildlife throughout the Missouri River basin.

4-055

4-055. Kortes Dam, North Platte River, Carbon County, Wyoming, 1946–1951. Wide World, 1949. P&P,LC-dig-ppmsca-17266.

This concrete gravity structure, located in a narrow gorge of the North Platte River, was the first dam built under the Pick-Sloan Missouri River Basin Project. Designed for power generation, it was located between Seminoe and Pathfinder dams to achieve the highest head between Seminoe tailwater and the Pathfinder reservoir. The powerhouse occupies the entire width of the canyon at the foot of the dam. Demand for power during the dam's construction led the Bureau of Reclamation to put the generators into service and build temporary transmission facilities even before the powerhouse was completed.

4-056

4-057

4-058

4-056. Garrison Dam construction showing intake structure, Missouri River, Riverdale Vic., McLean County, North Dakota, 1947–1954. Acme, 1951. P&P,LC-dig-ppmsca-17249.

Although dwarfed by Fort Peck Dam (4-048–4-053), Garrison is still one of the largest earthfill dams in the country. It extends more than 2 miles across the Missouri River and rises 210 feet above the riverbed, creating Lake Sakakawea for 180 miles upstream. Work on the dam's embankment took seven years to complete and required construction of the town of Riverdale to house a work force of four thousand people.

4-057. Embankment of Oahe Dam under construction, Missouri River, Pierre Vic., between Hughes and Stanley Counties, South Dakota, 1948–1962. United Press, 1954. P&P,LC-dig-ppmsca-17274.

Yet another massive earthen dam on the Missouri, the 9,300-foot-long Oahe Dam is a rolled earth structure with a crest height of 245 feet. The seven generators in the powerhouse provide electricity for much of the north-central United States, and its reservoir extends over 200 miles upstream to Bismark, North Dakota.

4-058. Canyon Ferry Dam under construction, Missouri River, Helena Vic., Lewis and Clark County, Montana, 1949–1954. Donald H. Demarest, photographer, 1951. P&P,LC-dig-ppmsca-17368.

This concrete gravity structure is located just downstream from the original Montana Power Company's Canyon Ferry Dam and power plant in the backwater of Hauser Lake. It was built for both power production and irrigation.

4-059. Kirwin Dam, North Fork Solomon River, Phillips County, Kansas, 1952–1955. United Press, 1955. P&P,LC-dig-ppmsca-17265.

This 12,646-foot-long earthen dam in north-central Kansas provided flood control and irrigation as part of the Missouri River Basin Project. The concrete overflow spillway is shown here.

4-060. Fort Randall Dam inlet works under construction, Missouri River, Pickstown, Charles Mix County, South Dakota, 1946–1956. Acme, 1951. P&P,LC-dig-ppmsca-17248.

Slightly smaller than Garrison Dam (4-076) this earthfill dam is 2 miles long and 160 feet high. As with Fort Peck (4-048–4-053) and Garrison dams, Fort Randall Dam required a new town site, Pickstown, to house a construction force of 3,500 workers.

4-059

4-060

5-001

ARKANSAS–TEXAS GULF–RIO GRANDE

ARKANSAS, RED, AND WHITE RIVERS

Arkansas, Canadian, Cimarron, Red, Salt, Prairie Dog, Washita, Clear Boggy, Blue, Kiamichi, and White Rivers

The Arkansas River is the southerly counterpart of the great Missouri. Rising in the Rocky Mountains, it flows 1,450 miles across the prairie to the Mississippi River, draining 195,000 square miles. The Canadian and Cimarron rivers are its main tributaries. The Arkansas has three distinct characters in its long path through central North America. At its headwaters, it runs as a steep mountain torrent, dropping 4,600 feet in 120 miles. At Cañon City, Colorado, it leaves the mountains and enters Royal Gorge, one of the deepest canyons in the United States. On its course through the prairies, it has wide shallow banks and occasionally floods but becomes a trickle during the summer because of the use of its water for irrigation.

5-001. Pensacola Dam, Neosho (Grand) River, Langley, Mayes County, Oklahoma, 1938–1940. Acme, 1940. P&P,LC-dig-ppm-sca-17252.

The State of Oklahoma was successful in obtaining federal support for this first large multipurpose dam in Oklahoma. A multiple arch dam, it spans a mile across the Grand River to impound the vast Lake O' the Cherokees.

Many nations of Native Americans lived near or along the Arkansas River, but the first Europeans to see the river were members of the Coronado expedition in 1541. In 1819, the Adams-Onís Treaty set the Arkansas as part of the frontier between the United States and Spanish Mexico, which it remained until the annexation of Texas and the Mexican-American War in 1846. Later, the Santa Fe Trail followed the Arkansas through much of Kansas. Important cities on the Arkansas include Wichita, Kansas; Tulsa, Oklahoma; and Little Rock, Arkansas. More than twenty-five dams on the river provide flood control, power, and irrigation. The McClellan-Kerr Arkansas River Navigation System (1946–1971) makes the river navigable to Tulsa, 450 miles upstream.

The Red River is the southernmost of the large tributaries of the Mississippi. It rises in several forks in the Llano Estacado (lower high plains) of Texas and flows eastward to the Atchafalaya and Mississippi rivers. In Texas, it flows rapidly through a canyon in semiarid plains, but later in its course it waters rich, red clay farmlands. Its major tributary, the Washita (Ouachita), joins the Red at Lake Texoma, a reservoir created by Denison Dam (1945; 5-049–5-054). South of Lake Texoma, the Clear Boggy, the Blue, and the Kiamichi rivers feed the Red clear water. Caddo people, with their distinctive mound-building culture, lived north of the river from pre-contact times. Spain and France debated the Red as the boundary separating New Spain and New France beginning in the early sixteenth century. Later, under the Adams-Onís Treaty of 1819, the Red became the southern border

5-002. Witchita village on Rush Creek, Red River Valley, Louisiana, 1852. James Ackerman, lithographer, 1854. P&P,LC-USZ62-11478.

5-002

5-003

5-003. Drilling rigs and the Red River bed, Wichita County, Texas. Homer T. Harden, photographer, 1919. P&P,LC-dig-pan-6a14791.

of the Louisiana Purchase. For many years, navigation was difficult on the lower course of the Red River due to floating trees that collected behind obstructions. One of these, the Great Raft, a 160-mile log jam built up through the centuries, was cleared from the river in the middle of the nineteenth century. The river is now navigable for small ships to above Natchitoches, Louisiana.

TEXAS GULF RIVERS

Sabine, Neches, Trinity, Brazos, Colorado of Texas, Guadalupe, San Antonio, and Nueces Rivers

The rivers of Texas run roughly parallel, draining the high plains of the Llano Estacado and the Edwards Plateau, cutting across the coastal plain, and emptying into the Gulf of Mexico. In the eighteenth century, the Spanish developed extensive irrigation systems along the San Antonio River as a part of their mission-building and colonization programs. As urbanization intensified in the twentieth century, efforts to cope with flash floods brought on by the region's heavy thunderstorms led to the construction of numerous flood control dams across these rivers, particularly along the Colorado and the Brazos. Demands for irrigation, municipal water supply, and hydroelectric power have led to additional dams.

RIO GRANDE BASIN

Chania, Puerco, Salado, and Pecos Rivers

The Rio Grande, known in Mexico as Río Bravo del Norte, rises in the San Juan Mountains of Colorado. The upper portion of the river flows along the sediment-filled basin of the Rio Grande Rift. East of El Paso, the river continues through desert until it reaches the subtropical lower Rio Grande valley, where its waters have been used extensively for agriculture. The river ends in a small sandy delta on the Gulf of Mexico.

Pueblos were thriving on the banks of the river and native people were irrigating the arid country when the Spanish explorer Coronado arrived in 1540. The Rio Grande valley at the river's mouth was developed agriculturally in the 1920s, but today the river there is often reduced to a trickle because of over-appropriation upstream. Use of the Rio Grande's water has been a subject of negotiation since 1906, when an international treaty aimed to "equitably distribute" the waters of the Rio Grande between the United States and Mexico. Subsequent interstate and international agreements led to the Rio Grande Compact of 1944, which further clarified allocation issues with respect to tributaries and established the International Boundary and Water Commission to manage the

5-004

5-004. Rio Grande River in flood. Unidentified photographer, 1952. P&P,LC-USZ62-42374.

5-005

5-005. Farmhands at a truck stop in Neches, Texas, on their way home from Mississippi to the Rio Grande valley. Russell Lee, photographer, 1939. P&P,LC-USF33-012466-M2.

resource. Today dams on the Rio Grande provide irrigation, flood control, and regulation of the river flow—namely, the Elephant Butte (1916), Caballo (1938), Amistad (1969), and Falcon (1954) dams. In dry years, the amount of water reaching Mexico is often less than what the treaty stipulates. Shifts in the riverbed have led to border disputes between the two countries, such as the 114-year controversy over the location of the border at El Paso, which was settled in 1968 when the river was channeled.

The Pecos River, a major tributary of the Rio Grande, has been dammed at Alamogordo, Avalon, and McMillan as a part of the Carlsbad Reclamation Project (1906) and at Red Bluff Dam a little further downstream. Long-standing interstate disputes about water use were settled in 1949, when a federal bill provided for a compact between New Mexico and Texas. In the heyday of ranching in west Texas, "west of the Pecos" was the term for the distinct and rugged region of the western tip of the state.

IRRIGATION DAMS

Acequias

The building of *acequias* (irrigation canals) was an important element in Spanish efforts to colonize Texas. Acequias had been widespread throughout Spain since the time of the Moorish conquest, and the early Spanish colonists brought with them sophisticated knowledge of how to construct large-scale irrigation systems.

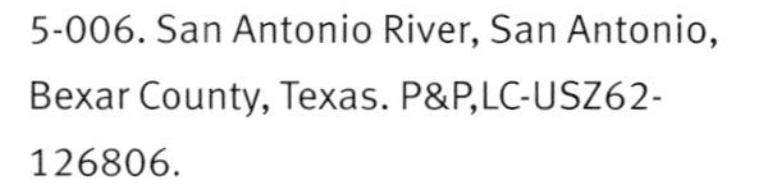
5-006. San Antonio River, San Antonio, Bexar County, Texas. P&P,LC-USZ62-126806.

5-006

The Franciscan missionaries developed an extensive network of acequias and dams for their missions on the San Antonio River. The main canal, the Acequia Madre de Valero (1718–1744), was followed by the Nuestra Señora de la Purísima Concepción de Acuña Mission acequia (1729), the San José y San Miguel de Aguayo Mission acequia (1730), the San Juan Capistrano Mission acequia (1731), and the San Francisco de la Espada Mission acequia (1731–1745). The last includes the stone Espada aqueduct, which spans an arroyo—it is claimed to be the only Spanish structure of its type still in use in the United States.

5-007

5-007. View from Velita Street toward San Fernando, San Antonio, Bexar County, Texas, 1836. H. Langkwitz, 1836. P&P,LC-USZ62-16310.

5-008

5-008. Acequia of Mission San Juan Capistrano, San Antonio, Bexar County, Texas, 1731. Unidentified photographer, 1905. P&P,LC-dig-ppmsca-17292.

As the missions gradually became secularized (recognized as civil communities with property ownership) at the end of the eighteenth century, city authorities became responsible for the distribution of water. After annexation by the United States, city control was discontinued and the acequias operated as community enterprises or mutual companies. With the expansion of San Antonio in the twentieth century, most of the canals were abandoned.

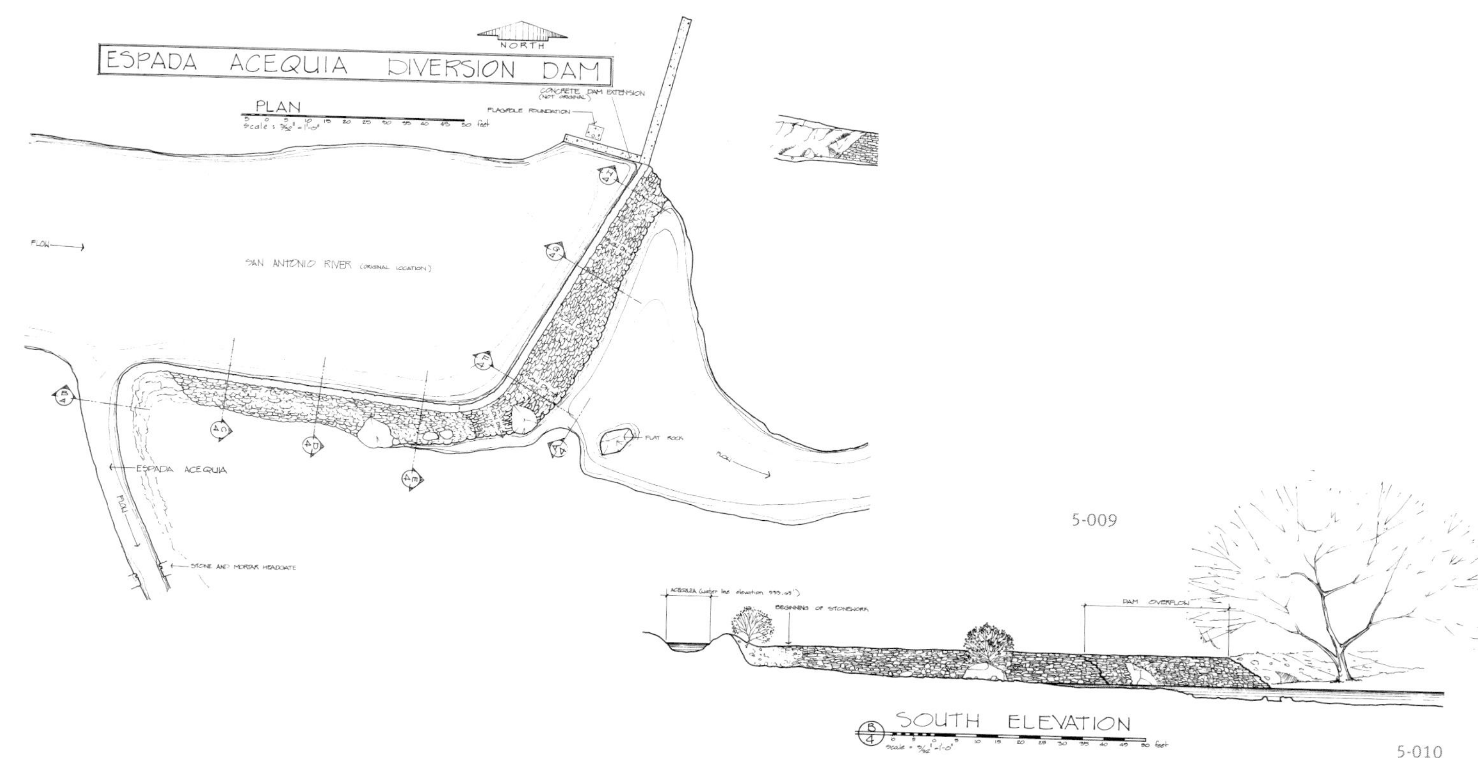

5-009. Plan, Espada Acequia Diversion Dam, San Antonio River, San Antonio, Bexar County, Texas, 1731–1745. Gary Rogers, delineator, 1973. P&P,HAER,TEX,15-SANT.V,4B.

This dam served the acequia of Mission San Francisco de la Espada in San Antonio.

5-010. South elevation, Espada Acequia Diversion Dam, San Antonio River, San Antonio, Bexar County, Texas, 1731–1745. Gary Rogers, delineator, 1973. P&P,HAER, TEX,15-SANT,V,4B.

5-011

5-011. Ruins of Espada Acequia Diversion Dam, San Antonio River, San Antonio, Bexar County, Texas, 1731–1745. Jet Lowe, photographer, 1983. P&P,HAER,TEX,15-SANT,V, 4B, no. 1.

Headwater Reservoir Dams

Although the Arkansas and the Rio Grande rivers run through arid plains and plateaus of the Southwest, their headwaters lie in the relatively water-rich Rocky Mountains. Many reservoir dams were built in these mountains and foothills to capture run-off and store it for agricultural use—often at some distance downstream.

5-012

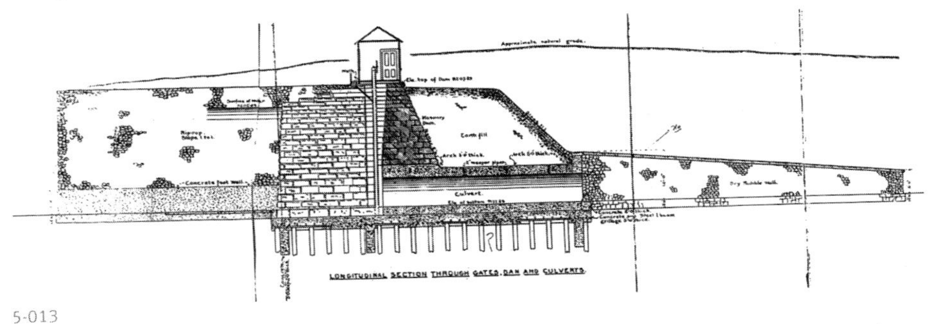

5-013

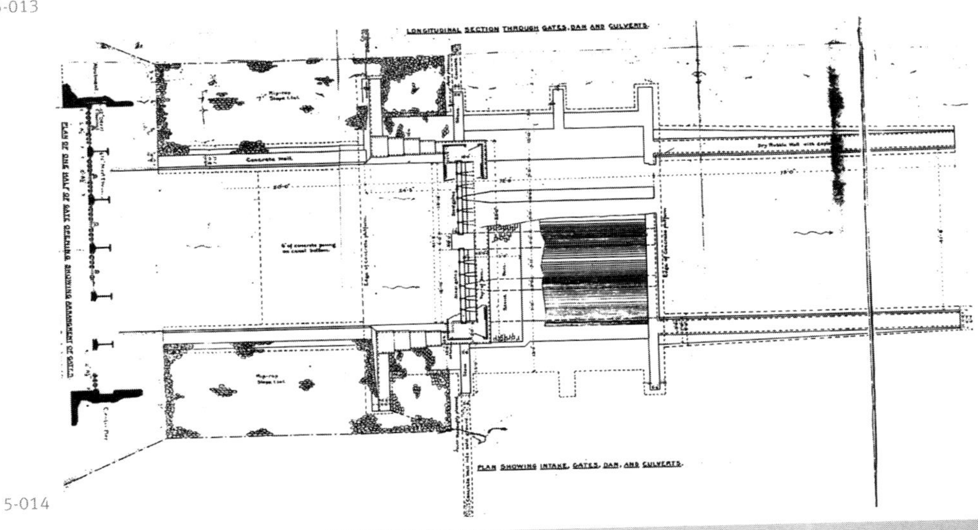

5-014

5-015

5-012. Upper Twin Lake, Lake County, Colorado. William Henry Jackson, photographer, ca. 1881–1889. P&P,LOT 12685,no. 20 (OSF).

5-013. Section, Twin Lakes Dam, Lake County, Colorado, 1898–1900. Twin Lakes Reservoir and Canal Company, delineator, 1899. P&P,HAER,COLO,34-TWLK.V,1-10, detail.

5-014. Plan, Twin Lakes Dam, Lake County, Colorado, 1898–1900. Twin Lakes Reservoir and Canal Company, delineator, 1899. P&P,HAER,COLO,34-TWLK.V,1-10,detail.

5-015. Twin Lakes Dam, Lake County, Colorado, 1898–1900, modified 1949–1951. Bureau of Reclamation. P&P,HAER, COLO,34-TWLK.V,1, no. 7.

This masonry and earthfill structure captured mountain run-off for irrigating the arid lower Arkansas River valley in eastern Colorado. It releases water into the Arkansas River to supply the Colorado Canal (1890–1891), which begins 150 miles downstream east of Pueblo and continues into the farmlands of Crowley County. The dam was superseded by a newer one downstream in 1980.

Carlsbad Irrigation District Dams

Irrigation in the Pecos River basin flourished under the Spanish land-grant colonization system and was continued after 1850 by the American settlers. These early systems were community ditches that diverted the river's flow without permanent diversion structures. Beginning in 1888, private investors built several dams and an extensive irrigation system of canals and flumes along the Pecos River, creating one of the largest nineteenth-century irrigation projects in the American West. Severe floods in 1893 and 1904 led them to sell the complex to the United States Reclamation Service. By 1907, the Reclamation Service had repaired the system, supplying 145 miles of ditches and irrigating 30,000 acres of alfalfa and cotton. Alamogordo Dam was added to the system in the 1930s, followed by Brantley Dam fifty years later.

5-016. Main Canal across the Pecos River, Carlsbad Irrigation District, Eddy County, New Mexico, 1890, destroyed 1902, rebuilt 1904. Unidentified photographer, ca. 1904. WC. P&P,LC-dig-ppmsca-17316.

A high, trestle-supported wooden flume, originally built in 1890 to continue the canal over the river, was destroyed by flooding in 1902. It was rebuilt as a massive concrete aqueduct.

5-016

5-017

5-017. Rock cut and Avalon Dam, Carlsbad Irrigation District, Pecos River, Eddy County, New Mexico, 1888. Unidentified photographer, ca. 1889. P&P,HAER,NM,8-CARL.V,1B, no. 50.

5-018

5-018. Spillway at Avalon Dam, Pecos Valley Irrigation and Improvement Company, Pecos River, Eddy County, New Mexico, 1888. HD1694 N6 C42 1987,p.163.

5-019

5-019. Remains of Avalon Dam after flood of October 1904, Pecos River, Eddy County, New Mexico, 1888, 1894. HD1694 N6 C42 1987,p.164.

5-020. Cross-section through old part of Avalon Dam, showing new construction, Carlsbad Irrigation District, Pecos River, Eddy County, New Mexico, 1888, 1894, rebuilt 1907, modified 1912. United States Reclamation Service, delineator, 1916. P&P,HAER,NM,8-CARL.V,1B-78,detail.

5-021. Site plan of Avalon Dam, Carlsbad Irrigation District, Pecos River, Eddy County, New Mexico, 1888, 1894, rebuilt 1907, modified 1912, 1936. Jim McDonald and Paula Albers, delineators, 1991. P&P,HAER,NM,8-CARL.V,1B,sheet 1.

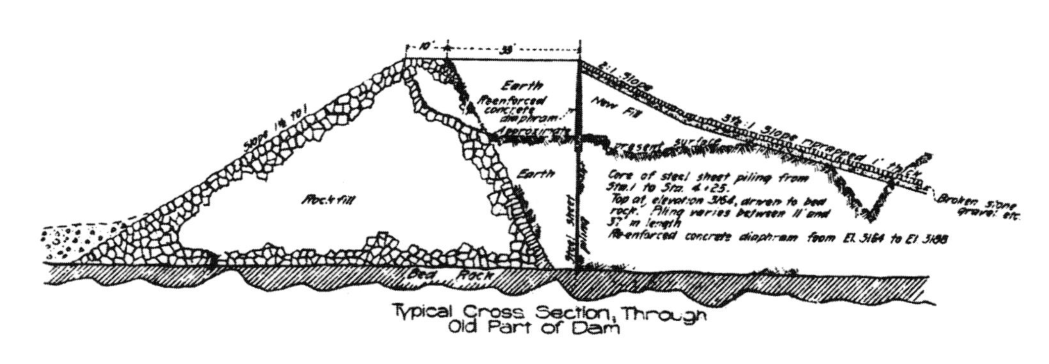

5-020

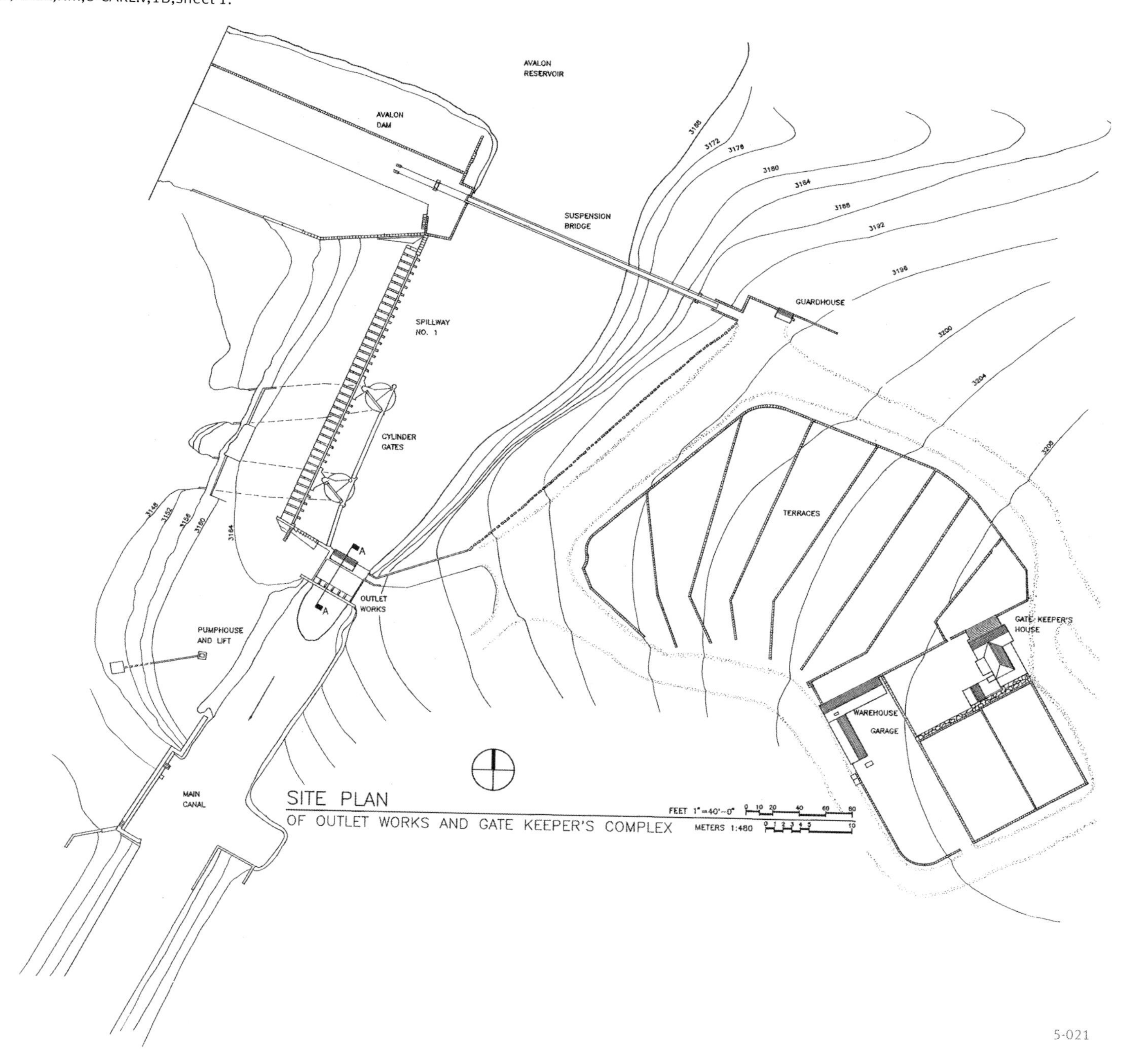

5-021

5-022

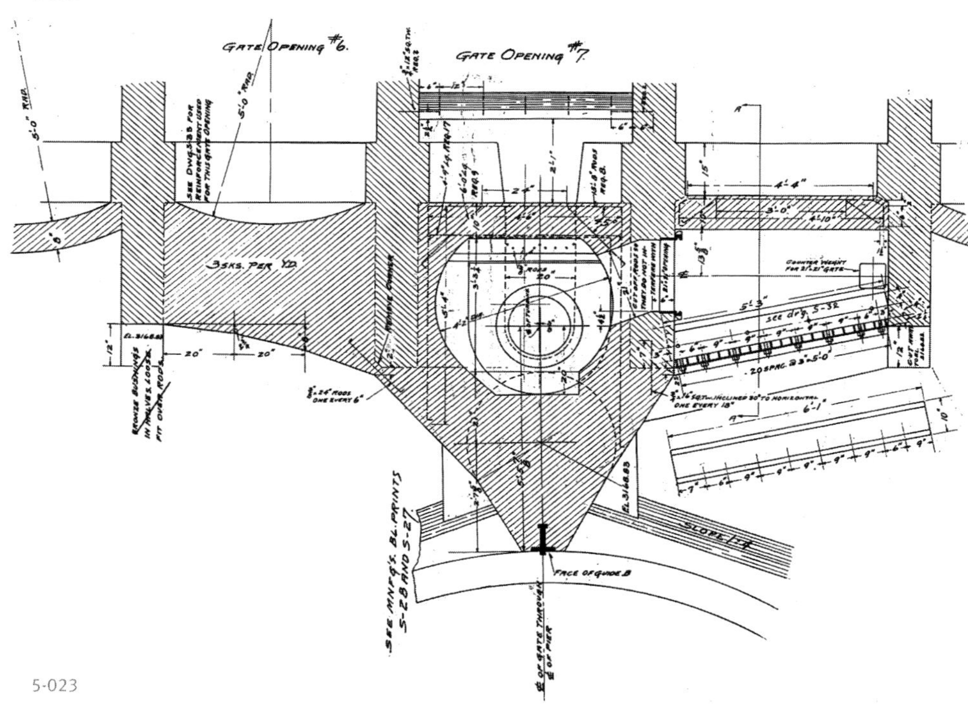

5-023

5-022. Water pouring into cylinder gates, Avalon Dam, Carlsbad Irrigation District, Pecos River, Eddy County, New Mexico, 1912. United States Reclamation Service. P&P,LC-dig-ppmsca-17332.

In addition to forming a small reservoir, Avalon Dam diverts water from the Pecos River into the main canal. The dam built in 1888 washed out five years later and again in 1904. The Reclamation Service rebuilt it in 1907 as a zoned earthfill structure and raised its height in subsequent years. There are three spillways and an outlet works. The unusual cylinder gates are also features of the Laguna Diversion Dam in Yuma, Arizona (see 6-098).

5-023. Plan of cylinder gate opening number 7, Avalon Dam, Carlsbad Irrigation District, Pecos River, Eddy County, New Mexico, 1912. United States Reclamation Service, delineator, 1916. P&P,HAER,NM,8-CARL.V, 1B-88,detail.

5-024

5-024. Construction of spillway, McMillan Dam, Carlsbad Irrigation District, Pecos River, Eddy County, New Mexico, 1888, 1894, rebuilt by United States Reclamation Service 1908. R.B.D., photographer, 1917. P&P,HAER,NM,8-CARL.V,1A, no. 49.

At the time of its first reconstruction after damage in 1894, McMillan Dam formed one of the largest artificial reservoirs in the world. Completely destroyed by flood in 1904, it was rebuilt by the Reclamation Service in 1908 as a zoned earthfill dam. In 1937, the crest was raised to accommodate silt accumulation. It was demolished in 1991 after the completion of Brantley Dam (5-030).

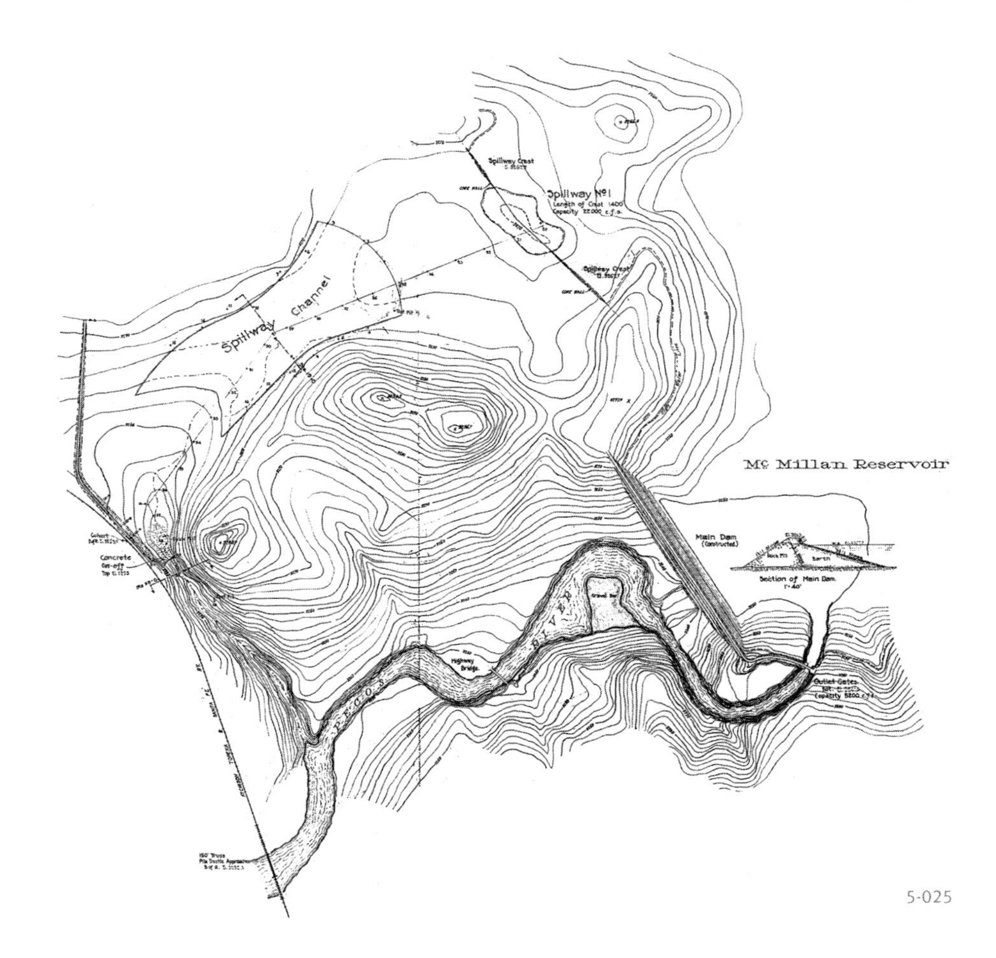

5-025

5-025. Plan of McMillan Dam, Carlsbad Irrigation District, Pecos River, Eddy County, New Mexico, 1888, 1894, rebuilt 1908. United States Reclamation Service, delineator, 1916. P&P,HAER,NM,8-CARL.V,1A-58,detail.

5-026

5-027

5-028

5-026. Headgate and reservoir, McMillan Dam, Carlsbad Irrigation District, Pecos River, Eddy County, New Mexico, 1888, 1894. Walter J. Lubken, photographer, 1906. P&P,HAER,NM,8-CARL.V,1A, no. 48.

5-027. Raising the gates at McMillan Dam, Carlsbad Irrigation District, Pecos River, Eddy County, New Mexico, 1888, 1894. Unidentified photographer, ca. 1895. P&P,HAER,NM,8-CARL.V,1A, no. 45.

5-028. CCC reconstruction of flood damage, McMillan Dam, Carlsbad Irrigation District, Pecos River, Eddy County, New Mexico, 1888, 1894, rebuilt 1908, modified 1937. Unidentified photographer, 1937. P&P,HAER,NM,8-CARL.V,1A, no. 51.

5-029

5-030

5-029. Outlet works, McMillan Dam, Carlsbad Irrigation District, Pecos River, Eddy County, New Mexico, 1908, modified 1937. Fred Quivik, photographer, 1990. P&P,HAER,NM,8-CARL.V,1A, no. 28.

5-030. Summer (Alamogordo) Dam, Carlsbad Irrigation District, Pecos River, Fort Summer Vic., De Baca County, New Mexico, 1936–1937, modified 1954–1956. Acme, 1938. P&P,LC-dig-ppmsca-17231.

Built by the Bureau of Reclamation as an additional unit in the Carlsbad Irrigation District, Alamogordo Dam is a zoned earthen structure 164 feet high and 3,084 feet long. In 1956, the dam was raised 16 feet and its spillway capacity increased. Its name was changed in 1974 to avoid confusion with the town of Alamogordo, New Mexico.

5-031. Brantley Dam, Carlsbad Irrigation District, Pecos River, Eddy County, New Mexico, 1983–1988. Fred Quivik, photographer, 1990. P&P,HAER,NM,8-CARL.V,1I, no. 2.

5-031

United States Reclamation Service Projects

By 1890, settlement and irrigation in the Rio Grande River valley had depleted the flow of the river, causing it to run dry at El Paso more frequently and for longer periods. When an American company obtained permission to dam the river, the Mexican government appealed the decision to the International Boundary Commission. Its decision confirmed that American water usage had significantly reduced the flow of the river at El Paso, and it recommended that the United States build a storage dam and guarantee Mexico one-half of the water supply.

With the passage of the Reclamation Act in 1902, the United States Reclamation Service began to explore potential dam sites on the Rio Grande, settling on Elephant Butte two years later. In 1906, the two countries agreed on a distribution of the Rio Grande for irrigation, with Mexico receiving its portion at the Acequia Madre, above the city of Juárez, and the United States paying for the dam and delivery of the water to the Mexican canal. Construction of Elephant Butte began shortly afterwards, making Elephant Butte the first dam to distribute waters across an international border. The dam is 301 feet high and 1,674 feet long, with one of the largest reservoirs in the world. After the completion of an additional storage facility (Caballo Dam, 25 miles downstream from Elephant Butte), a hydroelectric plant was added in 1940.

5-032. Foundation of Elephant Butte Dam riverbed, Rio Grande, Truth or Consequences, Sierra County, New Mexico, 1912–1916. Unidentified photographer, 1914. P&P,LC-dig-ppmsca-17373.

5-032

5-033. Elephant Butte Dam, Rio Grande, Truth or Consequences, Sierra County, New Mexico, 1912–1916. United States Reclamation Service, 1917. P&P,LC-dig-ppmsca-17303.

5-034. Crest of Elephant Butte Dam, Rio Grande, Truth or Consequences, Sierra County, New Mexico, 1912–1916. United States Reclamation Service, 1917. P&P,LC-dig-ppmsca-17302.

5-033

5-034

5-035

5-037

5-036

5-035. Upstream face of Elephant Butte Dam under construction, Rio Grande, Truth or Consequences, Sierra County, New Mexico, 1912–1916. George Grantham Bain Collection, ca. 1915. P&P,LC-DIG-ggbain-21700.

5-036. First use of spillway since 1916, Elephant Butte Dam, Rio Grande, Truth or Consequences, Sierra County, New Mexico, 1912–1916. Acme, 1941. P&P,LC-dig-ppmsca-17244.

5-037. Elephant Butte reservoir, Rio Grande, Truth or Consequences, Sierra County, New Mexico. HD1694 N6 C42 1987,p. 427.

5-038

5-038. Completing reconstruction of settling basin and concrete lining, Franklin Canal, Rio Grande, El Paso, El Paso County, Texas, 1914. Unidentified photographer, 1914. P&P,LOT 7592,no. 5.

The Franklin Canal supplies water to the upper El Paso River valley. Built in 1889–1890 by the El Paso Irrigation Company, it was acquired by the United States Reclamation Service in 1912 to become one of the Rio Grande Project's main canals. Additional works constructed in the Rio Grande River valley from 1914 to 1919 include the Mesilla Diversion Dam, Percha Diversion Dam, and Rincon Valley Canal.

5-039

5-039. Newly designed canal structure for Franklin Canal, Rio Grande, El Paso Valley, Texas, 1914. Unidentified photographer, 1914. P&P,LOT 7592,no. 6.

FLOOD CONTROL DAMS

As the Red and Arkansas rivers run into the Mississippi, they undergo the kind of intensive flooding that occurs throughout the Mississippi River basin. The rivers of the Texas Gulf region, on the other hand, run fast to the Gulf with intermittent and, at times, torrential flows, cutting canyons through the plateau and washes through the desert.

5-040. Flood damage, Clarendon, White River, Monroe County, Arkansas. Unidentified photographer, 1927. P&P,LC-USZ62-53837.

5-041. Dry bed of the Colorado River, Texas. Russell Lee, photographer, 1939. P&P,LC-USF 34-33187-D.

5-040

5-041

5-042

5-043

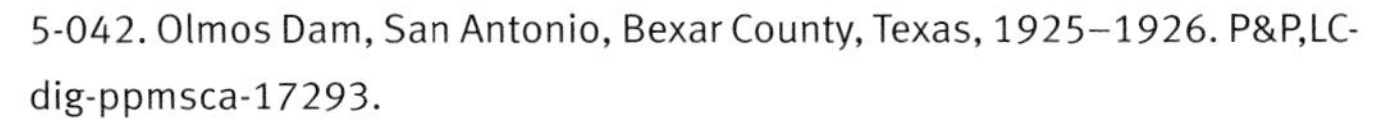

5-042. Olmos Dam, San Antonio, Bexar County, Texas, 1925–1926. P&P,LC-dig-ppmsca-17293.

After a catastrophic flood in 1921, the City of San Antonio adopted a flood control plan that involved the construction of Olmos Dam on the San Antonio River. The dam is fitted with outlet gates at its base, and the reservoir is kept empty and used as a municipal park.

5-043. Morris Sheppard Dam, Brazos River, Palo Pinto County, Texas, 1938–1941. Acme, 1941. P&P,LC-dig-ppmsca-17276.

Recurrent flooding of the Brazos River led to the first conservation legislation in Texas (in 1919) and the establishment of the Brazos River Conservation and Reclamation District ten years later, which became a model for similar legislation in other states. Originally called Possum Kingdom Dam, this structure was the first project of the district, funded through the Works Progress Administration and revenues from its power generation program. The dam provides flood control for Waco and the lower Brazos River valley and has contributed to irrigated farming and industrial development in the valley. The dam is a flat slab buttress construction 2,740 feet long and 190 feet high, with nine spillway gates in the center and a powerhouse.

5-044. Pensacola Dam, Neosho (Grand) River, Langley, Mayes County, Oklahoma, 1938–1940. Acme, 1940. P&P,LC-dig-ppmsca-17252.

The State of Oklahoma was successful in obtaining federal support for this first large multipurpose dam in Oklahoma. A multiple arch dam, it spans a mile across the Grand River to impound the vast Lake O' the Cherokees.

5-044

Highland Lakes Project

The largest flood control project in Texas was Highland Lakes Project. The initial impetus for this project was a half-completed hydroelectric dam in Burnet County, begun and then abandoned by a Chicago-based utility that went bankrupt with the onset of the Great Depression. The federal government agreed to fund completion of the project, provided it would be run by a public agency. The result was the formation in 1933 of the Colorado River Authority—modeled after the Tennessee Valley Authority—charged with water supply, flood control, electricity generation, and soil conservation programs in the lower Colorado River.

The period 1930–1950 saw the construction of the six dams of the Highland Lakes Project, creating a series of lakes on the Colorado River north of Austin. Two of them, Buchanan and Mansfield dams, served for water storage, while the others acted as "pass-through lakes." They were built in pairs. The first pair was constructed in the 1930s: Buchanan Dam and the smaller Inks Dam downstream. Over 2 miles long, Buchanan is the longest multiple arch dam in the nation. Ten years later, Wirtz Dam and Starcke Dam were built further downstream on the Lower Colorado. At the bottom of the chain of lakes and just north of the city of Austin, Mansfield Dam forms Lake Travis. Over a mile long and 266 feet high, it remains one of the largest masonry structures in the world.

5-045

5-046

5-045. Lake Bridgeport Dam, West Fork of the Trinity River, Bridgeport Vic., Wise County, Texas, 1929–1931, modernized 1969. Acme, 1941. P&P,LC-dig-ppmsca-17267.

In 1922, the periodic flooding of the Clear Fork and Trinity rivers caused significant damage to the city of Fort Worth, leading to the construction of two flood control dams, Lake Bridgeport and Eagle Mountain. Lake Bridgeport is a rolled earthfill dam, with outlet conduits in the base of the dam and a spillway of three 20-foot bays, two of which feature vertical lift gates. In 1969, the crest of the dam was raised 10 feet and a new 90-foot spillway was added, with eight roller lift gates.

5-046. Dry dams built for flood water control, Sandstone Creek, Upper Washita Soil Conservation District, Oklahoma, 1957. U.S. Soil Conservation Service, 1957. P&P,LC-dig-ppmsca-17311.

U. S. ARMY CORPS OF ENGINEERS DAMS

In the wake of the Flood Control Act of 1936, the U.S. Army Corps of Engineers established a Southwestern Division, charging it with responsibility for flood control from the Arkansas River to the Rio Grande. With the growing urbanization and industrialization of this region of the country, control of its flood-prone rivers became a national priority. And with the onset of World War II, hydropower generation did as well. Some of the major projects in this division included Conchas Dam, Denison Dam, and the McLellan-Kerr Navigation System on the Arkansas River.

5-047

5-047. Conchas Dam under construction, South Canadian River, Tucumcari Vic., San Miguel County, New Mexico, 1935–1938. AP, 1938. P&P,LC-dig-ppmsca-17240.

Conchas Dam was a federally funded project built for irrigation, flood control, power development, and municipal water supply in a sparsely settled region on the edge of the Dust Bowl. In 1926, New Mexico, Texas, Oklahoma, and Arkansas concluded a compact allocating the waters of the Canadian River; a few years later, Secretary of the Interior Harold Ickes commissioned a complete investigation of the watersheds of the Arkansas River. Working with the U.S. Army Corps of Engineers, the Arkansas Basin Committee recommended a number of projects in the basin, including this dam on the Canadian River. It is one of many Depression-era projects that supported irrigation and resettlement in the arid West. Built by the Corps of Engineers, Conchas Dam is a concrete gravity structure 235 feet high and 6,230 feet long.

5-048

5-048. Conchas Dam, South Canadian River, Tucumcari Vic., San Miguel County, New Mexico, 1935–1938. U.S. Army Corps of Engineers, 1939. HD1694 N6 C42 1987.

5-049

5-049. Intake structure, Denison Dam, Red River, Denison, Grayson County, Texas, 1942–1945. Wide World, 1943. P&P,LC-dig-ppmsca-17242.

At the time of its completion by the U.S. Army Corps of Engineers, this mile-long dam was the largest rolled earthfill dam in the world, and it influenced subsequent Corps projects in the West—particularly in terms of soil sampling and testing. It was built as a multipurpose dam during wartime, and its hydroelectric generation significantly contributed to industrial capacity in Texas and Oklahoma, while the sizable Lake Texoma has become a recreational and wildlife resource in the region.

5-050

5-050. Construction of water conduits to powerhouse, Denison Dam, Red River, Denison, Grayson County, Texas, 1942–1945. U.S. Army Corps of Engineers, 1940. P&P,LC-dig-ppmsca-17241.

5-051

5-051. Exterior of powerhouse, Denison Dam, Red River, Denison, Grayson County, Texas, 1942–1945. U.S. Army Corps of Engineers, 1944. P&P,LC-dig-ppmsca-17390.

5-052. Construction of draft tube forms, Denison Dam, Red River, Denison, Grayson County, Texas, 1942–1945. U.S. Army Corps of Engineers, 1943. P&P,LC-dig-ppmsca-17394.

5-053. Interior of powerhouse, Denison Dam, Red River, Denison, Grayson County, Texas, 1942–1945. U.S. Army Corps of Engineers, 1944. P&P,LC-dig-ppmsca-17393.

5-054. Interior of powerhouse, Denison Dam, Red River, Denison, Grayson County, Texas, 1942–1945. U.S. Army Corps of Engineers, 1944. P&P,LC-dig-ppmsca-17391.

5-052

5-053

5-054

INTERNATIONAL BOUNDARY AND WATER COMMISSION DAMS

Following the 1944 U.S.-Mexico Water Treaty, the International Boundary and Water Commission built several new dams to regulate and distribute the waters of the Rio Grande between the two countries. The first project of the commission was Falcon Dam, built 75 miles downstream from Laredo, Texas, and Nuevo Laredo, Tamaulipas. It remains the lowermost dam on the river. Designed primarily for water storage, it also controls floods and generates power to offset construction costs. The 5-mile-long earthfill dam created a large reservoir that covered hundreds of prehistoric archeological sites and three towns, (including the historic town of Guerrero, Tamaulipas) and displaced several thousand residents, requiring the construction of several new towns on both sides of the river.

Ten years later, Amistad Dam was built upstream, also as a very long (6-mile) earthfill dam, with a concrete gravity spillway section of sixteen gates. Mexico and the United States jointly operate both Falcon and Amistad dams. With the subsidence of Falcon reservoir, the ruins of historic Guerrero are once again visible.

5-055

5-055. Mexican and U.S. engineers collaborate on planning and construction of Falcon Dam, Rio Grande, Zapata Vic., Starr County, Texas, 1950–1954. Unidentified photographer, 1952. P&P,LC-dig-ppmsca-17379.

5-056. Falcon Dam, Rio Grande, Zapata Vic., Starr County, Texas, 1950–1954. AP, 1953. P&P,LC-dig-ppmsca-17245.

5-056

5-057. Family to be relocated by construction of Falcon Dam, Rio Grande, Zapata Vic., Starr County, Texas, 1950–1954. Unidentified photographer, 1952. P&P,LC-dig-ppmsca-17378.

5-058. Construction of Nuevo Guerrero, Tamaulipas, Mexico, Rio Grande, 1950–1954. Unidentified photographer, 1952. P&P,LC-dig-ppmsca-17377.

5-057

5-058

During the Civil War, the Union's attempt to capture Shreveport, Louisiana—known as the Red River military expedition—was dogged throughout the campaign by the river's low water level. On his retreat downstream, Admiral David Porter discovered that the river was so low his gunboats were trapped above the rapids at Alexandria. To save the fleet, Colonel Joseph Bailey of the Wisconsin Cavalry suggested damming the river to raise the water level and float the boats over shallow rapids. As a former lumberman and chief engineer, Bailey came up with a design that included a pair of wing dams across the river (one of trees and one of timber crib) that raised the water level to a depth sufficient for the ironclad ships. This took two weeks and was a spectacle that attracted sightseers from nearby Alexandria.

5-059. Admiral Porter's fleet passing Colonel Bailey's Dam with townspeople looking on, Red River, Alexandria, Rapides Parish, Louisiana, 1864. C. E. H. Bonwill, delineator, 1864. P&P,LC-USZ62-32153.

5-060. Sketch of the two breakwaters above Alexandria in the Red River, constructed by Lieut. Col. Bailey, to extricate the heavy iron-clads and transports of the Mississippi Squadron, under command of Rear Admiral D. D. Porter. F. H. Gerdes, delineator, 1864. G&M,G3992.R4S5 1864.G4.

5-059

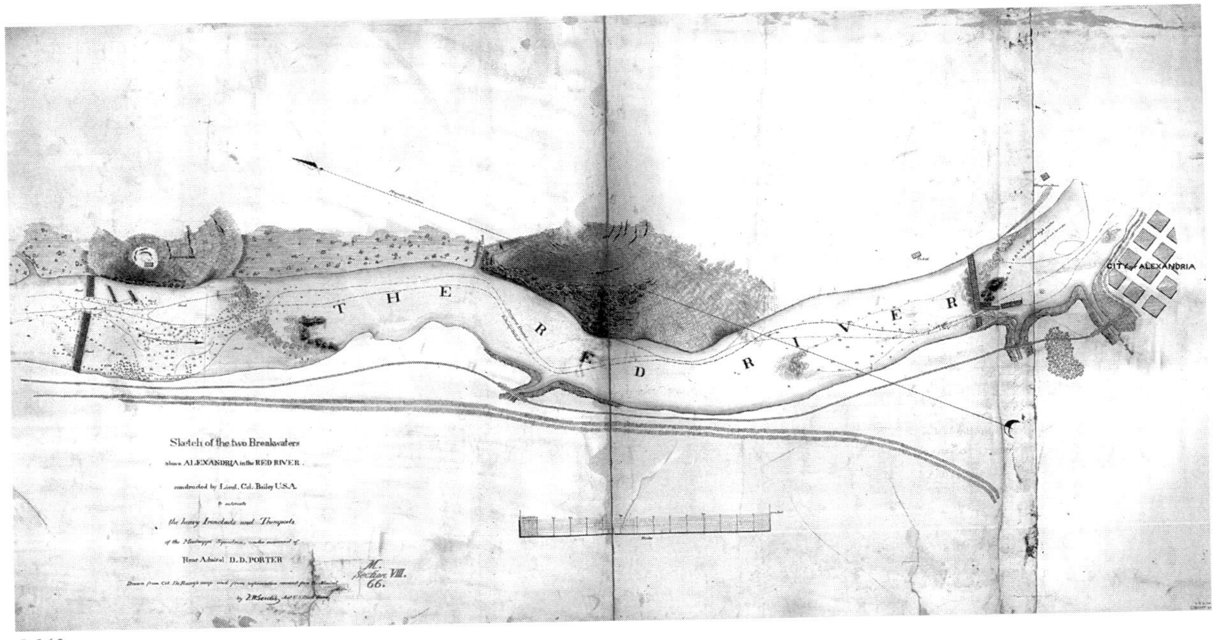

5-060

5-061. Tree dam section, Colonel Bailey's Dam, Red River, Alexandria, Rapides Parish, Louisiana, 1864. Col. Bailey's report to U.S. War Department, 1891–1895. P&P,HAER,LA,40-ALEX.V, 2, sheet no. 7.

5-062. Remains of tree dam, Colonel Bailey's Dam, Red River, Alexandria, Rapides Parish, Louisiana, 1864. George J. Castillo, photographer, 1984. P&P,HAER,LA,40-ALEX.V,2, no. 13.

5-061

5-062

COLORADO–GREAT BASIN

COLORADO RIVER BASIN

Gunnison, Dolores, Green, San Juan, Animas, Willcox Playa, Little Colorado, Gila, Salt, and Verde Rivers

The Colorado River drains the western slopes of the Rocky Mountains in the United States and part of Mexico, descending more than 10,000 feet as it flows to the Sea of Cortez. Three-quarters of the Colorado River basin is federal land and Indian reservations. Also called the Grand River in the nineteenth century, its major tributaries are the Green River of Utah and Wyoming, the Gunnison River of Colorado, the Dolores and San Juan rivers of New Mexico, and the Little Colorado and Gila rivers of Arizona. With John Wesley Powell's expeditions down the Colorado and George M. Wheeler's survey of the country west of the 100th meridian, photographers such as E. O. Beaman, Timothy O'Sullivan, John K. Hillers, and William Henry Jackson captured the grandiose landscapes of this sparsely inhabited and arid region, its multicolored sedimentary rocks carved by rivers into canyons and buttes.

6-001. Theodore Roosevelt Dam, Salt River, Gila County, Arizona, 1903–1911. E. E. Kunselman, photographer, 1916. P&P,LC-dig-ppmsca-17362.

Colorado's water appropriation laws controlled early water development projects in the river's upper reaches. Mining, irrigation, and large-scale diversion projects were contemporaneous with similar projects in eastern Colorado, Montana, and Idaho. Early Spanish irrigation systems such as the acequias were object-lessons to later settlers in the Colorado River basin, while Mormon farmers brought their own cooperative and energetic approach to irrigation projects in the Uinta Mountains east of Salt Lake City. The Colorado River basin is also home to some of the earliest irrigation projects in the country, projects that long predated the Spanish missions. Irrigation formed part of the

6-002

6-002. Grand Canyon, Colorado River, Arizona. John K. Hillers, photographer, 1872. P&P,LC-USZC4-8294.

everyday life of the Anasazi, Mogollon, and Hohokam Indians of the high deserts and arid tablelands of the Southwest. We can see some continuity in irrigation practices from these earliest agriculturalists, through the Spanish settlements and into subsequent waves of Anglo farmers.

The newly founded United States Reclamation Service built two major projects in this region—the Roosevelt Dam on the Salt River in Arizona (6-050–6-053), and the Truckee-Carson Project in Nevada (6-035–6-039). In 1922, the seven states in the Colorado's drainage basin signed the Colorado River Compact, an agreement on how to retain the river and divide it up. Six year later, Congress authorized construction of Boulder (now Hoover) Dam (6-09–6-105). This was a major engineering accomplishment of its time.

6-003. Colorado River basin, showing division into upper and lower basins. HD1695 W4 R68 2006,p. 241.

6-003

Shortly after the completion of Hoover Dam, planning began on Parker Dam to supply water to Los Angeles (6-106–6-107). Further downstream, Imperial Dam diverts water into the All-American Canal (6-108–6-111), turning 630,000 acres of California desert into farmland, yet creating problems with drainage and salinity. Overall, water from the lower Colorado River irrigates a million acres in the United States and half a million in Mexico, and it supplies urban populations in Arizona, California, and Nevada. Today the river rarely reaches the Gulf of California, as the Morelos Diversion Dam on the Mexican border sends what remains of the water to the cities of Mexicali and Tijuana.

Federal projects on the river have largely been carried out by the Bureau of Reclamation. These include the Colorado River Project, the Central Utah Project, the Colorado River Storage Project, and the Central Arizona Project—in all, nine major dams on the main stem of the river, and numerous additional dams on its tributaries. In the mid-1960s, widespread opposition to Glen Canyon Dam (6-116–6-117), just upstream from the Grand Canyon, helped shape current policies of water management and environmental protection.

6-004

6-004. Row crop irrigation in Yuma, Yuma County, Arizona. Ben D. Glaha, photographer, 1939. P&P,LC-USZ62-118075.

GREAT BASIN

Great Salt Lake and its tributaries; and the Humboldt, Truckee, and Carson rivers

The Great Salt Lake is the major remnant of Lake Bonneville, a large freshwater lake of the Pleistocene era that occupied much of western Utah. Located at the western margin of the Rocky Mountain's Wasatch Range, it receives water from the Bear, Weber, and Jordan rivers. With no outlet and steady evaporation, the lake became salty from accumulated minerals. This large body of water in an arid region attracted early attention. Native American cultures used the freshwater marshes and streams around the lake for hunting and fishing. When hearing the Frémont expedition (1843–1844) reports of a Great Salt Lake in the West, Mormon leaders in Nauvoo, Illinois, selected it as a destination, and the Mormons reached the Great Salt Lake Valley in 1847.

HIGH MOUNTAIN DAMS BUILT FOR INDUSTRY

Dams built for mining and milling on the western slopes of Colorado's Rocky Mountains were very much like those built on the eastern slopes. Generally timber-crib structures or earthen dams, they were rapidly erected in these remote locations to provide hydropower, to store water for hydraulic mining, or eventually (with adequate investment) to generate hydroelectricity in larger industrial operations like the Tacoma Project.

6-005. Terminal Dam, Tacoma Project, Cascade, Little Cascade and Elbert Creeks, La Plata County, Colorado, 1903–1905, demolished 1980. Unidentified photographer, 1905. P&P,HAER,COLO,33-TAC.V,4, no. 7.

6-005

6-006

6-007

6-006. Construction, Terminal Dam, Tacoma Project, Cascade, Little Cascade and Elbert Creeks, La Plata County, Colorado, 1903–1905, demolished 1980. Unidentified photographer, 1903. P&P,HAER,COLO,33-TAC.V,4-2.

6-007. Downstream face, Aspaas Dam, Tacoma Project, Cascade, Little Cascade and Elbert Creeks, La Plata County, Colorado, 1925–1926, demolished 1980. Unidentified photographer, 1925. P&P,HAER,COLO,33-TAC.V,2-5.

This hydroelectric development served mining operations in Silverton, 50 miles to the north, because the mountainous terrain had made coal deliveries prohibitively expensive. Modeled after the Ames hydroelectric facility near Telluride (developed by L. L. Nunn), the Tacoma project involves a number of diversion dams, tunnels, canals, and reservoirs, along with the powerhouse. The larger Terminal and smaller Aspaas dams were both timber-crib rock-filled structures.

6-008

6-009

6-010

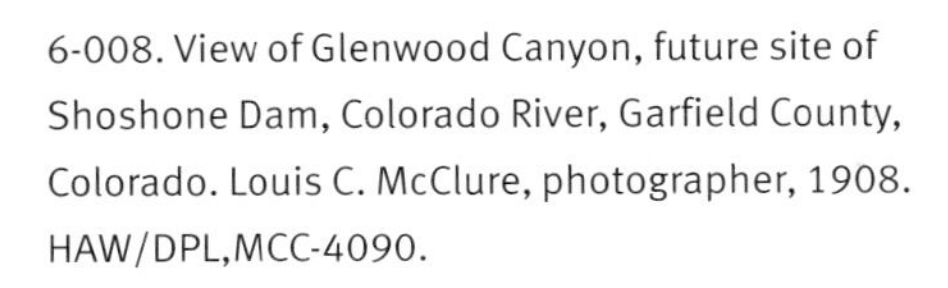

6-008. View of Glenwood Canyon, future site of Shoshone Dam, Colorado River, Garfield County, Colorado. Louis C. McClure, photographer, 1908. HAW/DPL,MCC-4090.

6-009. Shoshone Dam and Hydroelectric Plant, Colorado River, Garfield County, Colorado, 1905–1909. Harry M. Rhoads, photographer, ca. 1910. HAW/DPL,Rh-491.

Built by Central Colorado Power Company, Shoshone Dam is the oldest hydroelectric project in the state. Water from the dam is diverted into a tunnel in the canyon wall, dropping 287 feet into the power plant—at times, leaving the Colorado riverbed dry for the 2 miles between the dam and the power plant. In 1924, it was sold to the Public Service Company of Colorado.

6-010. Bear trap gates, Shoshone Dam, Colorado River, Garfield County, Colorado, 1905–1909. Louis C. McClure, photographer, 1910. HAW/DPL,MCC-3401.

6-011. Interior of Shoshone Hydroelectric Plant, Colorado River, Garfield County, Colorado, 1905–1909. Louis C. McClure, photographer, 1909. HAW/DPL,MCC-3391.

6-011

SMALL IRRIGATION DAMS

By far the most typical dam in the Colorado region and throughout the West is the small irrigation dam. Made of wood, stone, earth, or concrete, these dams were built by farmers and ranchers to store seasonal or intermittent flows, or as diversion structures to direct water into canals or reservoirs. Such dams and canals also have a long history in the Colorado River basin, as all Native Americans in the region relied on regular irrigation to grow their staple crops. Subsequent settlers established irrigation cooperatives or companies to construct dams and canals. In regions where settlers collaborated for the common benefit, such as the Mormon areas of Utah, they were adapted to supply collectively owned canals or small power plants.

6-012

6-012. Concho Dam, Concho Lake, Saint Johns Vic., Apache County, Arizona, ca. 1915. Russell Lee, photographer, 1940. P&P,LC-USF33-012943-M5.

6-013

6-013. Irrigation dam, Santa Clara River, Santa Clara, Washington County, Utah. Russell Lee, photographer, 1940. P&P, LC-USF33-012941-M2.

6-014. Raceway and conduit feeding a hydroelectric plant, Logan River, Cache County, Utah. Russell Lee, photographer, 1940. P&P,LC-USF34-037309-D.

6-015. Hydroelectric plant, Logan City, Cache County, Utah. Russell Lee, photographer, 1940. P&P,LC-USF34-037231-D.

6-016. Turbines in hydroelectric plant, Logan City, Cache County, Utah. Russell Lee, photographer, 1940. P&P,LC-USF34-037182-D.

6-014

6-015

6-016

6-017

6-017. Farm Security Administration cooperative irrigation pipeline, Saint George, Washington County, Utah. Russell Lee, photographer, 1940. P&P,LC-USF34-037880-D.

6-018. Bear River Canal Dam, Bear River, Box Elder County, Utah. Unidentified photographer, 1917. P&P,LC-dig-ppmsca-17361.

Private investors financed this timber-crib and stone diversion dam and its attendant canals to attract farmers to this region north of the Great Salt Lake. Purchased in 1902 by the Utah Idaho Sugar Company, the irrigation system contributed to settlement along the Bear River. The original diversion dam was replaced by Utah Power in 1927.

6-018

6-019. Uinta Mountains, Weber River valley, Morgan County, Utah. Russell Lee, photographer, 1940. P&P,LC-USF34-038968.

6-020. Historic Dams of the Upper Provo River, Summit and Wasatch counties, Utah, 1914–1935. Bonnie J. Halda, delineator, 1986. P&P,HAER,UTAH,22-KAM.V,1, no. 2.

A number of large Mormon-controlled irrigation companies built dams on the Upper Provo River to impound water within the Uinta and Wasatch-Cache National Forests. Although technologically simple, these small earthfill and masonry dams reflect the important role of irrigation in the settlement of Utah. The carefully built rock faces of some of these dams were mandated by the National Forest Service. These dams were later incorporated into the Central Utah Project.

6-019

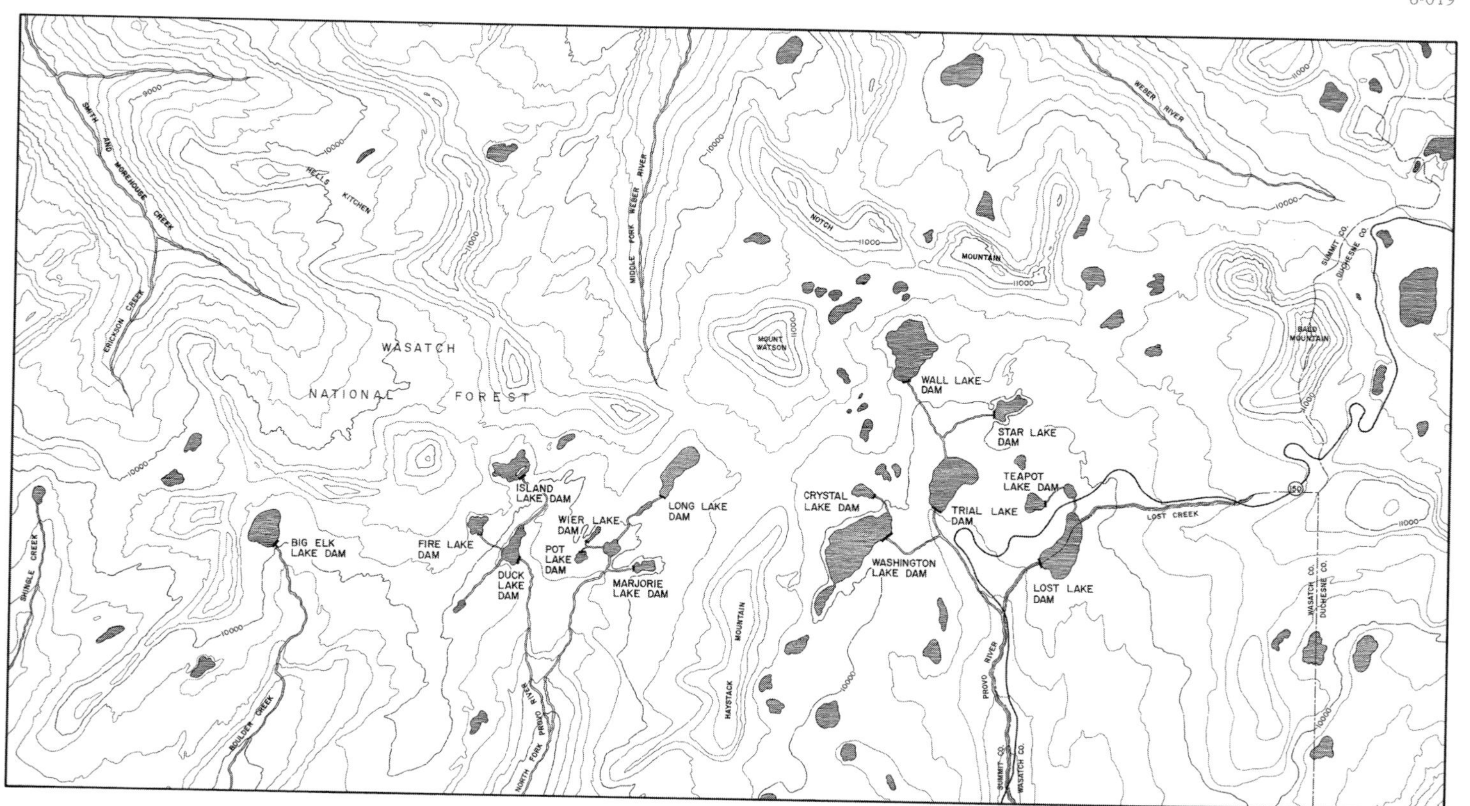

6-020

6-021

6-022

6-023

6-021. Construction, Trial Lake Dam, Kamas Vic., Summit County, Utah, 1912–1914. Unidentified photographer, 1912. P&P,HAER,UTAH,22-KAM.V,1-L, no. 11.

6-022. Downstream face, Washington Lake Dam, Kamas Vic., Summit County, Utah, 1913–1914. Unidentified photographer, 1913. P&P,HAER,UTAH,22-KAM.V,1-N, no. 11.

6-023. View of inclined outlet gate wheel, stem, and stem guide, Washington Lake Dam, Kamas Vic., Summit County, Utah, 1913–1914. Clayton B. Fraser and Robert W. Righter, photographers, 1985. P&P,HAER,UTAH,22-KAM.V,1-N, no. 7.

6-024. Marjorie Lake Dam, Kamas Vic., Summit County, Utah, 1935. Clayton B. Fraser and Robert W. Righter, photographers, 1985. P&P,HAER,UTAH,22-KAM.V,1-H, no. 3.

6-025. Historic Dams of the Uinta Mountains, Duschesne County, Utah, 1920–1951. Bonnie J. Halda, delineator, 1986. P&P,HAER,UTAH,7-MOHO.V,1.

The Uinta Mountains east of Salt Lake City form the headwaters of the Uinta, Strawberry, and Duchesne rivers, which drain into the Green River. The National Forest Service gave permission to Mormon irrigation companies to build dams to impound spring run-off that would otherwise move quickly through the Uinta basin. The waters stored behind these small earthfill or masonry dams would be released during the growing season. These dams, for the most part constructed between 1920 and 1930, are representative of early irrigation efforts in the Uinta basin. They were later incorporated into the Upalco Unit of the Central Utah Project.

The Green River, the largest tributary of the Colorado, rises near the Continental Divide and flows southerly 730 miles through deep canyons to the Colorado River in Canyonlands National Park. Its headwaters in the Uinta Mountains were developed by Mormon settlers in the nineteenth century. In the twentieth century, the Bureau of Reclamation's Colorado River Storage Project and Central Utah Project extensively developed the Green River basin for irrigation, mining, and hydroelectric power.

6-024

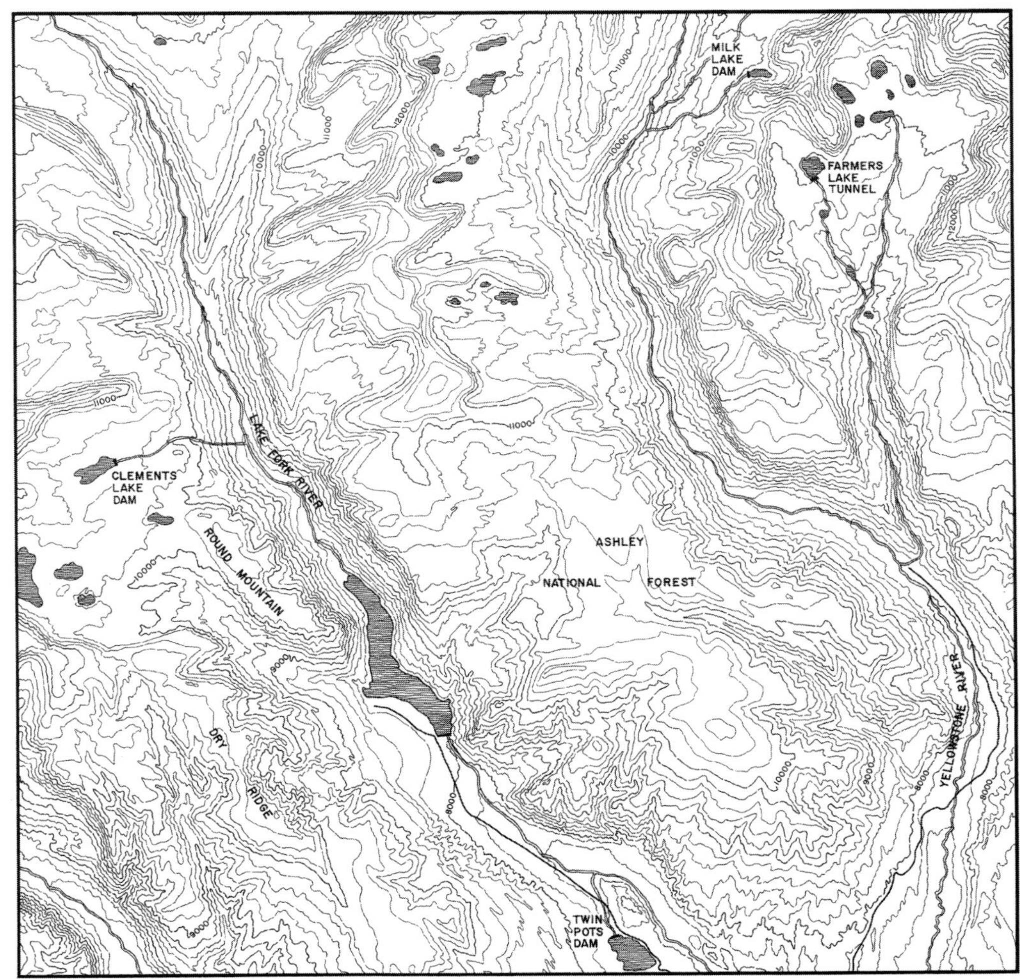

6-025

6-026

6-027

6-026. Upstream face, Five Point Lake Dam, Mountain Home Vic., Duchesne County, Utah, 1929. Clayton B. Fraser and Robert W. Righter, photographers, 1985. P&P,HAER,UTAH,7-MOHO.V,1-H, no. 1.

6-027. Downstream face, Milk Lake Dam, Mountain Home Vic., Duchesne County, Utah, 1935. Clayton B. Fraser and Robert W. Righter, photographers, 1985. P&P,HAER,UTAH,7-MOHO.V,1-K, no. 4.

6-028

6-028. Dam crest and outlet gate wheel, Milk Lake Dam, Mountain Home Vic., Duchesne County, Utah, 1935. Clayton B. Fraser and Robert W. Righter, photographers, 1985. P&P,HAER,UTAH,7-MOHO.V,1-K, no. 6.

6-029. Outlet and concrete collar, Five Point Lake Dam, Mountain Home Vic., Duchesne County, Utah, 1929. Clayton B. Fraser and Robert W. Righter, photographers, 1985. P&P,HAER,UTAH,7-MOHO.V,1-H, no. 6.

6-029

6-030

6-030. Roller used for earth compacting, Five Point Lake Dam, Mountain Home Vic., Duchesne County, Utah, 1929. Clayton B. Fraser and Robert W. Righter, photographers, 1985. P&P,HAER,UTAH,7-MOHO.V,1-H, no. 7.

6-031. Reservoir, Twin Pots Dam, Mountain Home Vic., Duchesne County, Utah, 1921, rebuilt 1931. Clayton B. Fraser and Robert W. Righter, photographers, 1985. P&P,HAER,UTAH,7-MOHO.V,1-M, no. 1.

6-031

EARLY DIVERSION PROJECTS

Colorado water diversion projects were enabled by the state's Doctrine of Prior Appropriation, which stated that the first users of a water supply maintained rights to that water as long as they used it beneficially. This made it possible to build canals and diversion projects across unsettled land without fearing that a later landowner could claim the water. The rapidly flowing rivers and dry climate of the West encouraged ambitious development schemes such as these.

Montezuma Valley Irrigation Company System

The system began with plans for a tunnel to divert water from the fast-flowing Dolores River to the adjacent Montezuma River valley to the south. Funded by local developers, it was one of the earliest large-scale irrigation projects in the Southwest. It is also notable for its mile-long tunnel, one of the first irrigation tunnels in the region. Other elements in the project include the Great Cut ditch, the Morton flume, and diversion dams for Narraguinnep Reservoir (1888, rebuilt 1907, modified 1956), Groundhog Reservoir, and Totten Lake (both built and failed in 1907 and rebuilt in subsequent years).

6-032. Morton flume, Montezuma Valley Irrigation Company System, Dolores River, Dolores Vic., Montezuma County, Colorado, 1885–1889. Unidentified photographer, 1951. P&P,HAER,COLO,42-DOL.V,5, no. 3.

6-032

6-033

6-034

A chief engineer and subsequent manager of the Montezuma Valley Irrigation Company, A. L. Fellowes, was appointed hydrographer for the U.S. Geological Survey from 1898 to 1902, when he mapped the route for the Gunnison Tunnel, which connected that river with the Uncompahgre Valley. In 1902, he joined the newly formed United States Reclamation Service.

6-033. Wooden headgate, Montezuma Valley Irrigation Company System, Dolores River, Dolores Vic., Montezuma County, Colorado, 1885–1889. Unidentified photographer, ca. 1897. P&P,HAER,COLO,42-DOL.V,5, no. 5.

6-034. Outlet, Dolores tunnel, Montezuma Valley Irrigation Company System, Dolores River, Dolores Vic., Montezuma County, Colorado, 1885–1889. Unidentified photographer, 1950. P&P,HAER,COLO,42-DOL.V,5, no. 1.

Gunnison Tunnel

Begun by private developers, the Gunnison Tunnel was taken over by the United States Reclamation Service in 1903 as one of its first five projects. Drilled through nearly 6 miles of rock, it brought water from the Black Canyon of the Gunnison River through a mountain range to the adjacent Uncompahgre River valley. Since 1967, the Bureau of Reclamation has built three dams upstream from the tunnel, impounding the Gunnison River for 40 miles.

6-035. Official opening of the Gunnison Tunnel, Uncompahgre Valley, Montrose, Montrose County, Colorado, 1905–1909. Almeron Newman, photographer, 1909. P&P,LC-dig-pan-6a02584.

6-035

FEDERAL RECLAMATION PROJECTS

It is difficult to overstate the scale and scope of federally built reclamation projects in the Colorado River basin. Beginning with the establishment of the United States Reclamation Service in 1902, dam building in this region continued unabated until the Bureau of Reclamation was restructured in the 1980s as a water management rather than dam-building agency.

Early reclamation projects built on the state of Colorado's experiences in large-scale water diversion and canal building—these include the Truckee-Carson Project, Laguna Diversion Dam, the Strawberry Valley Project, and the Grand Valley Diversion Dam. The best-known dam on the Colorado, of course, is Hoover Dam, completed in 1936 in Black Canyon by a consortium of investors that obtained the Bureau's backing for their site over a competing site at Marble Canyon. After the completion of Hoover Dam, irrigation and power generation went hand-in-hand with dam development along the Colorado River, and business interests in Southern California played a large role in influencing the goals of the federal agency. This period saw the construction of Parker and Imperial dams and their related aqueducts, which brought Colorado River water to metropolitan Southern California and agricultural valleys in the desert regions.

With the Colorado River Compact of 1922, the seven states that contain the river's watershed negotiated an agreement for the quantities of water to be allocated to each. But to actually capture and deliver the negotiated allocations to each state, a whole host of dams, reservoirs, diversion systems, and tunnels had to be built. A number of large federally funded water projects ensued, the most significant ones being the Central Arizona Project and Colorado River Project on the lower reaches of the Colorado River basin, and the Colorado–Big Thompson Project (1938–1947), Upper Colorado River Storage Project (1956–1978), and the Central Utah Project (1959–1992) on the upper Colorado. The Colorado–Big Thompson Project gathers snowmelt in Rocky Mountain National Park with four dams, pumps the water up to reservoirs, and transports it through a 13-mile-long tunnel across the Continental Divide. The water cascades down a number of diversion structures for irrigation and power production in arid northeastern Colorado. The Central Utah Project is a system of dams, tunnels, aqueducts, and canals (most notably Strawberry Dam) that collects run-off from the Uinta Mountains and conducts it across the Wasatch Range to the densely populated Salt Lake City region. This system ensures that Utah gets its share of water. The Colorado River Storage Project was undertaken to meet the requirement that a specified amount of water be supplied to the lower portion of the river, even in dry years. This involved the construction of a number of dams on the Colorado and its tributaries, including Glen Canyon Dam on the Colorado River in Arizona, Flaming Gorge Dam on the Green River in Utah, Navajo Dam on the San Juan River in New Mexico, and the Curecanti dams on the Gunnison River in Colorado. Finally, the Central Arizona Project involves pumping water from Lake Havasu up through a mountain tunnel into an aqueduct that supplies the cities of Phoenix and Tucson.

By the 1970s, spurred on by such massive projects as Glen Canyon Dam, ecological and environmental concerns began to make their way into legislation, such as the Environmental Protection Act (1970) and the Endangered Species Act (1973). After the oil crisis of 1973, responsibility for electricity generation moved from the Bureau of Reclamation to the Department of Energy. Between 1986 and 1992, the Bureau saw its construction budget cut by 50 percent, a de facto recognition that the country's dam-building days were over. Today the Bureau of Reclamation is concerned with managing these water resources for multiple goals that are responsive to political circumstances. Administering 45 percent of the surface water in the western United States, it is the largest wholesaler of water and the ninth-largest electric utility in the country.

Great Basin Projects

Located in the home state of Senator Francis Newlands, author of the Reclamation Act, the Truckee-Carson Project was one of the first five United States Reclamation Service projects—along with the Salt River (Arizona), Uncompahgre (Colorado), North Platte

6-036

6-036. Opening day ceremonies, Derby Diversion Dam, Truckee-Carson Project, Fallon, Churchill County, Nevada, 1903–1905. Black and Ferguson, 1905. P&P,LC-USZ62-97165.

(Nebraska), and Milk River (Montana) projects. The project includes the concrete gravity Derby Diversion Dam (1905) and the later Boca Dam (1939) on the Truckee River and its tributaries, and the earthfill Lahontan Dam (1915) on the Carson River. The rivers supply the Newlands irrigation project near Fallon, Nevada. The Truckee and Carson rivers rise in California's Sierra Nevada and flow eastward into desert sinks that are remnants of ancient Lake Lahontan. Mercury used in gold and silver mining has contributed to massive pollution of the Carson River.

6-037

6-037. Senator Francis Newlands, Representative Franklin Mondell of Wyoming, and L. H. Taylor, engineer for Nevada, stand on the newly completed Derby Diversion Dam, Truckee-Carson Project, Fallon, Churchill County, Nevada, 1903–1905. Unidentified photographer, 1905. HD1695 W4 R68 2006,p. 121,detail.

6-038

6-038. Lake Tahoe Dam, Truckee River, Truckee-Carson Project, Truckee Vic., Placer County, California, 1909–1913. Unidentified photographer, 1905. HD1695 W4 R68 2006,p. 148,detail.

6-039

6-040

6-039. Spillway through embankment of Lahontan Dam, Carson River, Truckee-Carson Project, Fallon Vic., Churchill County, Nevada, 1911–1915. HD1695 W4 R68 2006,p. 232,detail.

6-040. Rye Patch Dam, Humboldt River, Humboldt Project, Lovelock Vic., Pershing County, Nevada, 1935–1936 Acme, 1936. P&P,LC-dig-ppmsca-17279.

The Humboldt River rises in the Ruby Mountains of northeastern Nevada and meanders 280 miles in a southerly direction until it is vanishes into the desert sands. Tens of thousands of years ago, the freshwater springs at Rye Patch served as a watering hole for camels, bison, and elephants, and evidence is found for human settlement as far back as eight thousand years. In the westward migrations of the nineteenth century, Rye Patch was a welcome stop on the Humboldt Trail.

Although the settlers of the lower Humboldt River valley were initially promised to benefit from the Truckee-Carson Project, the United States Reclamation Service had underestimated the water requirements for desert farming, so it wasn't until the 1930s that the Humboldt Project was established to divert and regulate water supply to the valley. Rye Patch is an earthfill dam with a concrete spillway and outlet works. In 1971, the reservoir and the river below the dam became a state recreation area.

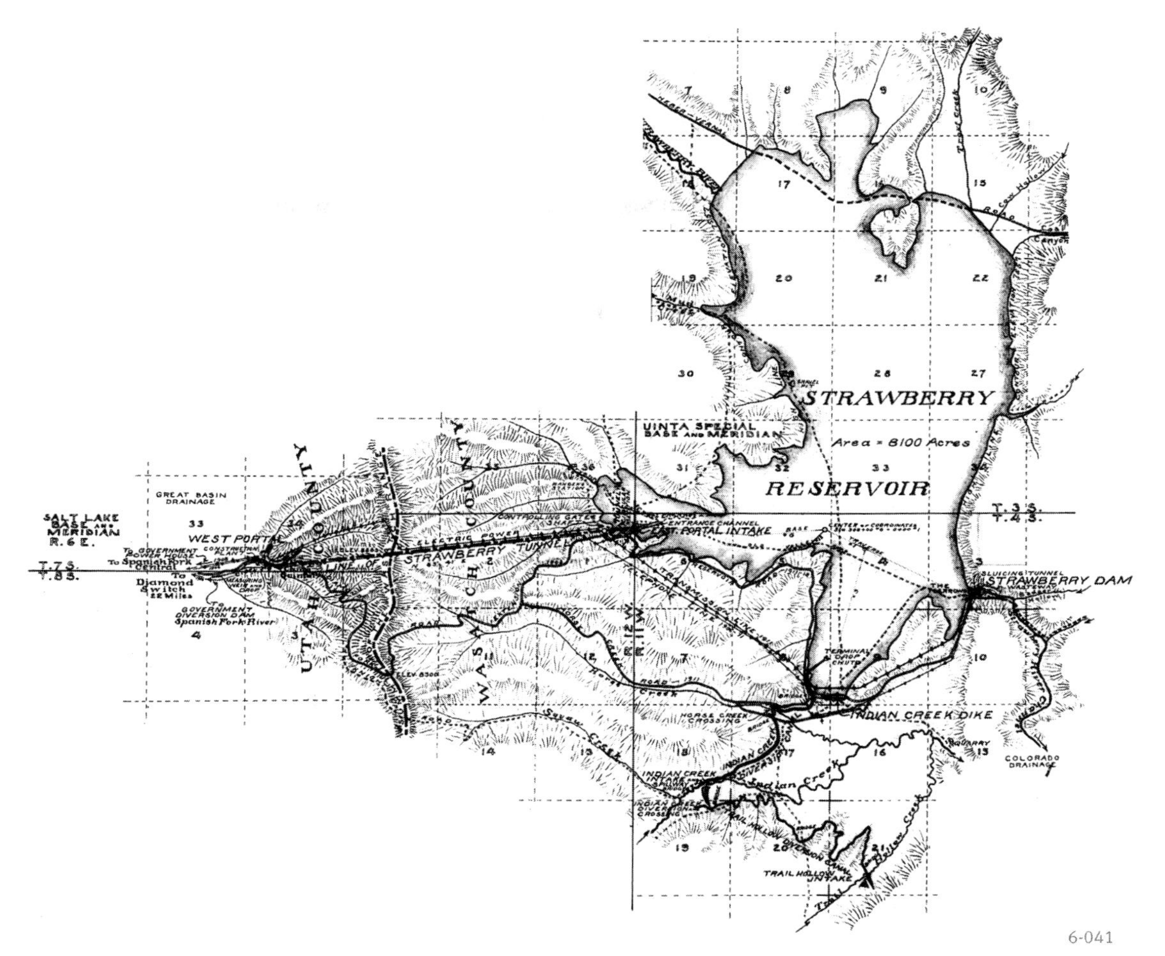

6-041

6-041. Map of reservoir, Strawberry Valley Project, Wasatch County, Utah, 1911-13. United States Reclamation Service, delineator, 1913. P&P,HAER,UTAH,25-PAYS,1-62,detail.

This early project of the United States Reclamation Service diverts the waters of Strawberry River through the Wasatch Range to a tributary of the Spanish Fork River in order to irrigate lands south of Utah Lake. Because the reservoir lands were part of the Uintah Indian Reservation, Congress removed them from reservation land to turn them over to the newly created United States Reclamation Service. The Strawberry Project promoted farming in southern Utah County, provided electrical power, and built roads. Dams in this project are Strawberry Dam (1911–1913), Spanish Fork Diversion Dam (1907–1908), and Indian Creek Dike (1911–1912). Other major features include Strawberry Tunnel (a 4-mile-long, concrete-lined tunnel), Strawberry Power Canal, three power plants, and the High Line (contour-following) Canal.

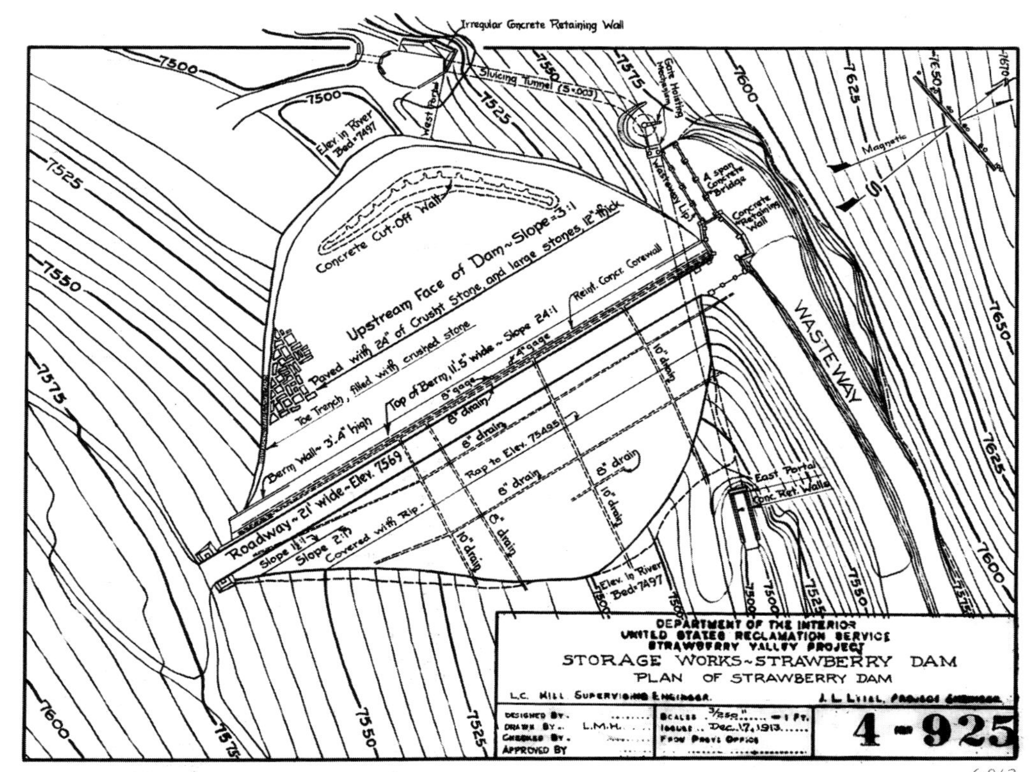

6-042

6-042. Plan, Strawberry Dam, Strawberry Valley Project, Wasatch County, Utah, 1911–1913. United States Reclamation Service, delineator, 1913. P&P,HAER,UTAH,25-PAYS,1-63.

6-043

6-046

6-043. Corewall excavation, Strawberry Dam, Strawberry Valley Project, Wasatch County, Utah, 1911–1913. Unidentified photographer, 1911. P&P,HAER,UTAH,25-PAYS,1, no. 3.

6-044. Fill nearing completion, Strawberry Dam, Strawberry Valley Project, Wasatch County, Utah, 1911–1913. Unidentified photographer, 1911. P&P,HAER,UTAH,25-PAYS,1, no. 5.

6-045. Upstream portal of tunnel, Strawberry Dam, Strawberry Valley Project, Wasatch County, Utah, 1911–1913. Unidentified photographer, 1911. P&P,HAER,UTAH,25-PAYS,1, no. 6.

6-046. Indian Creek Dike, Strawberry Valley Project, Wasatch County, Utah, 1911–1912. Unidentified photographer, 1912. P&P,HAER,UTAH,25-PAYS,1, no. 9.

6-044

6-045

6-047

6-047. Powerhouse turbine delivery, Strawberry Valley Project, Wasatch County, Utah, 1906–1917. Unidentified photographer, 1908. P&P,HAER, UTAH,25-PAYS,1, no. 25.

6-048. Construction, High Line Canal, Strawberry Valley Project, Wasatch County, Utah, 1906–1917. Unidentified photographer, 1915. P&P,HAER,UTAH,25-PAYS,1, no. 29.

6-048

Salt River Basin Reclamation Project

The Salt River Basin Reclamation Project encompasses a number of works built on the two tributaries of the Gila River, which flows into the Colorado. Long before Spanish or Anglo settlement, the Gila River valley was farmed using irrigation. Its tributaries, the Salt and Verde rivers, have long supported Native American civilizations and are marked with ruined cities and irrigation canals. These rivers meet near the modern-day city of Phoenix. One of the first five schemes undertaken under the Reclamation Act of 1902, the Salt River Project began with the construction of Roosevelt Dam east of Phoenix (1903–1911). Later additions include Mormon Flat (1923–1925), Horse Mesa (1924–1927) and Stewart Mountain (1930) dams on the Salt River, and Bartlett (1935–1939) and Horseshoe (1946) dams on the Verde. Coolidge and Painted Rock dams are the largest dams on the Gila River.

6-049. Downstream face, Roosevelt Dam, Salt River, Gila County, Arizona, 1903–1911. E. E. Kunselman, photographer, 1916. P&P,LC-USZ62-86217.

6-050. Roosevelt Dam, Salt River, Gila County, Arizona, 1903–1911. West Coast Art Company, 1913. P&P,PAN US GEOG-Arizona,no. 83 (E size).

The major feature of the Salt River Project, this dam marked the beginning of federal reclamation projects throughout the West. It served to bring water and power to the Salt River valley in Arizona (present-day Phoenix), making it one of the first multipurpose dams in the United States. Although it is arch shaped, Roosevelt is not a true arched dam, but a gravity structure that relies on its mass of concrete and masonry to resist the outward thrust of the retained water. At the time of its construction, it was the largest masonry arch dam in the world; it is now a national historic engineering landmark. The photographs show modifications to the spillway gates completed in 1939. In 1996, the structure was strengthened and the crest was raised 77 feet, adding 20 percent to the reservoir's storage capacity. A single-span, steel arch bridge was added over the dam at the same time.

6-049

6-050

6-051

6-051. Young man with fish, Roosevelt Dam, Salt River, Gila County, Arizona, 1903–1911. E. E. Kunselman, photographer, 1915. P&P, LC-dig-cph-3c05191.

6-052. Upstream face, Roosevelt Dam, Salt River, Gila County, Arizona, 1903–1911. E. E. Kunselman, photographer, 1915. P&P,LC-dig-cph-3c06345.

6-053. Upstream face showing modified spillway gates, Roosevelt Dam, Salt River, Gila County, Arizona, 1903–1911, modified 1936–1939. Unidentified photographer, ca. 1940. P&P,LC-USZ62-80238.

6-052

6-053

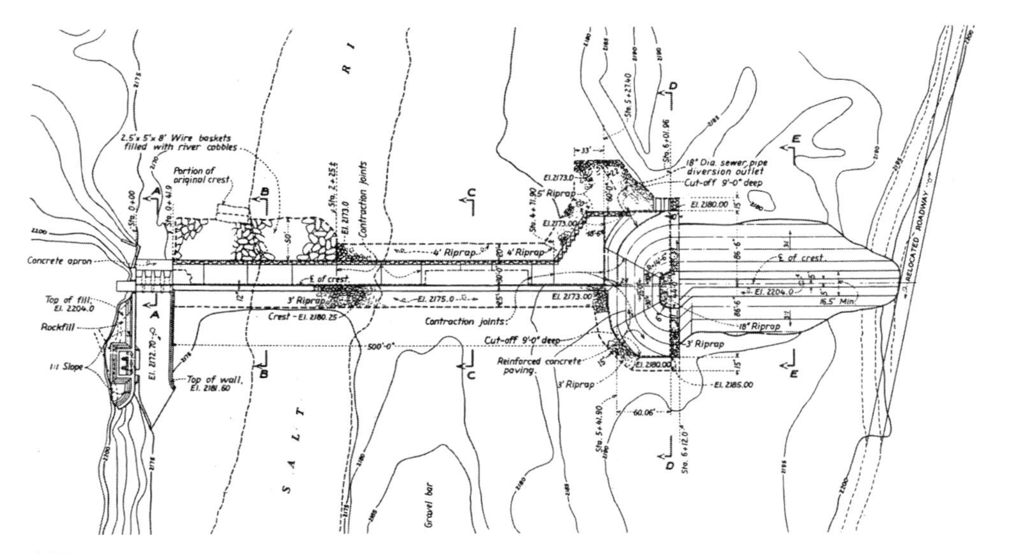

6-054

6-055

6-056

6-054. Plan, Roosevelt Diversion Dam, Salt River, Gila County, Arizona, 1904–1906, modified 1936–1937. Bureau of Reclamation, delineator, 1937. P&P,HAER,ARIZ,007-ROOS. V,1-78,detail.

The diversion dam and power canal provided hydropower for the construction of the Roosevelt Dam.

6-055. Roosevelt Diversion Dam, Salt River, Gila County, Arizona, 1904–1906, modified 1936–1937. Walter J. Lubken, photographer, 1906. P&P,HAER,ARIZ,007-ROOS.V,1, no. 26.

6-056. Gates at end of Cotton pressure pipe, Roosevelt Diversion Dam and Power Canal, Salt River, Gila County, Arizona, 1904–1906. Unidentified photographer, 1906. P&P,HAER,ARIZ,007-ROOS.V,1, no. 18.

6-057

6-058

6-057. Granite Reef Diversion Dam, Salt River, Mesa, Maricopa County, Arizona, 1908–1913. California Panorama Company, 1908. P&P,LC-dig-6a17238.

6-058. Granite Reef Diversion Dam, Salt River, Mesa, Maricopa County, Arizona, 1908–1913. West Coast Art Company, 1913. P&P,PAN US GEOG-Arizona,no. 86 (E size).

This concrete overflow dam diverts water into the Arizona and Grand canals, supplying water to Phoenix, Tempe, Mesa, and other areas of the Salt River valley. It lies downstream from Roosevelt, Horse Mesa, Mormon Flat, and Stewart Mountain dams on the Salt River.

6-059

6-059. Construction, Mormon Flat Dam, Salt River, Phoenix Vic., Maricopa County, Arizona, 1923–1925. Unidentified photographer, 1926. P&P,HAER,ARIZ,7-PHEN.V,4, no. 20.

Mormon Flat Dam was influenced by the elegant concrete multiple arch dams of Californian John Eastwood. An additional feature is the more pronounced curvature toward the base of the dam, where water loads are greatest. The Salt River Water Users Association built Mormon Flat and the similar Horse Mesa Dam as a part of the Salt River Project's hydroelectric expansion program in the 1920s, providing for continuous year-round power and avoiding fluctuations in winter months. A new spillway was added in 1938, and a new powerhouse and switchyard thirty years later.

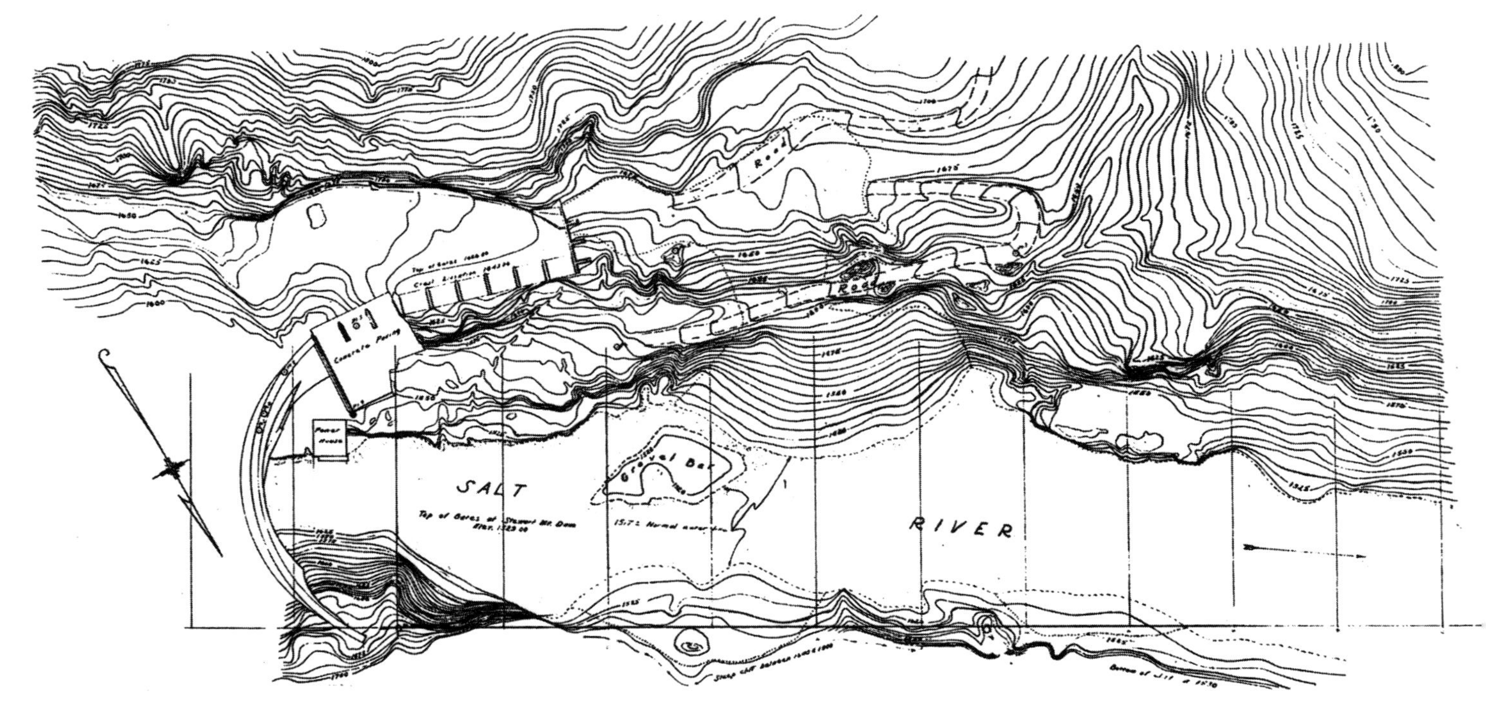

6-060

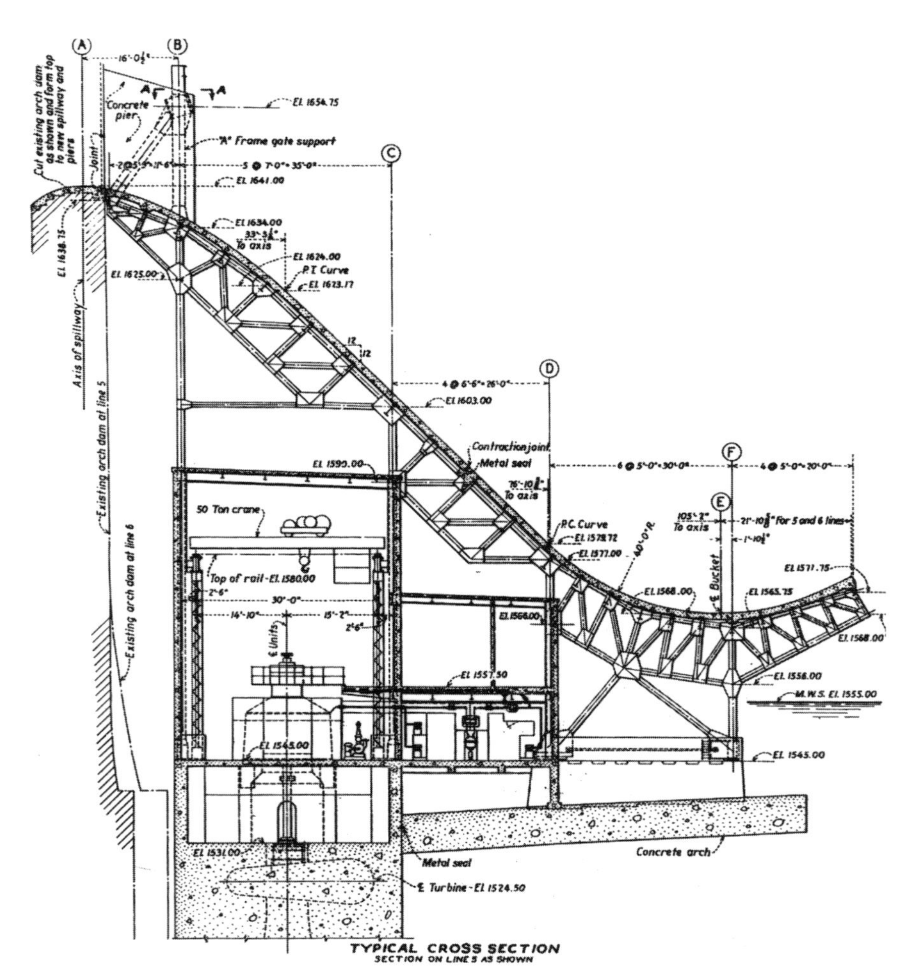

6-061

6-060. Site plan, Mormon Flat Dam, Salt River, Phoenix Vic., Maricopa County, Arizona, 1923–1925. Bureau of Reclamation, delineator, 1935. P&P,HAER,ARIZ,7-PHEN.V,4-37.

6-061. Section through spillway, Mormon Flat Dam, Salt River, Phoenix Vic., Maricopa County, Arizona, 1923–1925, modified 1936–1938. United States Bureau of Reclamation, delineator, 1936. P&P,HAER,ARIZ,7-PHEN.V,4-46,detail.

6-062. Installation of tainter gates, Mormon Flat Dam, Salt River, Phoenix Vic., Maricopa County, Arizona, 1923–1925. Unidentified photographer, 1925. P&P,HAER,ARIZ,7-PHEN.V,4, no. 14.

6-063. New spillway, Mormon Flat Dam with modified spillway, Salt River, Phoenix Vic., Maricopa County, Arizona, 1923–1925, modified 1936–1938. Ben D. Glaha, photographer, 1938. P&P,LC-dig-ppmsca-17397.

6-062

6-063

6-064

6-064. Dam site looking upstream, Horse Mesa Dam, Maricopa County, Arizona, 1924. Unidentified photographer, 1924. P&P,HAER,ARIZ,7-PHEN.V,3, no. 1.

6-065. Nearing completion, Horse Mesa Dam, Salt River, Phoenix Vic., Maricopa County, Arizona, 1924–1927. Unidentified photographer, 1927. P&P,HAER,ARIZ-7-PHEN.V,3, no. 22.

6-065

6-066. Design plan used in final trial, Horse Mesa Dam, Salt River, Phoenix Vic., Maricopa County, Arizona, 1924–1927. 1925. P&P,HAER,ARIZ,7-PHEN.V,3-43.

6-067. Dam crest, Horse Mesa Dam, Salt River, Phoenix Vic., Maricopa County, Arizona, 1924–1927. Unidentified photographer, 1927. P&P,HAER,ARIZ-7-PHEN.V,3, no. 20.

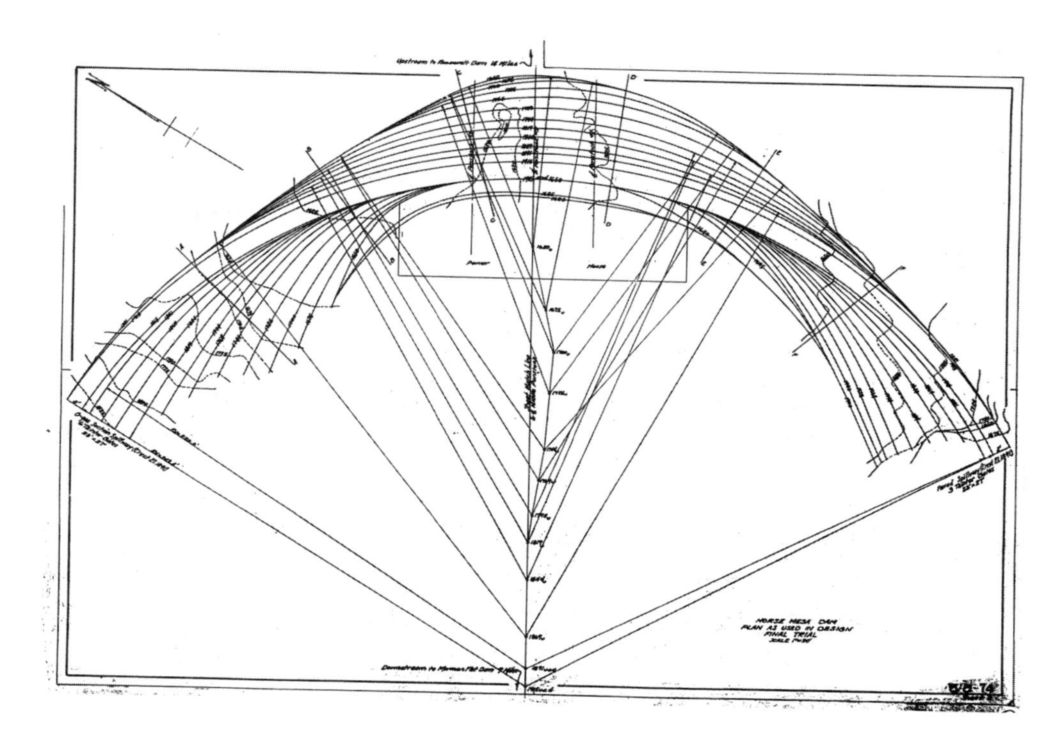

6-066

6-067

6-068

6-069

6-070

6-068. Installation of penstocks and scroll cases, Horse Mesa Dam, Salt River, Phoenix Vic., Maricopa County, Arizona, 1924–1927. Unidentified photographer, 1926. P&P,HAER,ARIZ,7-PHEN.V,3, no. 10.

6-069. North spillway, Horse Mesa Dam, Salt River, Phoenix Vic., Maricopa County, Arizona, 1924–1927. Mark Durben, photographer, 1988. P&P,HAER,ARIZ,7-PHEN.V,3, no. 31.

6-070. Workers with scroll cases, Horse Mesa Dam, Salt River, Phoenix Vic., Maricopa County, Arizona, 1924–1927. Unidentified photographer, 1926. P&P,HAER,ARIZ,7-PHEN.V,3, no. 9.

6-071

6-071. Bartlett Dam, Verde River, Phoenix Vic., Maricopa County, Arizona, 1936–1939. P&P,LC-dig-ppmsca-17399.

Jointly funded by the Bureaus of Reclamation and Indian Affairs, Bartlett was the highest multiple arch dam constructed in the United States at the time of its completion and the first built by the federal government. Its vaults are set between massive abutments that function as gravity dams.

6-072. Plan, Bartlett Dam, Verde River, Phoenix Vic., Maricopa County, Arizona, 1936–1939. Bureau of Reclamation, delineator, 1936. P&P,HAER,ARIZ,7-PHEN.V,2-47,detail.

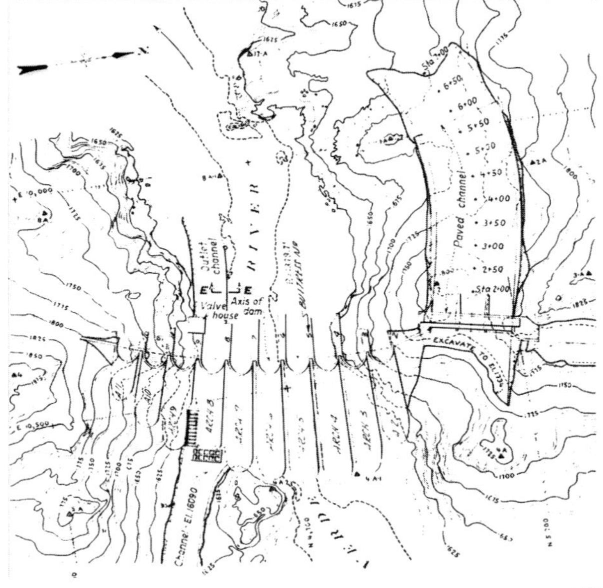

6-072

6-073

6-073. Construction, Bartlett Dam, Verde River, Phoenix Vic., Maricopa County, Arizona, 1936–1939. Unidentified photographer, 1939. P&P,HAER,ARIZ,7-PHEN.V,2, no. 27.

6-074

6-074. Upstream face, Bartlett Dam, Verde River, Phoenix Vic., Maricopa County, Arizona, 1936–1939. Bureau of Reclamation. P&P,LC-USZ62-80240.

6-075. Plan and section through slide gate outlets, Bartlett Dam, Verde River, Phoenix Vic., Maricopa County, Arizona, 1936–1939. Bureau of Reclamation, delineator, 1936. P&P,HAER,ARIZ,7-PHEN.V,2-69,detail.

6-076. Spillway gate and hoist chain, Bartlett Dam, Verde River, Phoenix Vic., Maricopa County, Arizona, 1936–1939. Unidentified photographer, 1939. P&P,HAER,ARIZ,7-PHEN.V,2, no. 30.

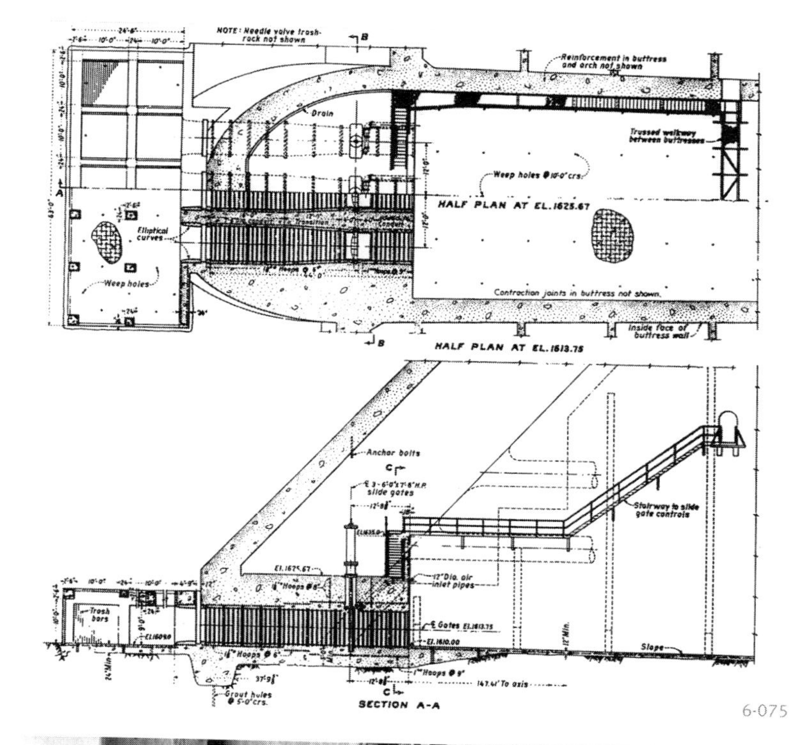

6-075

6-076

Indian Irrigation Service Projects

The United States Reclamation Service initiated the San Carlos Project to improve irrigation on the lands of the Gila Indian Reservation. While the Pima and Maricopa Indians along the Gila River had practiced irrigation for centuries, by the 1900s water usage by American settlers upstream began to seriously reduce the flow in the river, making the traditional canals useless. In 1913, the Indian Service assumed responsibility for the project and expanded it over an eight-year period to include a number of diversion dams, new and updated canals, wells, and other features. Several phases are visible on a Reclamation Service map of 1916, as the Sacaton, Blackwater, and Casa Blanca projects.

6-077

6-077. Coolidge Dam, San Carlos Project, Gila River, Peridot Vic., Pinal County, Arizona, 1925–1928. McCulloch Brothers, 1930. P&P,LOT 5155,G142.

This was the primary storage dam for the San Carlos Project. Built by the U.S. Indian Service (later the Bureau of Indian Affairs), the reinforced concrete, multiple dome design by Charles R. Olberg (1875–1938) is unique in the United States. Because funding for the dam was limited, the material-efficient multiple dome design was selected as the most cost-effective solution.

6-078. Boy scouts serving guests at dedication ceremonies. Coolidge Dam, San Carlos Project, Gila River, Peridot Vic., Pinal County, Arizona, 1925– 1928. Unidentified photographer, 1930. P&P,HAER,ARIZ,11-PERI.V,1, no. 38.

6-078

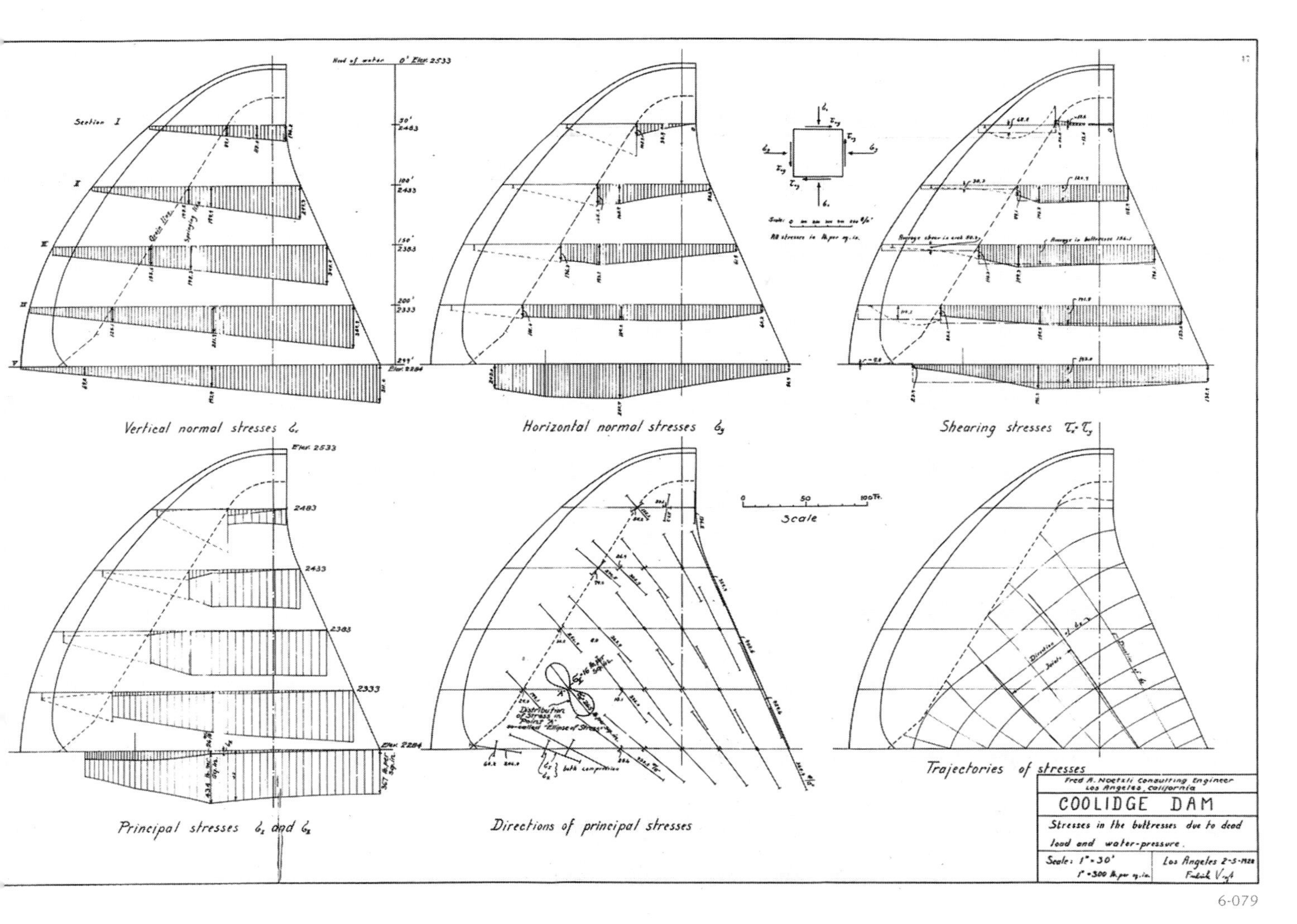

6-079

6-079. Stresses in the buttresses, Coolidge Dam, San Carlos Project, Gila River, Peridot Vic., Pinal County, Arizona, 1925–1928. Frederick Vogt, delineator, 1928. P&P,HAER,ARIZ,11-PERI.V,1, no. 72.

6-080

6-080. Upstream face, Coolidge Dam, San Carlos Project, Gila River, Peridot Vic., Pinal County, Arizona, 1925–1928. McCulloch Brothers, 1930. P&P,LOT 5155,G141.

6-081

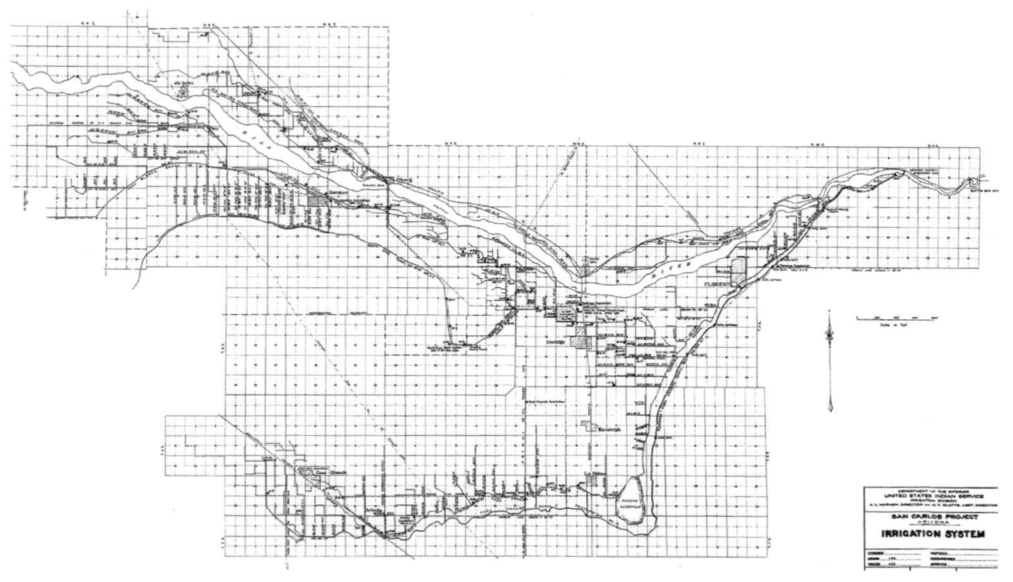

6-082

6-081. Downstream face, Coolidge Dam, San Carlos Project, Gila River, Peridot Vic., Pinal County, Arizona, 1925–1928. Cadenbach, 1928. P&P,HAER,ARIZ,11-PERI.V,1-34.

6-082. Map, San Carlos Project, Gila River, Pinal County, Arizona, 1939. U.S. Indian Service, delineator, ca. 1939. P&P,HAER, ARIZ,11-COOL,1, no. 3.

6-083. Plan, Ashurst-Hayden Dam, San Carlos Project, Gila River, Coolidge Vic., Pinal County, Arizona, 1921–1922. U.S. Indian Service, delineator, 1922. P&P,HAER,ARIZ, 11-COOL,1A-11.

The Ashurst-Hayden and Sacaton dams downstream from Coolidge Dam diverted the river to supply irrigation canals along the Gila Indian Reservation. These dam sites presented similar conditions on the Gila riverbed, of a deep filled-in canyon between solid rock abutments. The solution was to use a "floating" dam, made of a broad concrete slab anchored to the abutments. The Reclamation Service had previously used the design at Laguna Diversion Dam near Yuma (1905–1909; 6-099) and Granite Reef Dam on the Salt River (1906–1908; 6-067–6-068).

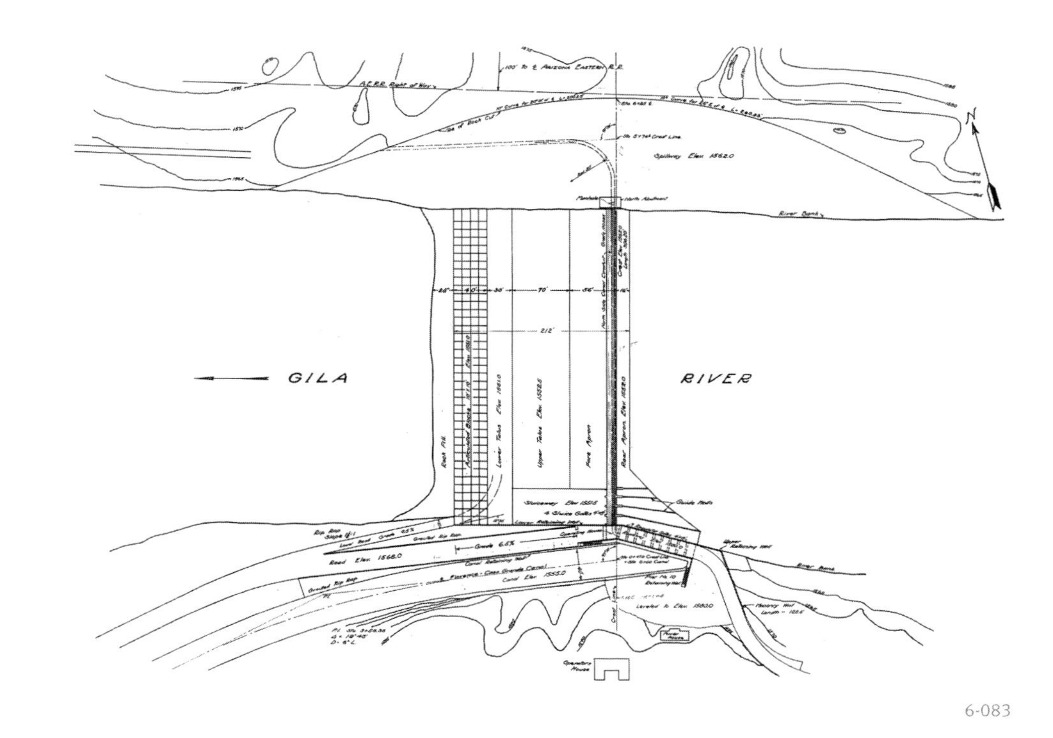

6-083

6-084. Downstream face, Ashurst-Hayden Dam, San Carlos Project, Gila River, Coolidge Vic., Pinal County, Arizona, 1921–1922. Bob Knotts, photographer, 1995. P&P,HAER,ARIZ,11-COOL,1A-1.

6-084

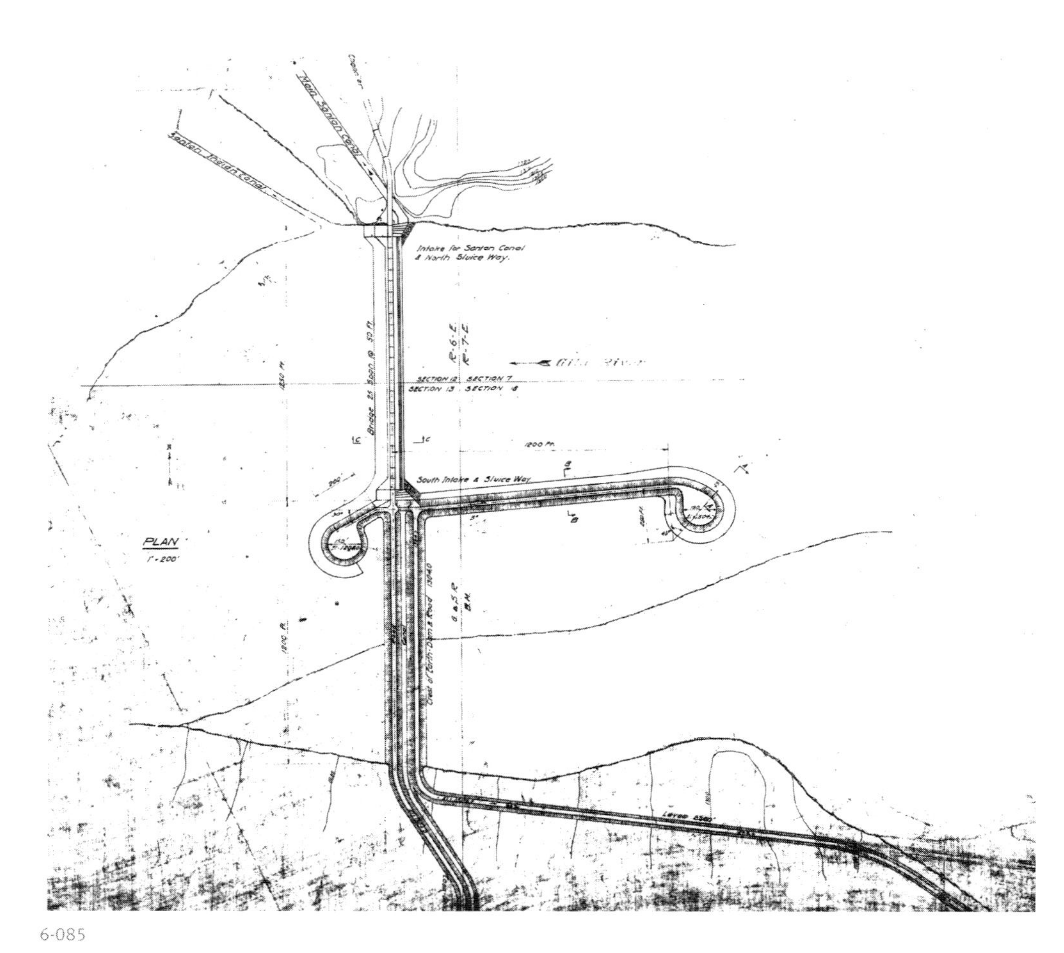

6-085

6-085. Plan, Sacaton Dam, San Carlos Project, Gila River, Pinal County, Arizona, 1923–1926. P&P,HAER,ARIZ,11-COOL, 1D, no. 26.

6-086

6-086. Sacaton Dam & Bridge from Quarry Hill. Downstream face, Sacaton Dam and Bridge, San Carlos Project, Gila River, Pinal County, Arizona, 1923–1926. Unidentified photographer, 1925. P&P,HAER,ARIZ, 1-COOL,1D-12.

Lower Colorado River Basin Projects

Laguna Diversion Dam

The Laguna Diversion Dam, built by the United States Reclamation Service as the main feature of an irrigation project diverting water from the Colorado, was the first government-built dam across that river. Preliminary investigations revealed deep alluvial deposits on the valley floor, making it prohibitively expensive to excavate to bedrock. Consequently, the "Indian weir," or floating dam system, was chosen, in which the construction rests on a silt base, and the design must take care to prevent undercutting. A 400-foot-wide strip across the riverbed was excavated to a depth of 25 feet, and three parallel concrete crest walls were constructed spanning the river. Loose rock was placed between these walls and a stone apron built on the downstream side, then the dam was covered with 18 inches of concrete. The weir is only 43 feet high, nearly two-thirds of which is below the riverbed. Both ends of the dam have concrete sluiceways fitted with iron roller gates on pillars. Turnouts upstream, equipped with flashboards (smaller gates), divert water into the canals. Because of cost overruns, the work was assumed by the government in 1907 and carried to completion in 1909 by force account under the supervision of the Reclamation Service.

6-087

6-087. Head gates, Laguna Diversion Dam, Colorado River, Yuma Vic., Yuma County, Arizona, 1905–1909. WC. P&P,LC-dig-ppmsca-17331.

Grand Valley Diversion Dam

In 1905, the Grand Valley Water Users Association was organized to cooperate with the newly formed United States Reclamation Service in developing a project for the region. The dam diverts water into a 55-mile high line canal to provide power and water to land along the Colorado River near Grand Junction. Six roller gates control flow over the crest; these were the first of their type used in the United States. The outbreak of World War I prevented the German patent-holders from providing the gates, so the Americans copied the design and built them. Further down the high line canal, the Orchard Mesa diversion siphons water under the Colorado River to the Grand Valley power plant and Orchard Mesa pumping plant, which lifts the water to irrigation canals.

6-089

6-088

6-088. Unirrigated land to come under government control, Grand Valley, Mesa County, Colorado. Unidentified photographer, ca. 1910. P&P,SSF-Irrigation-Colorado.

6-089. Orchard Mesa pumping plant, Grand Valley Project, Mesa County, Colorado. Louis C. McClure, photographer, 1911. HAW/DPL,MCC-1479.

Hoover (Boulder) Dam

This concrete arch-gravity dam, towering 730 feet over the Colorado River in the Nevada desert, captured the imagination of Depression-era America in a way that few projects, except perhaps the Empire State Building, could equal. Because of the immense size of the dam, no single contractor had the resources to make a qualified bid alone. The contract was awarded to a consortium of six general contractors who had joined forces to submit a bid. The Six Companies Incorporated were Morrison & Knudsen, Utah Construction, Pacific Bridge, J. F. Shea, Kaiser & Bechtel, and McDonald & Kahn. After completing Hoover Dam two years ahead of the allotted time, the Six Companies went on to build Grand Coulee Dam (see 7-076–7-085).

Before construction could begin on the dam, workmen known as high-scalers had to strip away loose rock from the walls of Black Canyon with jackhammers and dynamite. Four tunnels were built to divert the river's flow around the construction site. The contractors also had to lay 33 miles of railroad, build 7 miles of highway, and construct a transmission line to bring power from San Bernardino, 222 miles away. Because they did not complete the planned town site of Boulder City before beginning work on the dam, the work force settled in several squatter camps in miserable conditions. Discontent with these camps and with working conditions at the dam, brought on by the accelerated construction schedule, led to a strike in the first year of work. It was violently suppressed, and Boulder City was hastily completed.

6-090. Black Canyon, Colorado River, Clark County, Nevada, 1930–1936. P&P,LC-USZ62-114352.

6-090

6-091. Distant view, Hoover (Boulder) Dam, Colorado River, Boulder City, between Clark County, Nevada, and Mohave County, Arizona, 1930–1936. Ben D. Glaha, photographer, 1938. P&P,LC-dig-cph-3b07810.

6-092. Diversion tunnel outlets under construction, Hoover (Boulder) Dam, Colorado River, Boulder City, between Clark County, Nevada, and Mohave County, Arizona, 1930–1936. Ben D. Glaha, photographer, 1932. P&P,LC-dig-cph-3c14274.

6-091

6-092

Boulder Dam was an extraordinarily large dam, requiring unprecedented amounts of concrete in its construction—roughly 3.75 million cubic yards. The scale of the concrete work posed problems with overheating and uncontrolled cracking as the concrete set. The solution was to build the dam as a series of blocks to dissipate heat, and to limit concrete placement in any block to 5 feet every three days. Coils of steel pipe embedded in each block circulated chilled water as the concrete set. The blocks were keyed together to interlock and grouted as each section was complete.

Boulder Dam's Art Deco styling by the architect Gordon Kaufmann (1888–1949) provided a modern aesthetic for this spectacular project—one of the first modern-styled dams in the country. It was an appropriate decorative scheme for such an undertaking, both as an example of technological modernity and somehow particularly fitting for this project in the desert—so close to the burgeoning metropolis of Los Angeles, and later Las Vegas.

6-093. Drillers excavating dam foundation, Hoover Dam, Colorado River, Boulder City, Clark County, Nevada, 1930–1936. Bureau of Reclamation, 1933. P&P,LC-USZ62-117351.

6-093

6-094

6-096

6-095

6-094. Pouring concrete, Hoover Dam, Colorado River, Boulder City, between Clark County, Nevada, and Mohave County, Arizona, 1930–1936. Ben D. Glaha, photographer, 1934. P&P,LC-dig-cph-3b36025.

6-095. Ground crew, Hoover Dam, Colorado River, Boulder City, between Clark County, Nevada, and Mohave County, Arizona, 1930–1936. Ben D. Glaha, photographer, 1934. P&P,LC-USZ62-7245.

6-096. Construction, Hoover Dam, Colorado River, Boulder City, Clark County, Nevada, 1930–1936. Ben D. Glaha, photographer, 1934. P&P,LC-USZ62-60072.

6-097

6-098

6-097. Placing concrete in spillway, Hoover Dam, Colorado River, Boulder City, between Clark County, Nevada, and Mohave County, Arizona, 1930–1936. Bureau of Reclamation, 1933. P&P,LC-dig-ppmsca-17340.

6-098. Concrete mixing plant, Hoover Dam, Colorado River, Boulder City, between Clark County, Nevada, and Mohave County, Arizona, 1930–1936. Ben D. Glaha, photographer, 1934. P&P,LC-dig-cph-3c14357.

6-099

6-099. Interior of the Babcock and Wilcox plant at Hoover Dam, Colorado River, Boulder City, between Clark County, Nevada, and Mohave County, Arizona, 1930–1936. Ben D. Glaha, photographer, 1934. P&P,LC-USZ62-114364.

6-100

6-100. Intake towers at low reservoir level, Hoover Dam, Colorado River, Boulder City, between Clark County, Nevada, and Mohave County, Arizona, 1930–1936. Ben D. Glaha, photographer, 1934. P&P,LC-USZ62-114273.

6-101. View from northwest rim, Hoover Dam, Colorado River, Boulder City, between Clark County, Nevada, and Mohave County, Arizona, 1930–1936. Unidentified photographer, 1938. P&P,LC-USZ62-22446.

6-102. Spillway, Hoover Dam, Colorado River, Boulder City, between Clark County, Nevada, and Mohave County, Arizona, 1930–1936. Bureau of Reclamation. P&P,SSF-Dams-Arizona and Nevada-Hoover,BC-5502.

6-101

6-102

6-103

6-104

6-103. Transmission lines, Hoover Dam, Colorado River, Boulder City, between Clark County, Nevada, and Mohave County, Arizona, 1930–1936. Bureau of Reclamation, 1941. P&P,LC-dig-ppmsca-17403.

6-104. Reservoir at night, Hoover Dam, Colorado River, Boulder City, between Clark County, Nevada, and Mohave County, Arizona, 1930–1936. Bureau of Reclamation, 1940. P&P,LC-dig-cph-3b07811.

Parker Dam

Although the Bureau of Reclamation built this project, the Metropolitan Water District of Southern California financed it as a part of the Colorado River Aqueduct (1932–1941) that pumps Colorado River water to metropolitan consumers in the Los Angeles area. Located 100 miles downstream from Hoover Dam, Parker provides a reservoir for this aqueduct, which is a series of tunnels, lifts (pumping stations), and pipelines that carry water across the desert ranges to the Los Angeles basin. Later aqueducts served by Parker Dam include the San Diego (1947) and the Central Arizona Project (1985) aqueducts.

6-105. Downstream face, Parker Dam, Colorado River, Parker, between Mohave County, Arizona, and San Bernardino County, California, 1936–1939. Ben D. Glaha, photographer, 1941. P&P,LC-dig-ppmsca-17337.

6-106. Iron Mountain Lift and Colorado Aqueduct, Parker Dam, Colorado River, Parker, between Mohave County, Arizona, and San Bernardino County, California, 1936–1939. Spence Air, 1939. P&P,LC-dig-ppmsca-17369.

6-105

6-106

Imperial Dam

Imperial Dam is the Bureau of Reclamation's major irrigation diversion structure on the lower Colorado River. The dam diverts water from the Colorado River to the All-American Canal (so named because it runs entirely within the United States) and the Gila Main Canal, irrigating more than a million acres of land in the United States and Mexico. A slab and buttress–type structure 3,475 feet long, it raises the water surface elevation of the Colorado 23 feet. The Imperial reservoir has been largely silted up, forming a shallow lake with well-defined channels to the canals. Desilting basins lie downstream of each of the head works.

6-107

6-107. View upstream from the All-American Canal toward Imperial Dam, Colorado River, Yuma County, Arizona, and Imperial County, California, 1935–1938. Ben D. Glaha, photographer, 1939. P&P,LC-dig-cph-3a37935.

6-108. Roller gates to the All-American Canal, Imperial Dam, Colorado River, Yuma County, Arizona, and Imperial County, California, 1935–1938. Ben D. Glaha, photographer, 1939. P&P,LC-USZ62-60068.

6-109. "Natural and artificial rivers intersect," wash cut by the Colorado Flood of 1905 bridged by the siphon of the All-American Canal, Imperial County, California, 1939. Acme, 1939. P&P,LC-dig-ppmsca-17312.

6-108

6-109

6-110. "Man-made river in a desert sea," All-American Canal nearing completion, Imperial County, California, 1938. Acme, 1938. P&P,LC-dig-cph-3b36456.

6-110

Davis Dam

Davis Dam, located 67 miles downstream from Hoover Dam, creates Lake Mohave out of the Colorado River between these two structures. It regulates Hoover Dam releases to meet downstream uses, including the annual delivery of 1.5 million acre-feet of water to Mexico, in accordance with the 1944 treaty. The zoned earthfill structure is 200 feet high and 1,600 feet long.

6-111. Davis Dam, Colorado River, Bullhead City Vic., Mohave County, Arizona, 1942–1950. Acme, 1950. P&P,NYWTS-Subj.-Dams-Davis-Arizona.

6-111

Colorado River Storage Project

The Colorado River Storage Project, authorized in 1956, stores water for the states of the Upper Colorado River basin for irrigation, flood control, and hydroelectricity generation. It is made up of four storage units: Glen Canyon on the Colorado, Flaming Gorge on the Green, Navajo on the San Juan, and the Curecanti (Aspinall) complex of three dams on the Gunnison. While innovations in dam-building technology enabled the Bureau of Reclamation to construct ever-larger dams in more remote locations, protests arose over the ecological impacts of these projects spanning white-water rivers in the scenic, still-undeveloped wilderness. In 1968, municipal and industrial uses, the maintenance of water quality, recreation, and the improvement of fish and wildlife habitats were given priority over electricity production. Revenues from power generation repay project costs and subsidize irrigation in the region.

Flaming Gorge Dam

Located in the remote Uinta Mountain range of northeastern Utah, this concrete arch dam built by the Bureau of Reclamation was the first dam completed in the Colorado River Storage Project.

John Wesley Powell, in his 1869 expedition down the Green River, named Flaming Gorge for the brilliantly colored sandstone of its walls. The gorge is now flooded for

6-112

6-113

6-112. Flaming Gorge, Green River, Utah. UPI, 1961. P&P,LC-dig-ppmsca-17246.

6-113. Flaming Gorge Dam under construction, Green River, Dutch John Vic., Daggett County, Utah, 1957–1962, modified 1978, 1984. UPI, 1961. P&P,LC-dig-ppmsca-17246.

91 miles upstream. The dam has significantly affected the Green River ecosystem—producing colder, sediment-free water that has endangered several native fish species, and preventing annual spring floods, with the consequent reduction of beaches and sand bars downstream. The Bureau of Reclamation no longer operates the dam as a peaking plant, but it attempts to reduce dramatic daily fluctuations.

Navajo Dam

The completion of the 3,650-foot-long and 420-foot-high earthfill Navajo Dam had significant effects on the riverine habitat of the San Juan River, decimating fish and other animal populations. Modifications to the dam's current operation now simulate natural flows in an attempt to minimize the damage.

6-114. Navajo Dam, San Juan River, San Juan and Rio Arriba Counties, New Mexico, 1958-63. Bureau of Reclamation, delineator, 1958. P&P,LC-dig-ppmsca-17271.

6-114

Glen Canyon Dam

When sites were being considered for the first major Colorado River dam, Marble Canyon was in competition with Black Canyon, the site ultimately chosen for Hoover Dam. It offered solid bedrock foundations, a large and sparsely populated basin, and a nearby source of construction-grade rock and sand. Thirty years later, Glen Canyon Dam was finally authorized on this site, as the major feature in the massive Colorado River Storage Project. Designed to store water for dams on the lower Colorado and to generate hydroelectricity, it was the largest construction contract in the history of the Bureau of Reclamation and the third-highest dam in the world. The 1,560-foot-long and 710-foot-high concrete arch dam created Lake Powell out of 180 miles of the Colorado River.

6-115

6-115. Construction of Glen Canyon Dam, Colorado River, Page, Coconino County, Arizona, 1957–1964. Robert Kornfeld, photographer, 1963. P&P,NYWTS-Subj.-Dams-Glen Canyon Dam-Arizona.

Glen Canyon Dam is also located upstream from the Grand Canyon. Within two decades of its completion, concerns arose about the dam's effects on the canyon, as it suppressed annual floods, captured nutritive sediments, affected water temperatures, and damaged spawning grounds. Rising public outcry and numerous research projects on the dam's impact led to passage of the Grand Canyon Protection Act of 1992, which required Glen Canyon Dam to be operated for the protection of the canyon's ecosystems rather than for maximizing hydropower generation. Currently, the Bureau of Reclamation attempts to simulate spring floods with annual releases of water. Dry years have led to a significant drop in reservoir levels and an associated drop in hydroelectric production, effectively making the dam obsolete. The controversy surrounding the construction of Glen Canyon Dam played a key role in galvanizing the environmental movement against large dams.

6-116. Downstream face of Glen Canyon Dam, Colorado River, Page, Coconino County, Arizona, 1957–1964. Unidentified photographer, 1965. P&P,LC-dig-ppmsca-17251.

6-116

ANOTHER GREAT BASIN DAM

Mountain Dell Dam is an early example of multiple arch dam construction, built as a water reservoir for Salt Lake City. The city engineer put out bid packages for three competing dam designs: a curved gravity dam, an Ambursen flat-slab dam, and a multiple arch dam by John Eastwood, who had developed the design for mountainous sites in California (see 8-018–8-041). The low bids for the Eastwood design came in at 30 percent less than the alternatives; in addition, the design was felt to offer less risk of taxing the relatively weak foundations at the dam site. The dam was raised 40 feet in 1925.

6-117

6-117. Mountain Dell Dam, Parley's Creek, Salt Lake City Vic., Utah, 1914–1917, modified 1924–1925. Jack E. Boucher, photographer, 1971. P&P,HAER,UTAH,18-SALCI,22, no. 2.

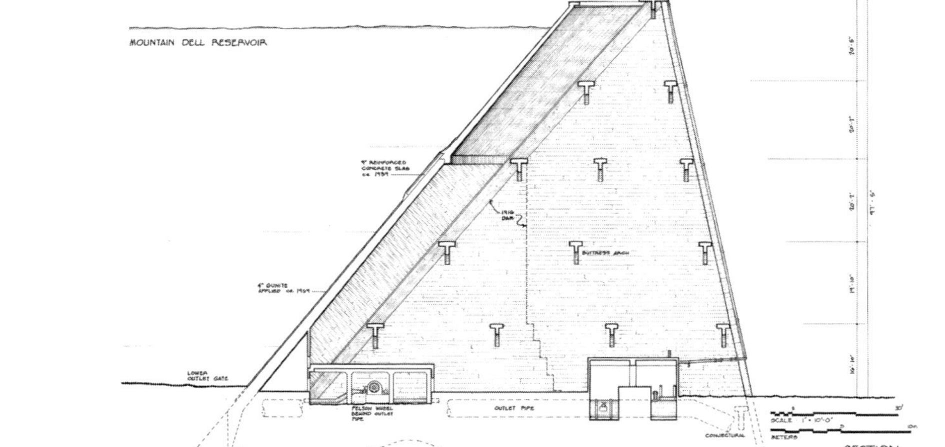

6-118

6-118. Section, Mountain Dell Dam, Parley's Creek, Salt Lake City Vic., Utah, 1914–1924. Charles Madsen, delineator, 1971. P&P, HAER,UTAH,18-SALCI,22.

6-119

6-119. Crest, Mountain Dell Dam, Parley's Creek, Salt Lake City Vic., Utah, 1914–1924. Jack E. Boucher, photographer, 1971. P&P,HAER,UTAH,18-SALCI,22, no. 5.

6-120. Upstream face with empty reservoir, Mountain Dell Dam, Parley's Creek, Salt Lake City Vic., Utah, 1914–1924. Jack E. Boucher, photographer, 1971. P&P,HAER,UTAH,18-SALCI,22, no. 18.

6-121. Downstream face showing buttresses, Mountain Dell Dam, Parley's Creek, Salt Lake City Vic., Utah, 1914–1924. Jack E. Boucher, photographer, 1971. P&P,HAER,UTAH,18-SALCI,22, no. 3.

6-120

6-121

COLUMBIA

COLUMBIA RIVER BASIN

Kootenai, Clark Fork, Pend Oreille, Spokane, Okanogan, Yakima, Snake, Henrys Fork, Boise, Salmon, Clearwater, and Willamette Rivers

The Columbia River drains 250,000 square miles in the Pacific Northwest. Its headwaters are in the Canadian Rockies, and it winds a circuitous route through the Rockies, then high plateaus, lava beds, and the Cascade Range before it enters the Pacific Ocean through a wide estuary. The river has been the focus for the life and ways of Native peoples in this region for thousands of years, serving as a site for economic, social, and cultural exchange centered on the salmon catch.

The Columbia's westernmost tributary—the Willamette River—was the principal destination on the Oregon Trail. Rapidly settled after the California gold rush, the valley has long been the most densely populated part of Oregon. The Snake River is the Columbia's chief tributary, flowing from the Continental Divide into the Columbia's watershed and serving as a route for westward immigrants along its course. Rising near Yellowstone, it flows through Jackson Lake in the Grand Teton Mountains, cascades down several notable falls in Idaho—where it supports that

7-001. Steel workers on wing dam of Grand Coulee Dam, Columbia Basin Reclamation Project, Columbia River, between Douglas and Grant Counties, Washington, 1933–1942. Bureau of Reclamation, 1942. P&P, LC-USZ62-67862.

state's agricultural and industrial enterprises—and then continues north through Hell's Canyon, where it is contained by several massive hydroelectric dams before it joins the Columbia River. The Snake's many hydroelectric plants are a major source of electricity in the Northwest, and its watershed provides irrigation for the Bureau of Reclamation's Minidoka, Boise, Palisades, and Owyhee projects.

In 1932, federal dam building began in earnest in the Columbia River basin, and today there are over 250 reservoirs and 150 hydroelectric projects in the basin. There are six federal and five nonfederal dams on the main stem of the Columbia River. The Grand Coulee and Chief Joseph dams are on the river's upper course. Priest Rapids, Wanapum, Rock Island,

7-002. Great Falls of the Snake River, Idaho Territory. Chromolithograph of original painting by Thomas Moran, 1876. P&P,LC-USZC4-3252.

7-002

Rocky Reaches, and Wells dams further downstream are among the largest commercially built hydroelectric facilities in the country. Bonneville, The Dalles, John Day, and McNary dams on the lower reaches of the river create a 300-mile navigation channel from the Pacific Ocean to the Columbia's confluence with the Snake. While the dams provide flood control, municipal water supply, and navigation, the development of hydroelectric power has had a significant effect on the economy of the Pacific Northwest, contributing to its military and industrial development (i.e., Hanford nuclear reservation, Kaiser shipbuilding, Boeing aircraft). In addition, the dams provide water for the Columbia Basin Irrigation Project in central Washington state, transforming the naturally arid region into an agricultural center.

7-003. The Columbia River passing over the Grand Coulee Dam. F. B. Pomeroy, photographer, 1942. P&P,SSF-Dams-Wash-Grand Coulee-1942.

7-003

HYDROELECTRIC DAMS BUILT BY INDUSTRY AND IRRIGATION INTERESTS

Located between the Mormon agriculturalists of Utah and the mining enterprises of western Montana, southern Idaho was an important area for the development of hydroelectric dams. With the discovery of gold in the 1860s, the Boise basin became a center for placer gold mining. Hydraulic mining operations on the Boise and Payette rivers entailed exten-

7-004. Placer mining at Twin Springs, Boise Vic., Ada County, Idaho. Horace C. Myers, photographer, 1901. P&P,LC-dig-ppmsca-17299.

7-004

sive construction of flumes and siphons to carry water to the mine sites and massive amounts of earth moving—both technologies that were used in dam construction. Mining also required plenty of power to process the ores. A number of spectacular falls on the Snake River (American Falls, Twin Falls, Shoshone Falls, Swan Falls) provided opportunities to harness hydroelectric power. While mining enterprises were the first to exploit this power source, dams on the eastern reaches of the Snake appeared shortly thereafter, to support irrigation in this agricultural region and provide electricity to its growing cities.

7-005. Upper Boise Hydraulic Mining Company flume, Boise Vic., Ada County, Idaho. W. E. Pierce & Company, ca. 1898. P&P,LC-dig-ppmsca-17300.

7-005

7-006

7-006. De Lamar quartz mine and mill, Silver City, Owyhee County, Idaho. W. E. Pierce & Company, ca. 1898. P&P,LC-dig-ppmsca-17301.

7-007

7-007. Swan Falls Dam, Snake River, Kuna Vic., Owyhee and Ada Counties, Idaho, 1900–1901, altered 1907, 1910, 1913, 1918–1921, 1936, 1944. Duane W. Garrett, photographer, 1984. P&P,HAER,ID,37-KU.V,1-2.

One of the earliest hydroelectric dams in the Pacific Northwest and the first in Idaho, Swan Falls Dam supplied power for mining operations in Silver City, Idaho. An early example of the use of long-distance transmission lines (after Niagara Falls–Buffalo and Oregon City–Portland), power from Swan Falls Dam was also sent north to support irrigation and municipalities in the Boise River valley.

The dam comprises two sections separated by a lava island. A concrete gravity dam spans the left channel, and a timber-crib spillway dam on the right. Pits and raceways for the turbines were excavated into the rock. When the Idaho Power Company acquired the dam in 1916, it was used as a training facility for new engineers because of the variety of manually operated equipment dating from several periods at the site.

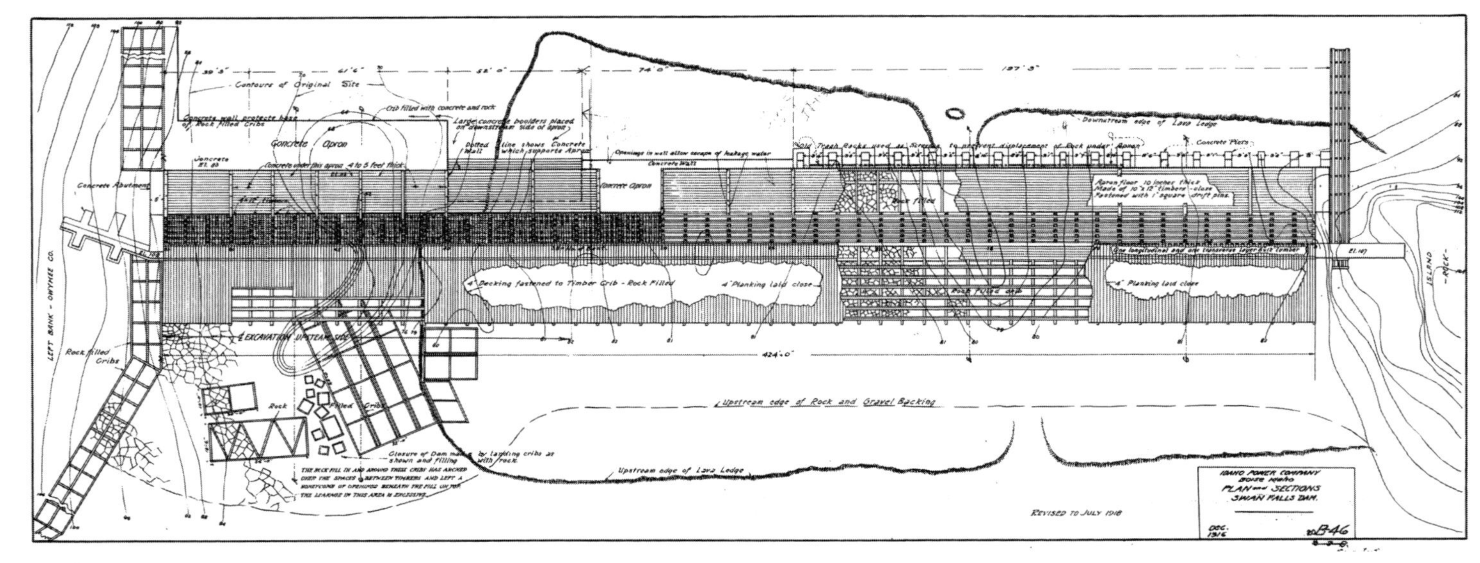

7-008

7-010

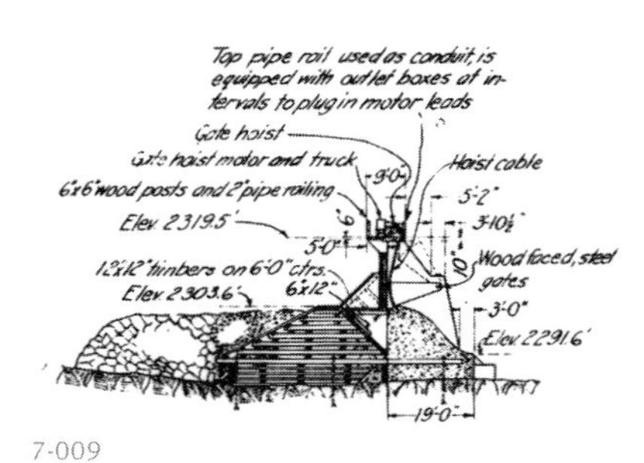

7-009

7-008. Plan of timber-crib left channel showing spillway, Swan Falls Dam, Snake River, Kuna Vic., Owyhee and Ada Counties, Idaho, 1900–1901, altered 1907, 1910, 1913, 1918. Idaho Power Company, delineator, 1916, revised 1918. P&P,HAER,ID,37-KU.V,1-100.

7-009. Cross-section of dam through timber-crib spillway, Swan Falls Dam, Snake River, Kuna Vic., Owyhee and Ada Counties, Idaho, 1900–1901, altered 1907, 1910, 1913, 1918–1921. C.W.S. of Idaho Power Company, delineator, 1922. P&P,HAER, ID,37-KU.V,1-105,detail.

7-010. Spillway and gates, Swan Falls Dam, Snake River, Kuna Vic., Owyhee and Ada Counties, Idaho, 1900–1901, altered 1907, 1910, 1913, 1918–1921, 1936, 1944. Michael L. Cordell, photographer, 1984. P&P,HAER,ID,37-KU.V,1, no. 19.

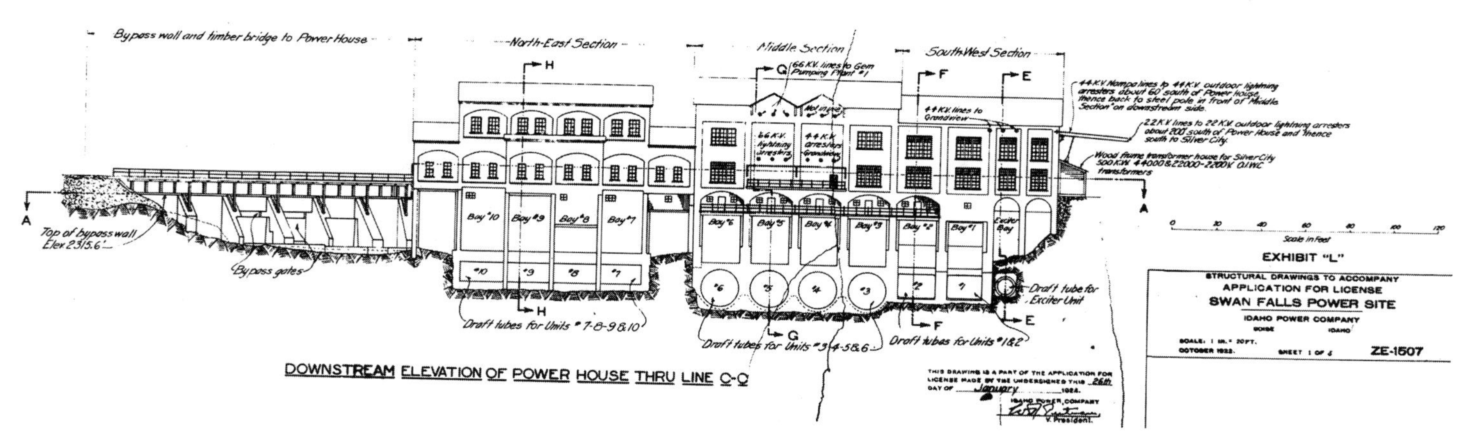

7-011

7-011. Downstream elevation of powerhouse, Swan Falls Dam, Snake River, Kuna Vic., Owyhee and Ada Counties, Idaho, 1900–1901, altered 1907, 1910, 1913, 1918–1921. C.W.S. of Idaho Power Company, delineator, 1922. P&P,HAER,ID,37-KU.V,1-101,detail.

7-012. Main floor of powerhouse, Swan Falls Dam, Snake River, Kuna Vic., Owyhee and Ada Counties, Idaho, 1900–1901, altered 1907, 1910, 1913, 1918–1921. Michael L. Cordell, photographer, 1984. P&P,HAER,ID,37-KU.V,1, no. 52.

7-012

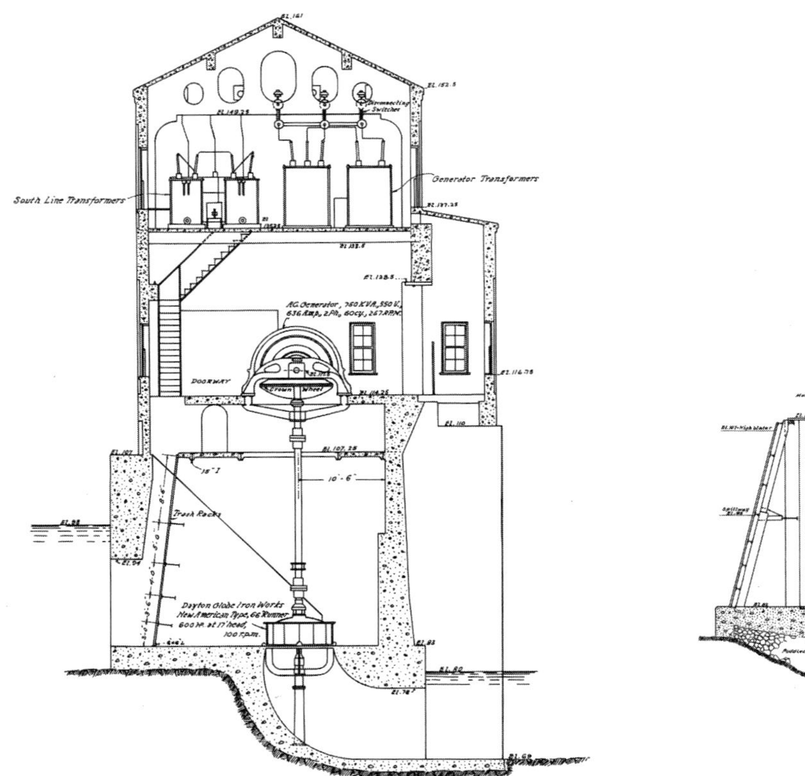

7-013

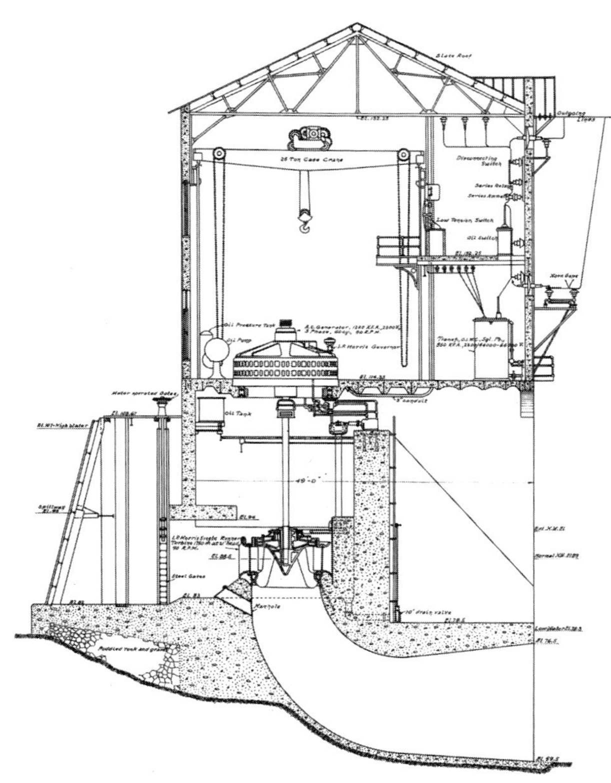

7-014

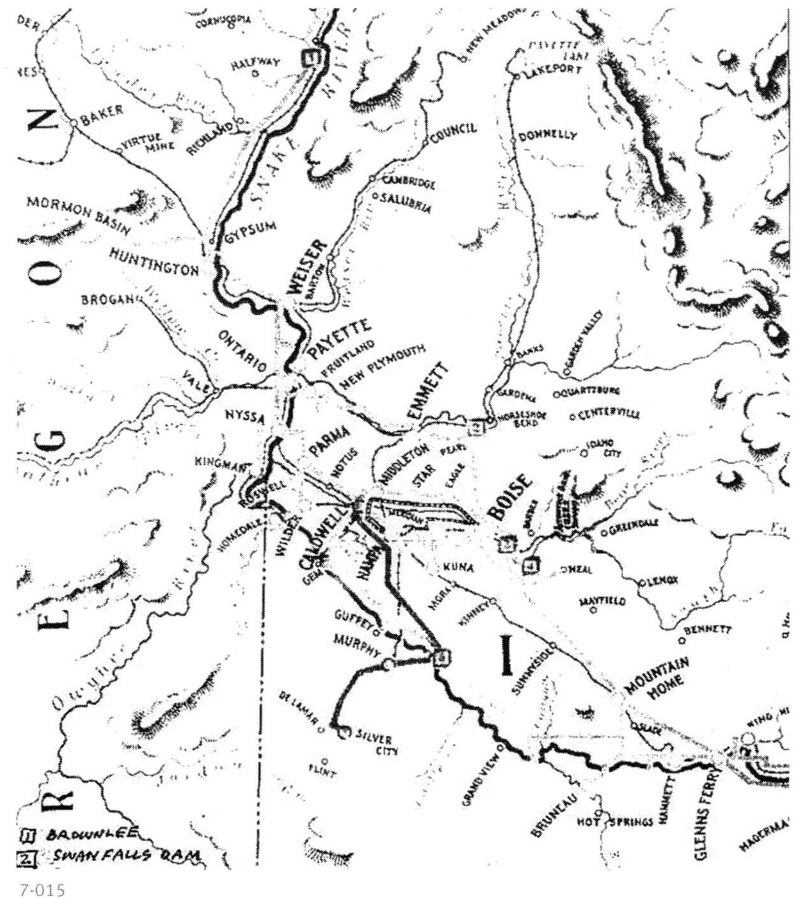

7-015

7-013. Powerhouse installation in 1907, Swan Falls Dam, Snake River, Kuna Vic., Owyhee and Ada Counties, Idaho, 1900–1901, altered 1907. 1914. P&P,HAER,ID,37-KU.V,1-97.

7-014. Powerhouse installation in 1913, Swan Falls Dam, Snake River, Kuna Vic., Owyhee and Ada Counties, Idaho, 1900–1901, altered 1907, 1910, 1913. 1914. P&P,HAER,ID,37-KU.V,1-99.

7-015. Hydroelectric power stations and transmission lines, Boise Valley, Idaho. ALAD,118067.

7-016

7-016. American Falls Dam and Hydroelectric Plants, Snake River, American Falls, Power County, Idaho, Island Plant 1901–1902; West Side Plant 1907; Dam and East Side Plant 1913. F. C. Bohlson, photographer, 1921. P&P,HAER,ID,39-AMFA,1-6,detail.

The Pocatallo region, located in southeastern Idaho between Butte, Montana, to the north and Salt Lake City to the south, lay at the intersection of a number of major rail lines that followed the old westward trails. The numerous falls of the Snake River, as it passes through this region, were developed early on for hydroelectric power and irrigation. One example is the American Falls plant, significant as an example of early low-head hydroelectric power in the West and for its long-distance power transmission. Purchased by Idaho Power Company in 1916, the power plants and dam were removed from service when the Bureau of Reclamation rebuilt the American Falls Dam in 1927.

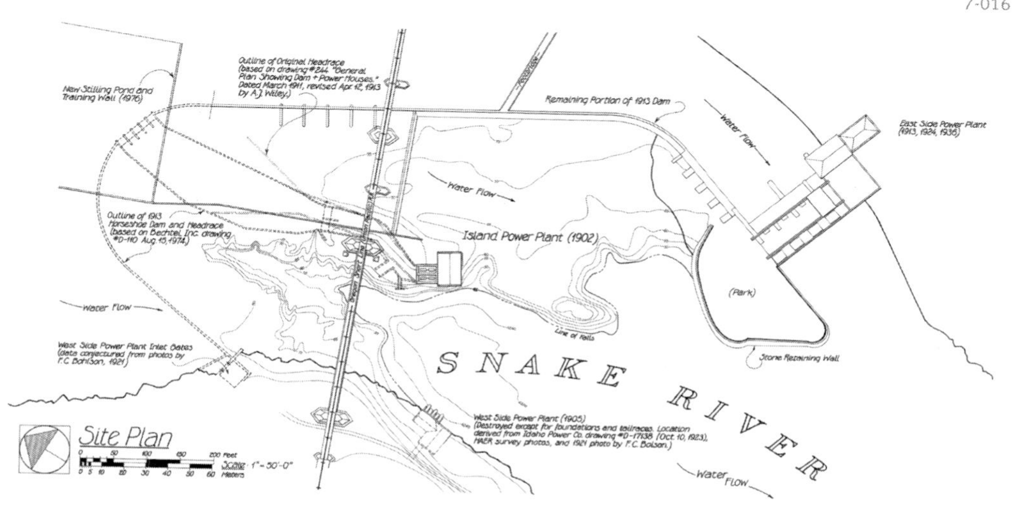

7-017

7-017. Plan of American Falls Dam and Hydroelectric Plants, Snake River, American Falls, Power County, Idaho, 1901–1902; West Side Plant 1907; Dam and East Side Plant 1913, 1924–1927. Bob Levy, Charles G. Poor, Janet Hochull, delineators, 1977; redrawn Richard K. Anderson Jr., 1978, 1983. P&P,HAER,ID,39-AMFA,1,sheet 1.

7-018

7-018. American Falls Island Power Plant, Snake River, American Falls, Power County, Idaho, 1901–1902. William P. Melville, photographer, 1976. P&P,HAER,ID,39-AMFA,1, no. 26.

7-019

7-019. Shoshone Falls Diversion Dam and Hydroelectric Plant, Snake River, Twin Falls Vic., Idaho, 1901–1907. Clarence E. Bisbee, photographer, ca. 1913. P&P,LC-dig-ppmsca-17306.

The Snake River drops a spectacular 212 feet over a horseshoe rim at Shoshone Falls. While early promoters of the region recognized the tourist potential of the falls, irrigation interests succeeded in diverting most of the Snake River for agricultural purposes. Shoshone Falls is the site of a hydroelectric plant built from 1901 to 1907 involving a diversion dam, rock-cut tunnels, and a power house. This image shows the diversion dam.

7-020. Shoshone Falls Diversion Dam and Hydroelectric Plant, Snake River, Twin Falls Vic., Idaho, 1901–1907. Amos Studio, 1910. P&P,LC-dig-cph-3c01964.

The chute and powerhouse are visible on the far left below the falls.

7-020

7-021

7-021. Milner Dam, Snake River; Murtaugh, Twin Falls County, Idaho, 1903–1905. William H. Eaton, photographer, 1985. P&P,HAER,ID,27-TWIF.V,1, no. 138.

Milner Dam was built under the Carey Act of 1894, which made federal lands available to the states so they could work with private canal companies to finance water delivery. The Carey Act was superseded by the Reclamation Act of 1902, which directly funded reclamation projects in sixteen western states with the proceeds from the sale of federal lands.

Construction of Milner Dam began with a temporary rock-fill crib dam that diverted water to a hydroelectric plant to provide power. The dam consists of three earthfill embankments with wooden cores linking two islands. Tunnels excavated through the solid rock of the islands diverted water from the river channel, allowing construction of earthfill embankments. A continuous spillway of ninety-nine gates spans the crest of the south island. At the close of construction, the tunnels were filled and the water levels allowed to rise across both islands and the three dams. A 160-mile siphon and canal system was built in tandem with the dams.

7-022. Plan and section, Milner Dam, Snake River, Murtaugh, Twin Falls County, Idaho, 1903–1905. Twin Falls Canal Company, delineator. P&P,HAER,ID,27-TWIF.V,1-192 and 183.

7-023. Workers in tunnel during construction of Milner Dam, Snake River, Murtaugh, Twin Falls County, Idaho, 1903–1905. Unidentified photographer, ca. 1903. P&P,HAER,ID,27-TWIF.V,1, no. 168.

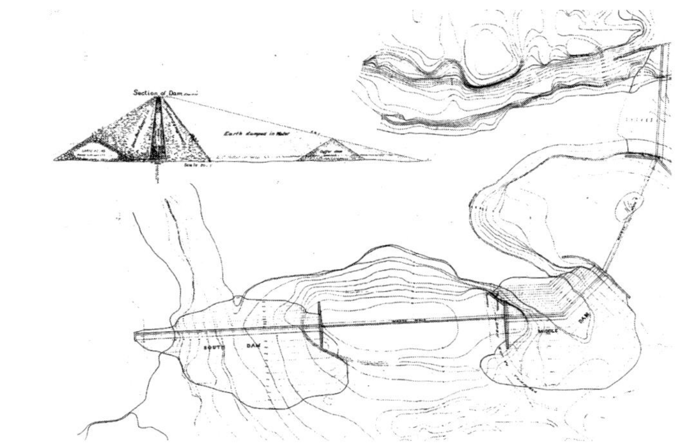

7-022

7-023

7-024

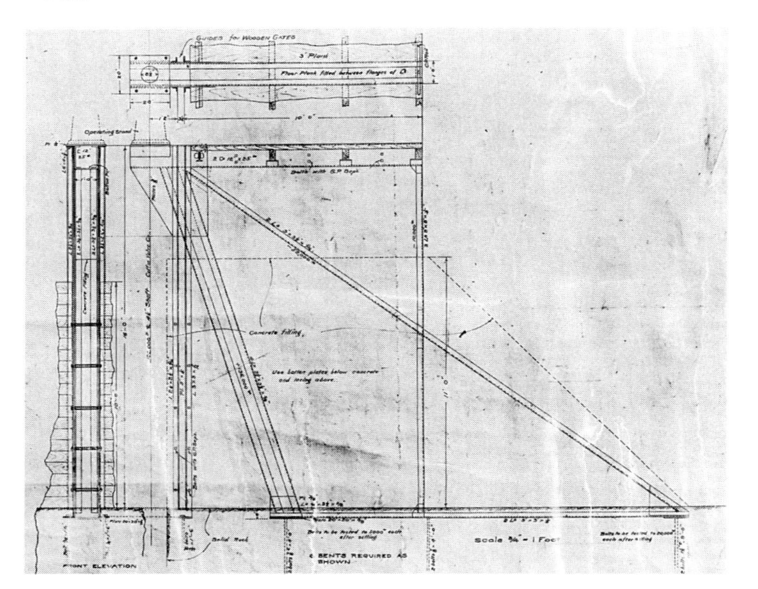

7-025

7-026

7-027

7-028

7-024. Approach to tunnel, Milner Dam, Snake River, Murtaugh, Twin Falls County, Idaho, 1903–1905. Unidentified photographer, ca. 1904. P&P,HAER,ID,27-TWIF.V,1, no. 182.

7-025. Spillway gates, Milner Dam, Snake River, Murtaugh, Twin Falls County, Idaho, 1903–1905. Twin Falls Canal Company, delineator. P&P,HAER,ID,27-TWIF.V,1-191.

7-026. Milner Dam, Snake River, Murtaugh, Twin Falls County, Idaho, 1903–1905. C. R. Savage, photographer, 1905. P&P,HAER,ID,27-TWIF.V,I-68.

7-027. Rock Creek Siphon south of Kimberly, Twin Falls County, Idaho. C. R. Savage, photographer. P&P,HAER,ID,27-TWIF.V,I, no. 85.

7-028. Dry Creek Headgates on Main Canal, south of Murtaugh, Twin Falls County, Idaho. C. R. Savage, photographer. P&P,HAER,ID,27-TWIF.V,I, no. 82.

7-029

7-030

7-029. Construction, Thompson Falls Hydroelectric Dam, Clark Fork River, Thompson Falls, Sanders County, Montana, 1913–1917. E. J. Frazier, photographer, 1913. P&P,LC-dig-pan-6a07361.

The Clark Fork River originates at the Continental Divide in the Anaconda-Butte region. From the 1880s, until settling ponds were built in 1955, effluents from mining and smelting operations in this area found their way into the Clark Fork River drainage. The first dam on the river is Milltown Dam, built in 1907–1908 at the confluence of the Blackfoot and Clark Fork rivers several miles upstream from Missoula. Waste from copper mines and milling facilities collected behind the dam, leading to its removal in 1983. The second major dam on the river, at Thompson Falls, provided hydropower to mines in Idaho. The image shows the dam under construction, with the dewatered section to the left.

7-030. Forebay channel and foreman's bungalow, Thompson Falls Hydroelectric Dam, Clark Fork River, Sanders County, Montana, 1913–1917. Unidentified photographer, 1916. P&P,HAER,MONT,45-THOFA,3A, no. 21.

7-031

7-032

7-031. Enloe Dam, original powerhouse and penstock under construction, Similkameen River, Okanogan County, Washington, 1903–1906, rebuilt 1919–1920. Unidentified photographer, ca, 1920. P&P,HAER,WASH,24-ORO.V,1, no. 20.

Located at a 33-foot-high vertical drop along the Similkameen River, this site attracted early interest in hydroelectric generation. A timber-crib dam and power plant built in 1903–1906 supplied power to towns, farmers, and mines. Purchased by the Okanogan Valley Power Company in 1913, it was rebuilt as a concrete gravity dam several years later. The penstocks lead several hundred feet downstream to surge tanks before dropping into the powerhouse. The second penstock was added in 1923.

7-032. Concrete abutment, Enloe Dam, Similkameen River, Okanogan County, Washington, 1919–1920, modified 1923. Harvey S. Rice, photographer, 1990. P&P,HAER,WASH,24-ORO.V,1, no. 14.

7-033

7-033. Second penstock under construction and remains of demolished powerhouse, Enloe Dam, Similkameen River, Okanogan County, Washington, 1919–1920, modified 1923. Unidentified photographer, 1923. P&P,HAER,WASH,24-ORO.V,1, no. 21.

7-034. Penstock intake controls, Enloe Dam, Similkameen River, Okanogan County, Washington, 1919–1920, modified 1923. Harvey S. Rice, photographer, 1990. P&P,HAER,WASH,24-ORO.V,1, no. 115.

7-034

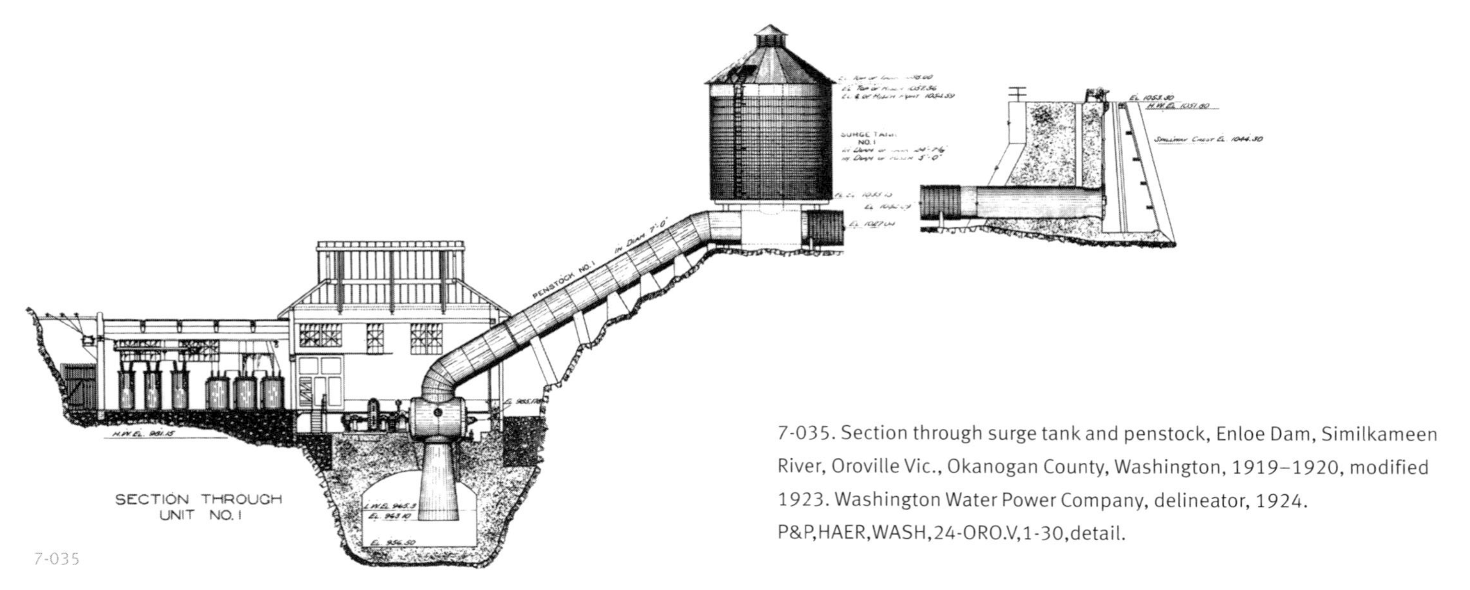

7-035

7-035. Section through surge tank and penstock, Enloe Dam, Similkameen River, Oroville Vic., Okanogan County, Washington, 1919–1920, modified 1923. Washington Water Power Company, delineator, 1924. P&P,HAER,WASH,24-ORO.V,1-30,detail.

7-036

7-036. Anyox Dam, Anyox Creek, Alice Arm, British Columbia, Canada, 1922–1924, abandoned 1935. Unidentified photographer, ca. 1923. P&P,HAER,CAL,19-LITRO.V, 1, no. 19.

This notable multiple arch dam by John Eastwood (and his only Canadian dam) was built for the Granby Company, a copper-mining enterprise in northwestern British Columbia. From 1912 to 1935, the Granby Company extracted 750 million pounds of copper, 140,000 ounces of gold, and 8 million ounces of silver from this region, making it one of the largest and most productive mines in the British Empire. The associated townsite, Anyox, grew to house 2,700 people before its collapse and abandonment during the Great Depression.

7-037. Rock Island Dam, Columbia River, Wenatchee Vic., between Chelan and Douglas Counties, Washington, 1930–1933. Wide World, 1932. P&P,LC-dig-ppmsca-17278.

Built by the Puget Sound Power & Light Company, Rock Island Dam was the first dam to span the Columbia River. The 1,424-foot-long spillway is divided by a central fishway into east and west sections, of fourteen and seventeen gates respectively, with the eastern section connecting to the powerhouse. The powerhouse has been expanded several times since construction.

7-037

7-038

7-039

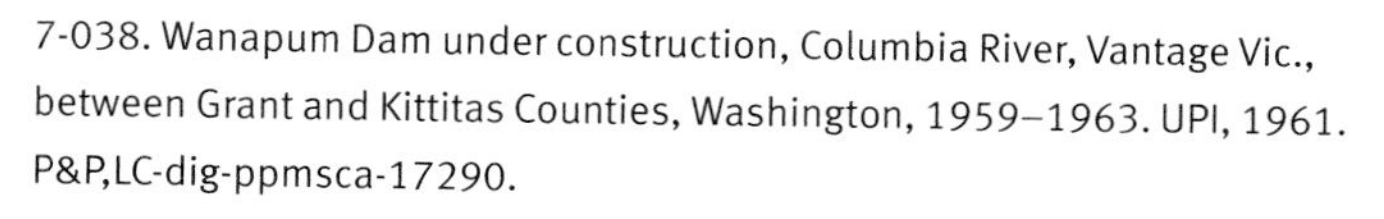

7-038. Wanapum Dam under construction, Columbia River, Vantage Vic., between Grant and Kittitas Counties, Washington, 1959–1963. UPI, 1961. P&P,LC-dig-ppmsca-17290.

Wanapum Dam was built in conjunction with Priest Rapids Dam for power generation by the Grant County Public Utilities district. Although the dam was fitted with fish ladders, modifications made to the turbines in recent years facilitate the survival of salmon moving through the system.

7-039. Turbine cone for Wanapum Dam, Columbia River, Vantage Vic., between Grant and Kittitas Counties, Washington, 1959–1963. UPI, 1961. P&P,LC-dig-ppmsca-17289.

7-040

7-041

7-040. Brownlee Dam, Snake River, Hells Canyon, Idaho, 1955–1959. Chapin Photo Shop, 1957. P&P,LC-dig-ppmsca-17254.

In 1947, Idaho Power applied for a license to construct three dams in and near Hells Canyon on the Snake River, the deepest canyon on the North American continent. This application led to a national debate in Congress and the media, with one side advocating public power and federal control of the project, while the other supported private development. In 1955, the Federal Power Commission licensed Idaho Power to construct the three projects: Hells Canyon, Oxbow, and Brownlee Dams.

7-041. Noxon Rapids Dam, Clark Fork River, Sanders County, Montana, 1959. Unidentified photographer, 1959. P&P,LC-dig-ppmsca-17273.

Built by the Washington Water Power Company of Spokane, the concrete gravity Noxon Rapids Dam is one of the largest hydropower projects in Montana, supplying power to the entire Northwest. The dam severely affected the spawning habitat for trout and salmon in the Clark Fork River.

FEDERAL RECLAMATION PROJECTS

After the establishment of the United States Reclamation Service in 1902, two of its first projects were in southern Idaho—the Boise Irrigation Project to the west, and the Minidoka project on the eastern reaches of the Snake River between Pocatello and Twin Falls. In both of these arid and high-altitude regions, agriculture required a dependable supply of water. Irrigation projects in Washington state began shortly thereafter. By 1905, the Reclamation Service had begun work on the Okanogan Project in the valley of the same name near the Canadian border; and the Yakima Project in arid southeastern Washington, where demands for irrigation had been spurred by the advent of the railroad. Early projects in Oregon included the Umatilla Project near the Washington border and the Klamath Basin Project, with its Lost River Diversion Dam, on the Californian border. The Bureau of Reclamation continued to build dams in this region over subsequent decades, culminating in the massive Columbia Basin Reclamation Project and its Grand Coulee Dam.

Minidoka Project

The Minidoka Project contains seven dams and thousands of miles of canals and laterals that irrigate over one million acres in Idaho and Wyoming. Major dams in the project include Minidoka (1904–1906), Jackson Lake (1910–1911), and the rebuilt American Falls (1925–1927).

7-042

7-042. Minidoka Dam, Minidoka Project, Snake River, Minidoka Vic., between Minidoka and Cassia Counties, Idaho, 1904–1906. United States Reclamation Service. TG23 J33 1988,p. 289.

The earth- and rock-fill Minidoka Dam stretches 4,475 feet across the Snake River, incorporating a 2,400-foot-long concrete gravity spillway into its design to accommodate the tremendous flow of the Snake. As the first hydroelectric project of the Reclamation Service, it set a precedent for subsequent federal power-generating facilities, such as the Grand Coulee Dam on the Columbia River (7-078–7-088).

7-043

7-043. Minidoka Dam powerplant, Minidoka Project, Snake River, Minidoka Vic., between Minidoka and Cassia Counties, Idaho, 1904–1906, powerplant added and spillway raised 1909, 1910–1911. United States Reclamation Service, 1911. HD1695 W4 R68 2006,p. 85.

7-044

7-044. Jackson Lake Dam, Minidoka Project, South Fork of Snake River, Moran Vic., Teton County, Wyoming, 1907, 1910–1911, modified 1916. P&P,stereo-1s01723.

The first Jackson Lake Dam, built in 1907, was a log crib dam at the outlet of the natural lake. When this failed in 1910, it was replaced by an earthen dam, spillway, and other improvements that raised the height of the lake nearly 40 feet. The earthen dam is 4,920 feet long, incorporating a concrete gravity spillway of 222 feet.

Boise Irrigation Project

Irrigation had been established in the Boise River valley from the earliest days of its mining settlement but was restricted to lands immediately adjacent to the river. With the formation of a water users organization, the United States Reclamation Service agreed to develop a diversion dam on the Boise River, a main-stem canal, and the Arrowrock reservoir dam. Anderson Ranch Dam was added to the project during World War II.

7-045. Boise River Diversion Dam, Boise Irrigation Project, Boise River, Boise Vic., Ada County, Idaho, 1908–1910. United States Reclamation Service, 1909. P&P,HAER,ID,1-BOISE.V,1-A, no. 35.

7-046. Dam, powerhouse, and canal, Boise River Diversion Dam, Boise Irrigation Project, Boise River, Boise Vic., Ada County, Idaho, 1908–1910, powerhouse 1912. United States Reclamation Service, 1912. P&P,HAER,ID,1-BOISE.V,1-A, no. 38.

7-045

7-046

7-047

7-047. Canal gates, Boise River Diversion Dam, Boise Irrigation Project, Boise River, Boise Vic., Ada County, Idaho, 1908–1910, powerhouse 1912. Clayton B. Fraser, photographer, 1989. P&P,HAER,ID,1-BOISE.V,1-A, no. 9.

7-048. Arrowrock Dam, Boise Irrigation Project, Boise River, Ada County, Idaho, 1911–1915. Unidentified photographer, 1915. HD1695 W4 R68 2006,p. 126, detail.

Built as a reservoir dam for irrigation storage, the concrete arch Arrowrock Dam, with a crest height of 350 feet, was the highest dam in the world until Owyhee (7-072–7-073) was completed in Oregon seventeen years later. The Reclamation Service employed a number of innovative practices in its construction, such as the use of a pipe grid to transport cement pneumatically and an overhead cableway to deliver concrete to any point on the construction site; both of these were later used at Hoover Dam (6-092–6-105). Other firsts include experimentation with the concrete mixture to reduce heat buildup, the use of thermometers as heat gauges, and construction joints that varied with the dam thickness.

7-048

7-049. Water released through upper set of needle valves, Arrowrock Dam, Boise Irrigation Project, Boise River, Ada County, Idaho, 1911–1915. Bain News Service, 1915. P&P,LC-dig-cph-3c22832.

The needle valve, designed by Reclamation engineer O. H. Ensign in 1906, was used here as in other early high-head installations of the United States Reclamation Service. A modified cylinder gate valve mounted on the upstream face of the dam, it was operated by a controlled release of water from the reservoir. In 1937, the dam crest was raised, and in 2004 its needle valves were replaced with clamshell gates that allowed releases at any reservoir level and could be maintained without dewatering.

7-049

7-050. Anderson Ranch Dam, South Fork of Boise River, Pine Vic., Elmore County, Idaho, 1941–1950. Acme photograph of drawing, 1945. P&P,LC-dig-ppmsca-17232.

Located 42 miles upstream from Arrowrock, Anderson Ranch Dam provides water storage, power generation, and silt control on the Boise River system. Another Herculean endeavor by the United States Bureau of Reclamation, this 456-foot-high earthfill dam was the highest embankment dam in the world at the time of its construction. It was constructed in the face of manpower and material shortages during World War II, taking ten years to complete. The power plant serves the Bureau of Reclamation's Southern Idaho Power System, with surplus power going to the Bonneville Power Administration.

7-050

7-051

7-052

7-053

7-051. Anderson Ranch Dam under construction, South Fork of Boise River, Pine Vic., Elmore County, Idaho, 1941–1950. Wide World, 1945. P&P,LC-dig-ppmsca-17330.

7-052. Cherry orchards at Emmett, Gem County, Idaho. Russell Lee, photographer, 1941. P&P,LC-USF34-039674-D.

7-053. Japanese-American farm worker at camp under FSA administration. Rupert, Minidoka County, Idaho. Russell Lee, photographer, 1942. P&P,LC-USF34-073925-E.

Japanese-Americans interned at Hunt, Idaho, formed part of the work force.

Okanogan Project

The Okanogan Project was the first United States Reclamation Service project in Washington State. Its main elements are two storage dams, the hydraulic earthfill Conconully Dam (built 1907–1910) and the later Salmon Lake Dam (1919–1921), as well as Salmon Creek Diversion Dam downstream at the canal headworks. Project canals and laterals serve five thousand acres along the Okanogan River.

7-054

7-054. Salmon Creek Diversion Dam, Okanogan Project, Salmon Creek, Okanogan County, Washington, 1906. United States Reclamation Service, ca. 1908. P&P,Unprocessed items,HAER No. WA-68.

Salmon Creek Diversion Dam is a typical concrete irrigation dam. Located several miles downstream from Conconully Dam, it is a 6-foot-high diversion weir that spans the creek, sending its water into a canal system. The ogee, or curved, crest of the weir allows overflow back into Salmon Creek, but usually the entire stream is diverted during irrigation season. The dam was rebuilt in 1999 to improve fish passage.

Yakima Reclamation Project

Eastern Washington State is high desert, sunny, and arid. Agriculture in the Yakima Valley received an impetus with the arrival of the Northern Pacific Railroad in 1886, which became the first large private investor in irrigation. By 1900, demand for water had outstripped the capacity of the Yakima River, and it was clear that if agriculture were to expand, water storage was necessary. After the foundation of the United States Reclamation Service in 1902, the government purchased Northern Pacific Railroad facilities and sought to create an integrated system for 175 miles of the Yakima River valley. With the addition of new dams every few years, the Yakima Project came to include six reservoir dams–Bumping Lake (1910), Kachess (1912), Clear Creek (1914), Keechelus (1917), Tieton (1925), and Cle Elum (1933)—as well as a number of diversion dams, power and pumping plants, canals, and laterals, most of which have been replaced with closed pipes. Opening dry lands to settlement, the project spurred the development of irrigated agriculture throughout eastern and central Washington.

Keechelus Dam is representative of earthen-dam engineering and construction of the early twentieth century. Tieton, also an earthfill dam, set a world record for height at the time of its completion in 1925. Its designers, Frank Weymouth and Frank Crowe, went on to play key roles in other Bureau of Reclamation projects.

7-055

7-055. Temporary crib dam at Lake Keechelus, Yakima Reclamation Project, Kittitas County, Washington, 1906–1907. Unidentified photographer, 1907. P&P,Unprocessed items,HAER No. WA-80.

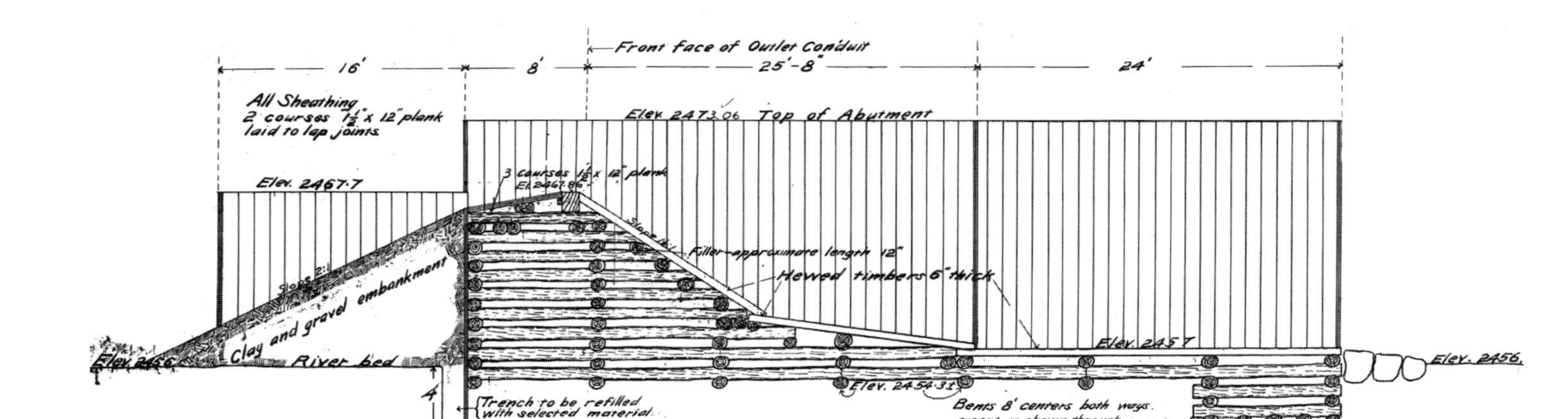

7-056

7-056. Construction drawing of temporary crib dam at Lake Keechelus, Yakima Reclamation Project, Kittitas County, Washington, 1906–1907. United States Reclamation Service, 1906. P&P,Unprocessed items,HAER No. WA-80.

7-057. Plan of dam site, Keechelus Dam, Yakima Reclamation Project, Easton Vic., Kittitas County, Washington, 1913–1917. United States Reclamation Service, 1911. P&P,Unprocessed items,HAER No. WA-80.

7-058. Aerial view, Keechelus Dam, Yakima Reclamation Project, Easton Vic., Kittitas County, Washington, 1913–1917. P&P, Unprocessed items,HAER No. WA-80.

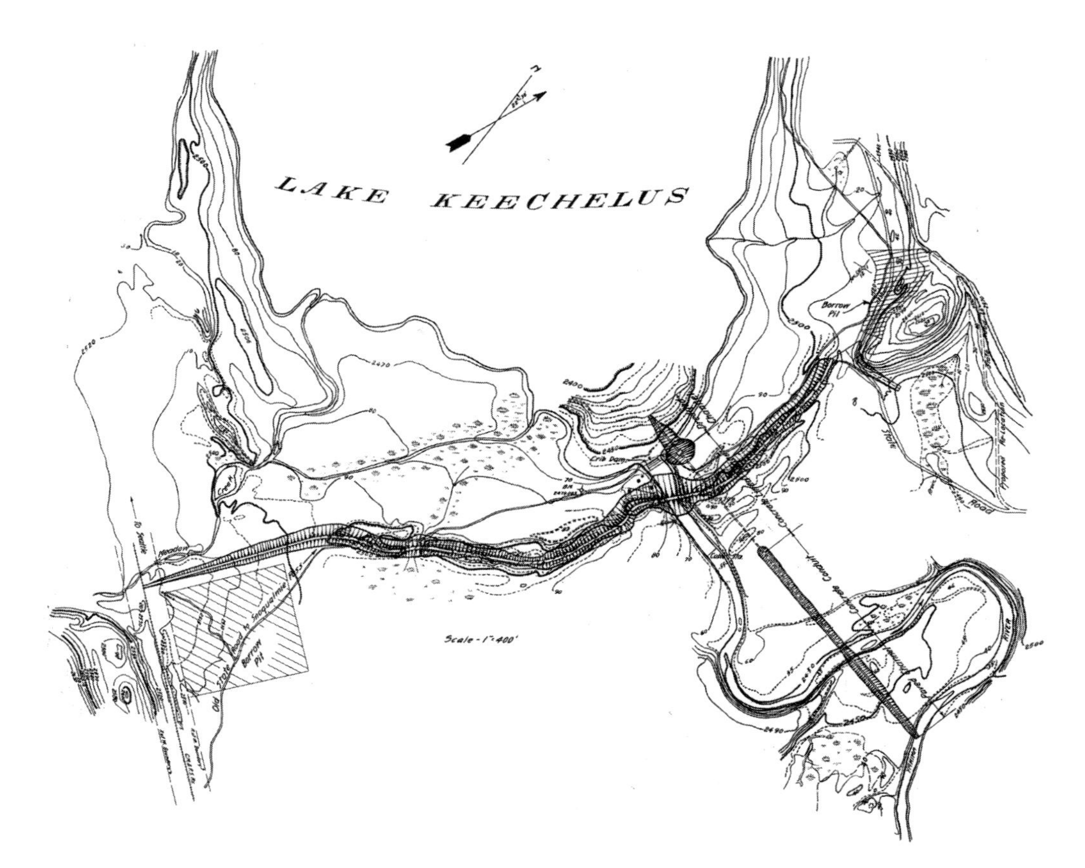

7-057

7-058

7-059

7-060

7-061

7-059. Embankment with foundation and rock in place on left, Keechelus Dam, Yakima Reclamation Project, Easton Vic., Kittitas County, Washington, 1913–1917. Unidentified photographer, 1914. P&P,Unprocessed items,HAER No. WA-80.

7-060. Bucyrus steam shovel, Keechelus Dam, Yakima Reclamation Project, Easton Vic., Kittitas County, Washington, 1913–1917. Unidentified photographer, 1916. P&P,Unprocessed items,HAER No. WA-80.

7-061. Upper slope, Keechelus Dam, Yakima Reclamation Project, Easton Vic., Kittitas County, Washington, 1913–1917. Unidentified photographer, 1914. P&P,Unprocessed items,HAER No. WA-80.

7-062

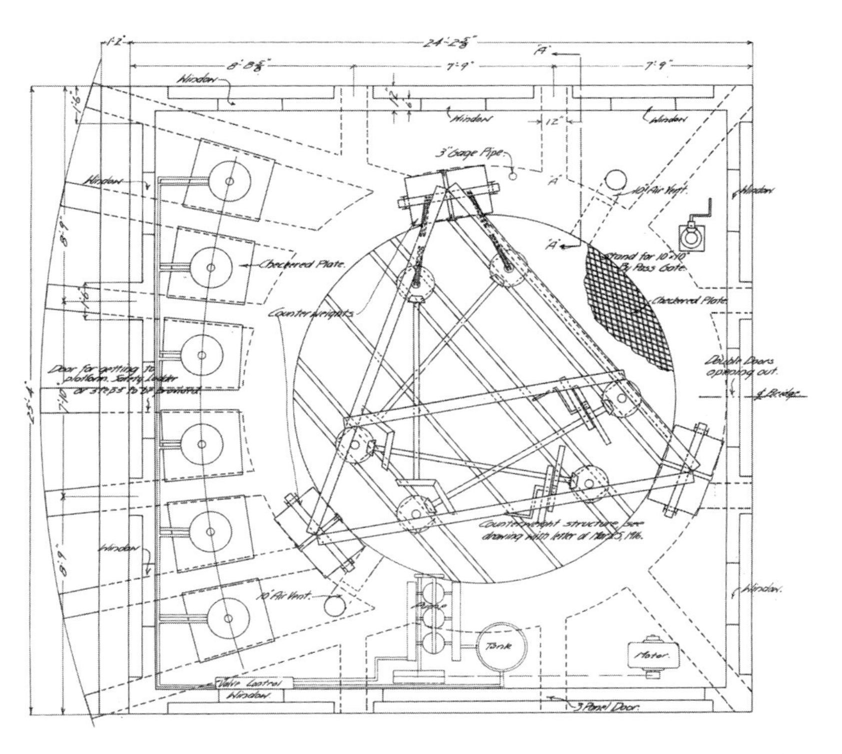

7-063

7-062. Gate tower, Keechelus Dam, Yakima Reclamation Project, Easton Vic., Kittitas County, Washington, 1913–1917. Clayton B. Fraser, photographer, 2001. P&P, Unprocessed items,WA-80-D.

7-063. Plan of gate tower, Keechelus Dam, Yakima Reclamation Project, Easton Vic., Kittitas County, Washington, 1913–1917. United States Reclamation Service, 1916. P&P,Unprocessed items,WA-80-D.

7-064. Inside flume under construction, Keechelus Dam, Yakima Reclamation Project, Easton Vic., Kittitas County, Washington, 1913–1917. United States Reclamation Service, 1916. P&P, Unprocessed items,HAER No. WA-80.

7-064

7-065

7-066

7-065. Outlet channel and stilling basin, Keechelus Dam, Yakima Reclamation Project, Easton Vic., Kittitas County, Washington, 1913–1917. Clayton B. Fraser, photographer, 1992. P&P,Unprocessed items,WA-80-A, no. 3.

7-066. Spillway forms, Keechelus Dam, Yakima Reclamation Project, Easton Vic., Kittitas County, Washington, 1913–1917. Unidentified photographer, 1916. P&P,Unprocessed items,WA-80-C.

7-067

7-067. Spillway and concrete channel, Keechelus Dam, Yakima Reclamation Project, Easton Vic., Kittitas County, Washington, 1913–1917. J. S. Moore, photographer, 1938. P&P,Unprocessed items,WA-80-C.

7-068. Concrete channel, Keechelus Dam, Yakima Reclamation Project, Easton Vic., Kittitas County, Washington, 1913–1917. Clayton B. Fraser, photographer, 2001. P&P,Unprocessed items,WA-80-C, no. 1.

7-068

7-069

7-070

7-069. Tieton Dam, Yakima Reclamation Project, Naches Vic., Yakima County, Washington, 1917–1925. United States Reclamation Service. P&P,HAER,WASH,39-NACH.V,2, no. 34.

7-070. Needle valve, Tieton Dam, Yakima Reclamation Project, Naches Vic., Yakima County, Washington, 1917–1925. Glade Walker, photographer, 1987. P&P,HAER,WASH,39-NACH.V,2, no. 7.

7-071. Kittitas Division Main Canal, Yakima Reclamation Project, Cle Elum Vic., Kittitas County, Washington, 1911–1931. A. A. Whitmore, photographer, 1931. P&P,LC-USZ62-37572.

7-071

Other Reclamation Projects

7-072. Lost River Diversion Dam, Klamath Basin Project, Lost River and Klamath River, Klamath Falls Vic., Klamath County, Oregon, 1911–1912. Unidentified photographer, 1913. P&P,HAER,ORE,18-KLAFA,1B, no. 9.

The earliest dams built for the Klamath Basin Project of southern Oregon were the earth- and rock-fill Clear Lake Dam (1908–1910) on California's Lost River, followed by this diversion dam downstream, which sent water from the Lost River into a canal leading to the Klamath River in order to reduce flooding of agricultural lands in the Tule Lake basin. The Lost River Diversion Dam is a horseshoe-shaped, multiple arch concrete structure with earth embankment wings. In the 1920s, additional dams were added to the project, including Gerber and Link River dams, and the Anderson-Rose, Malone, and Miller Diversion dams.

7-072

7-073

7-074

7-073. Owyhee Dam, Owyhee Project, Owyhee River, Malheur County, Oregon, 1928–1932. Raymond Merritt, photographer, 1949. P&P,SSF-Dams-Oregon-Owyhee-1949.

Owyhee Dam served as the storage reservoir for the Owyhee Project in arid southeastern Oregon on the Idaho border. The concrete arch dam with a crest height of 417 feet set a world record for height at the time of its construction. The Bureau of Reclamation used Owyhee Dam as a testing ground for the new construction techniques and engineering innovations it was developing for Hoover Dam, a project of an unprecedented scale that would be 300 feet higher than Owyhee. These included the "trial load" method of design and the circulation of cool river water through setting concrete to control excess heat.

7-074. Morning Glory Spillway of the Owyhee Reservoir, Owyhee Project, Malheur County, Oregon, 1928–1932. Russell Lee, photographer, 1941. P&P,LC-USF34-038939-D.

7-075. Pre-cast concrete flume on C Canal, Owyhee Project, Malheur County, Oregon, 1928–1932. Bureau of Reclamation, 1941. P&P,LC-dig-cph-3a00758.

7-076. Opening the gate valve of the Malheur siphon, Vale-Owyhee Project, Malheur County, Oregon, 1928–1932. Bureau of Reclamation, 1936. P&P,LC-dig-ppmsca-17347.

7-075

7-076

FEDERAL MULTIPURPOSE DAMS IN THE COLUMBIA RIVER BASIN

The federal government's first two projects on the Columbia River date from the 1930s. The Bureau of Reclamation built the Grand Coulee Dam for irrigation, and the U.S. Army Corps of Engineers built the Bonneville Dam to generate electricity and facilitate navigation on the river. Yet the scale of these projects was so large, and the effects they had on transforming the Pacific Northwest so great, that one cannot see them in isolation. Rather, they became the principal features of a massive federal enterprise involving power generation, federal investment in western lands and industries, and modernization of an entire region.

In 1937, after extensive public debate, the government established the Bonneville Power Administration to produce power from the two dams, build transmission systems, and distribute power at wholesale rates throughout the Pacific Northwest. Over the

7-077. Grand Coulee, Columbia River, site of present dam. Sarony, Major & Knapp lithograph, ca. 1850. P&P,LC-USZ62-31256.

7-077

subsequent decades, the federal government built four more dams on the main stem of the river, and linked up all the hydroelectric projects on the Columbia and its tributaries to the Columbia River Power System, making it the largest hydroelectric system on the continent. It included Bureau of Reclamation dams on the Snake and others on the Rogue River in Oregon, and it also extended across national borders—in 1964, the Columbia River Treaty arranged for three storage dams on the Canadian reaches of the river: Keenleyside (1968), Mica (1973), and Revelstoke (1984) dams. These store water to meet hydroelectric demands in the Columbia River projects downstream.

Federal dam-building on the Columbia River is comparable in scale and ambition to federal work in the Tennessee Valley Authority (see 2-079a–2-118). Both projects were dogged by battles over the role of the government in power generation, distribution, and regulation—known as public power. Both were multipurpose projects that served as instruments of modernization in economically depressed or undeveloped regions. Both played key roles in arming the country during World War II and in the nuclear programs that followed. Both displaced long-settled riverine communities—farmers in the

7-078. Jackhammer crew, Grand Coulee Dam, Columbia Basin Reclamation Project, Columbia River, between Douglas and Grant Counties, Washington, 1933–1942. Bureau of Reclamation, ca. 1935. P&P,LC-dig-ppmsca-17402.

7-078

Tennessee Valley and Native American fishing sites, homes, and towns on the Columbia River. There were nonetheless significant differences. While regional planning, conservation, and resettlement programs were central features of the TVA, they had little presence in the Pacific Northwest. And the Columbia River had a visible presence of Native Americans, who had always lived in close relation to the river.

Grand Coulee Dam

The Grand Coulee Dam is the centerpiece of the Columbia Basin Reclamation Project, which reaches southward from the dam 120 miles to the confluence of the Columbia and Snake rivers, and touches eight counties in Washington State. Designed to irrigate a million acres of land, it also, through the Bonneville Power Administration, supplies power throughout the western United States, making Washington State an industrial power in aerospace, nuclear processing, and shipbuilding. It was built by Consolidated Builders Incorporated (CBI), a consortium of construction companies including Henry Kaiser, of Kaiser Shipbuilding, and the Six Companies (responsible for Hoover Dam).

7-079

7-079. Casing sections, Grand Coulee Dam, Columbia Basin Reclamation Project, Columbia River, between Douglas and Grant Counties, Washington, 1933–1942. Bureau of Reclamation. P&P,LC-dig-ppmsca-17400.

7-080. Riggers en route to powerhouse on the right, Grand Coulee Dam, Columbia Basin Reclamation Project, Columbia River, between Douglas and Grant Counties, Washington, 1933–1942. F. B. Pomeroy, photographer, 1942. P&P,LC-dig-ppmsca-17342.

7-081. Construction, Grand Coulee Dam, Columbia Basin Reclamation Project, Columbia River, between Douglas and Grant Counties, Washington, 1933–1942. Unidentified photographer, 1937. HD1695 W4 R68 2006,p. 338,detail.

7-080

7-081

7-082

7-083

7-084

7-082. Scrollcase for generating unit, Grand Coulee Dam, Columbia Basin Reclamation Project, Columbia River, between Douglas and Grant Counties, Washington, 1933–1942. Bureau of Reclamation, 1942. P&P,LC-dig-ppmsca-17392.

7-083. 108,000 kW generator for the Grand Coulee Dam, under construction at the Westinghouse Electric Company in East Pittsburgh, Allegheny County, Pennsylvania. Westinghouse, 1941. P&P,LC-dig-ppmsca-17317.

7-084. Water over the spillway, Grand Coulee Dam, Columbia Basin Reclamation Project, Columbia River, between Douglas and Grant Counties, Washington, 1933–1942. Bureau of Reclamation, 1942. P&P, FSA/OWI,CB-6629.

7-085. Men riding a skip to repair face of dam, Grand Coulee Dam, Columbia Basin Reclamation Project, Columbia River, between Douglas and Grant Counties, Washington, 1933–1942. F. B. Pomeroy, photographer, 1942. P&P,LC-dig-ppmsca-17341.

7-086. Grand Coulee Dam, Columbia Basin Reclamation Project, Columbia River, between Douglas and Grant Counties, Washington, 1933–1942. Charles A. Libby & Son, 1942. P&P,LC-dig-pan-6a12962.

7-085

7-086

Bonneville Dam

Built by the U.S. Army Corps of Engineers under Roosevelt's New Deal, Bonneville Dam was the first of eight large federal lock and dam structures in the Columbia River basin. It was the principal feature of the Bonneville Power Administration, established in 1937 to transmit and sell electricity through a system that eventually linked all the dams on the Columbia River. Energy production at Bonneville Dam alone is over one million kilowatts, enough to power half a million homes for a year. Much of this power is sold to California utility companies. Constructed in the Columbia River gorge, the Bonneville Dam is a concrete gravity dam 1,230 feet long, with eighteen movable-crest steel gates and a lock system.

7-087

7-089

7-088

7-087. Bonneville Dam with fish ladder in foreground, Columbia River, between Multnomah County, Oregon, and Skamania County, Washington, 1934–1937. Russell Lee, photographer, 1941. P&P,LC-dig-cph-3a00759.

7-088. Inside powerhouse, Bonneville Dam, Columbia River, between Multonomah County, Oregon, and Skamania County, Washington, 1934–1937. Bureau of Reclamation, 1943. P&P,LC-dig-ppmsca-17395.

7-089. Workers installing steel reinforcing for buttress, Bonneville Dam, Columbia River, between Multnomah County, Oregon, and Skamania County, Washington, 1934–1937. Public Works Administration. P&P,LC-dig-ppmsca-17346.

Hungry Horse Dam

This Bureau of Reclamation project, built during wartime, provided additional upstream water storage for electricity generation at Grand Coulee and Bonneville dams. The 564-foot-high dam is a variable-thickness concrete arch structure with a crest length of 2,115 feet and one of the highest morning glory spillways in the world.

7-090. Drilling holes for dynamite, Hungry Horse Dam, South Fork of Flathead River, Flathead County, Montana, 1948–1953. Acme, 1949. P&P,LC-USZ62-26042.

7-090

7-091. Hungry Horse Dam nearing completion, South Fork of Flathead River, Flathead County, Montana, 1948–1953. UPI, 1952. P&P,LC-dig-ppmsca-17260.

7-092. Penstock pipes, Hungry Horse Dam, South Fork of Flathead River, Flathead County, Montana, 1948–1953. Acme, 1950. P&P,LC-dig-ppmsca-17261.

7-093. Spiral casing sections, Hungry Horse Dam, South Fork of Flathead River, Flathead County, Montana, 1948–1953. Acme, 1950. P&P,LC-dig-ppmsca-17262.

7-092

Chief Joseph Dam

Authorized after severe floods in 1948, Chief Joseph Dam was built by the U.S. Army Corps of Engineers 50 miles downstream from the Grand Coulee Dam. Sixteen hydropower turbines were installed in 1958 and eleven were added twenty years later, making it the second-largest hydropower producer in the country, after the Grand Coulee. It produces 2.6 million kilowatts, enough power to supply the Seattle metropolitan area. The concrete gravity dam is over a mile long and 236 feet high. The intake structure behind the powerhouse is half that length and 150 feet high, serving the twenty seven generating units. There is no fish ladder at this dam, preventing salmon from migrating further up the Columbia.

7-094. Chief Joseph Dam, Columbia River, Bridgeport Vic., Douglas County, Washington, 1949–1955. UPI,1956. P&P,LC-dig-ppmsca-17238.

7-094

7-095

7-095. Idaho Power and Transportation Company Dam, Snake River, Idaho Falls, Bonneville County, Idaho, ca. 1900. Frederick J. Bandholtz, photographer, 1909. P&P,PAN US GEOG-Idaho no. 22 (E size).

Built by Idaho Power and Light Company 4 miles downstream from the city of Idaho Falls, this dam and plant were sold to Utah Power and Light in 1913, and then purchased by the city of Idaho Falls in 1937 as part of its municipal utility. The dam sits just above a 20-foot drop in the Snake River and two other dams.

MUNICIPAL HYDROELECTRIC DAMS

Seattle's Hydroelectric Dams

Public power had an early start in Seattle. In 1902, city residents approved a hydroelectric development at the falls of the Cedar River, which on its completion in 1905 became the nation's first municipally-owned hydroelectric project. The plant performed so well, and demand for municipal power rose so dramatically, that the Seattle City Council soon decided to create a separate lighting department, known as Seattle City Light. Under the superintendency of the self-taught engineer J. D. Ross (1872–1939), Seattle City Light engaged in construction of three dams on the Skagit River, a tributary of Puget Sound.

7-096. Diablo Dam, Skagit River, Newhalem Vic., Whatcom County, Washington, 1927–1929. Jet Lowe, photographer, 1989. P&P,HAER,WASH,37-NEHA.V,1-F-1.

7-096

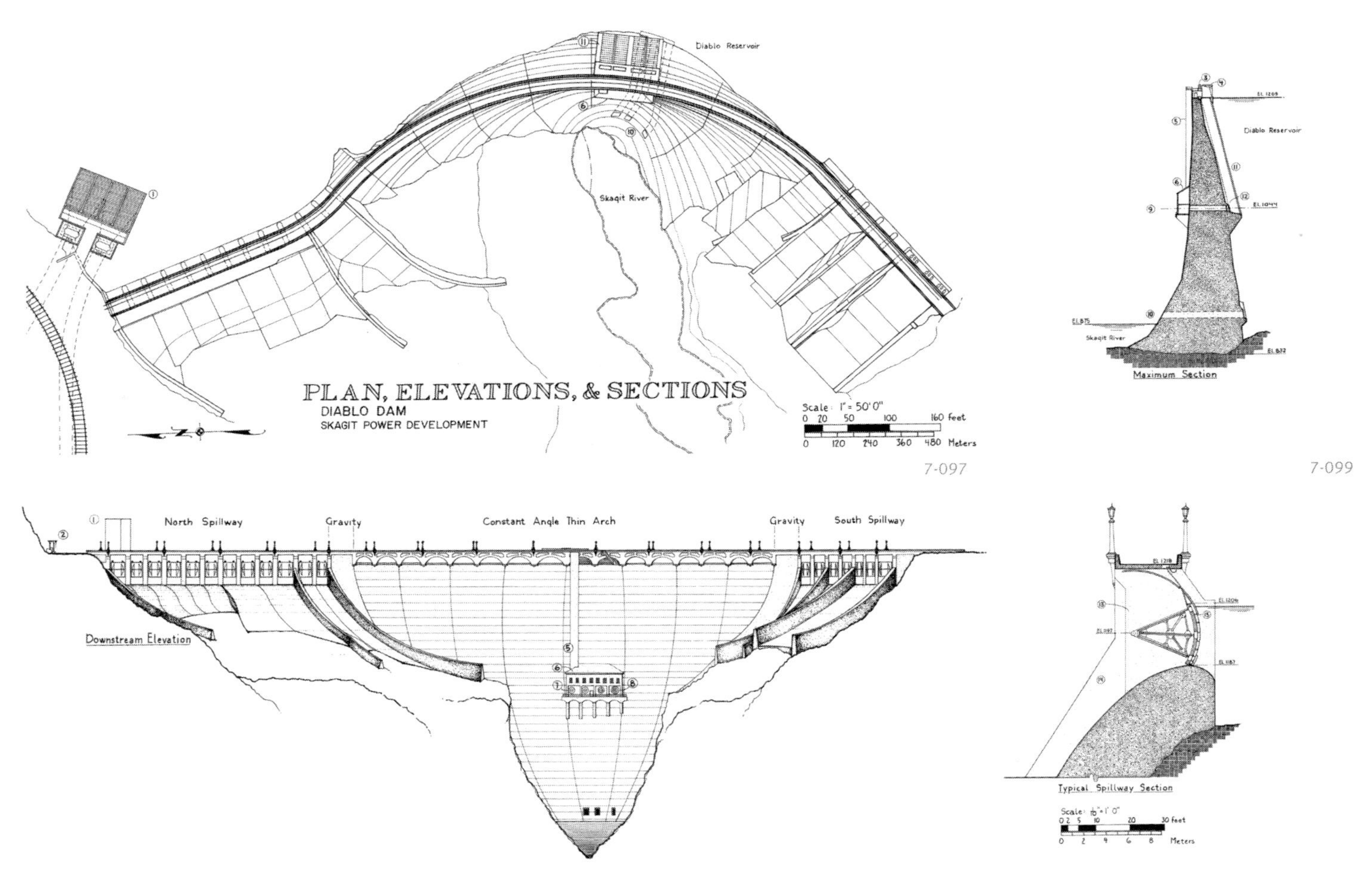

7-097

7-099

7-098

7-100

7-097. Plan, Diablo Dam, Skagit River, Newhalem Vic., Whatcom County, Washington, 1927–1929. Dione De Martelaere, delineator, 1990. P&P,HAER, WASH,37-NEHA.V,1-F-,sheet no. 1,detail.

7-098. Downstream elevation, Diablo Dam, Skagit River, Newhalem Vic., Whatcom County, Washington, 1927–1929. Dione De Martelaere, delineator, 1990. P&P,HAER,WASH,37-NEHA.V,1-F-,sheet no. 1,detail.

7-099. Maximum section, Diablo Dam, Skagit River, Newhalem Vic., Whatcom County, Washington, 1927– 1929. Dione De Martelaere, delineator, 1990. P&P,HAER,WASH,37-NEHA.V,1-F-,sheet no. 1,detail.

7-100. Spillway section, Diablo Dam, Skagit River, Newhalem Vic., Whatcom County, Washington, 1927–1929. Dione De Martelaere, delineator, 1990. P&P,HAER,WASH,37-NEHA.V,1-F-,sheet no. 1,detail.

7-101. Needle valves, Diablo Dam, Skagit River, Newhalem Vic., Whatcom County, Washington, 1927–1929. Jet Lowe, photographer, 1989. P&P,HAER,WASH,37-NEHA.V,1-F, no. 5.

7-101

7-102

7-103

7-102. Ross (Ruby) Dam under construction, Skagit River, Newhalem Vic., Whatcom County, Washington, 1937–1940. Bureau of Reclamation, 1940. P&P,LC-dig-ppmsca-17389.

7-103. Projected view of completed dam, Ross Dam, Skagit River, Newhalem Vic., Whatcom County, Washington, 1937–1940, modified 1943–1949. Bureau of Reclamation, delineator. P&P,LC-dig-ppmsca-17388.

7-104. Gorge High Dam, Skagit River, Newhalem Vic., Whatcom County, Washington, 1954–1961 (replaced original Gorge Dam of 1919–1924). Jet Lowe, photographer, 1989. P&P,HAER,WASH, 37-NEHA.V,1-C, no. 3.

7-105. Plan, Gorge High Dam, Skagit River, Newhalem Vic., Whatcom County, Washington, 1954–1961. Dale O. Waldron, delineator, 1990. P&P,HAER,WASH,37-NEHA.V,1-C-sheet no. 1, detail.

7-106. Outlet section, Gorge High Dam, Skagit River, Newhalem Vic., Whatcom County, Washington, 1954–1961. Dale O. Waldron, delineator, 1990. P&P,HAER, WASH,37-NEHA.V,1-C-sheet no. 1,detail.

7-104

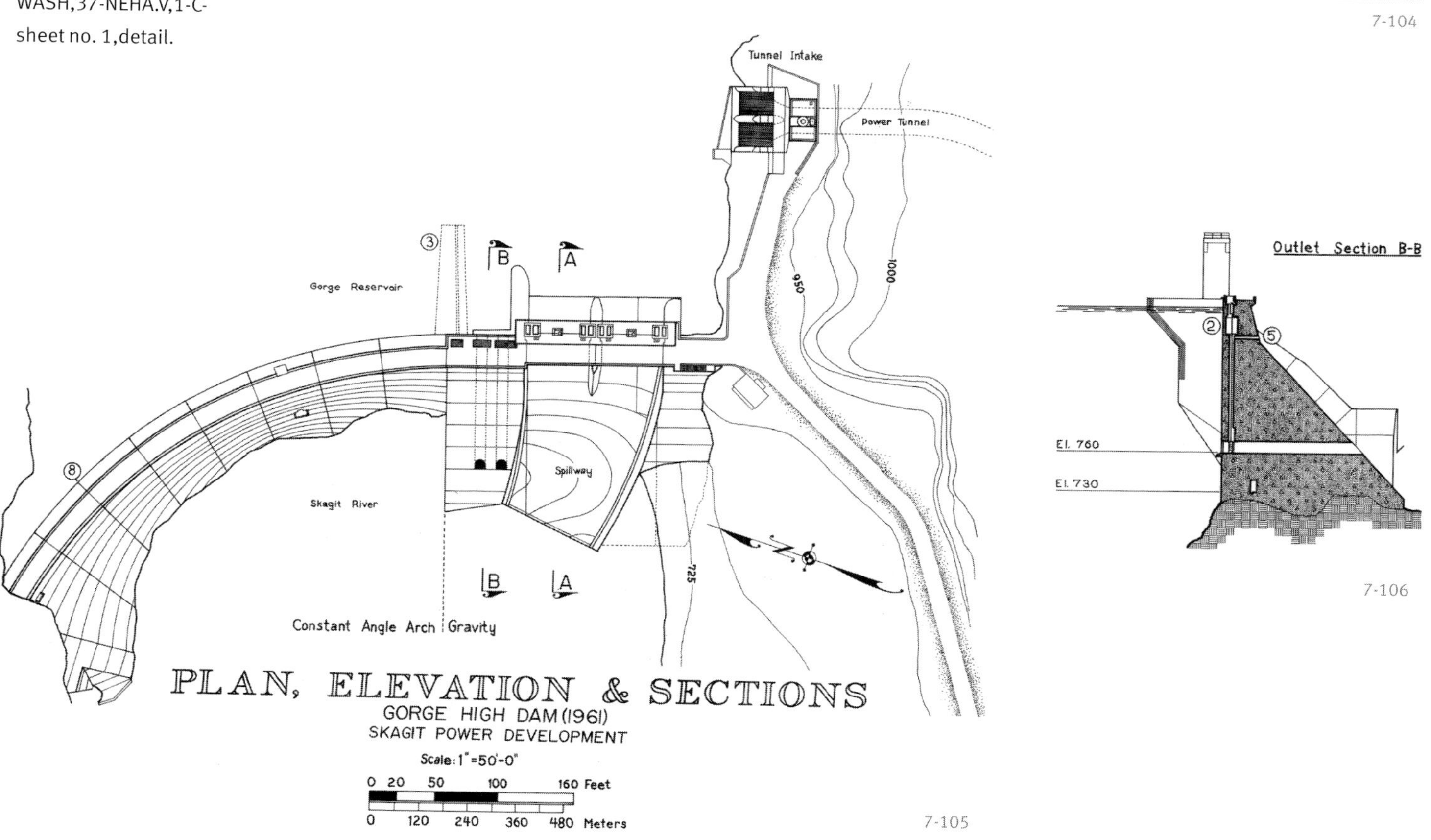

7-105

7-106

Tacoma's Hydroelectric Dams

Following Seattle's hydroelectric developments on the Skagit River, the city of Tacoma built the first La Grande dam on the Nisqually River in 1912, replacing it in the 1930s with a new La Grande Dam and the Alder Dam.

7-107. Excavated west abutment, Alder Dam, Nisqually River, Alder Vic., Pierce County, Washington, 1942–1945. Tacoma Power Company, 1942. P&P,LC-dig-ppmsca-17383.

7-108. Concrete sections, Alder Dam, Nisqually River, Alder Vic., Pierce County, Washington, 1942–1945. Tacoma Power Company, 1943. P&P,LC-dig-ppmsca-17384.

7-107

7-108

7-109

7-109. Upstream face, Alder Dam, Nisqually River, Alder Vic., Pierce County, Washington, 1942–1945. Tacoma Power Company, 1944. P&P,LC-dig-ppmsca-17381.

7-110. Downstream face of blocks 31 & 32, Alder Dam, Nisqually River, Alder Vic., Pierce County, Washington, 1942–1945. Tacoma Power Company, 1944. P&P,LC-dig-ppmsca-17382.

7-110

7-111

7-112

7-113

7-111. Upstream face of west abutment, La Grande Dam, Nisqually River, La Grande Vic., Pierce County, Washington, 1942–1945. Tacoma Power Company, 1943. P&P,LC-dig-ppmsca-17386.

7-112. Upstream face of east abutment, La Grande Dam, Nisqually River, La Grande Vic., Pierce County, Washington, 1942–1945. Tacoma Power Company, 1943. P&P,LC-dig-ppmsca-17385.

7-113. Waterwheel case for La Grande Dam, Nisqually River, La Grande Vic., Pierce County, Washington, 1942–1945. Pelton Water Wheel Company. P&P,LC-dig-ppmsca-17387.

7-114. Spillway, La Grande Dam, Nisqually River, La Grande Vic., Pierce County, Washington, 1942–1945. D. L. Glenn, photographer, 1943. P&P,LC-dig-ppmsca-17380.

7-114

CALIFORNIA

CENTRAL VALLEY

Sacramento, Pit, Feather, McCloud, American, Stanislaus, Eel, Yuba, San Joaquin, Mokelumne, Tuolumne, Merced, Fresno, Kings, and Kern Rivers

Bordering the Pacific Ocean for a thousand miles, California straddles a long central valley, bounded on one side by the Sierra Nevada and on the other by a coastal range. Two rivers drain this valley, the Sacramento flowing southward and the San Joaquin northward, to meet at a large delta that opens into the San Francisco Bay. With the highest mountains and deepest valleys in the continental United States, year-round snow cover and parched deserts, rich agricultural bottom lands and a moderate growing climate, the state is a bit like a nation unto itself.

When colonized in the eighteenth century by Spanish missionaries, California had one of the most densely settled native populations in North America. With the huge inflow of Americans looking for gold after 1849, its population grew rapidly. Innovations in dam building are legion in California. First, placer mining required large-scale water diversions and hydraulic technology for mining and processing ore, which directly led to a number of innovations in hydroelectric

8-001. Littlerock Dam, Little Rock Creek, Palmdale Vic., Los Angeles County, California, 1922–1924. Jet Lowe, photographer, 1981. P&P,HAER,CAL,19-LITRO.V,1, no. 22.

power generation and dam design. At the same time, municipalities were scrambling to secure water for their growing populations—most notably, San Francisco's Hetch Hetchy reservoir in Yosemite Park and Los Angeles's capture of the Owens River and, later, Colorado River water. Finally, among the state's thirstiest users of water were the agriculturalists of the Central Valley, who needed secure water supplies from the adjacent mountains to fully take advantage of the region's year-round growing climate.

As the state's municipalities, agribusiness, and industries grew, California's seemingly insatiable demand for water ran into conflict with its development of park lands, the Sierra Nevada offering both. California was one of the first states to experience popular resistance to dam building — in the public outcry over the Hetch Hetchy project constructed in Yosemite, the country's first national park, and fought by the Sierra Club, one of its first conservation societies.

The state is criss-crossed by aqueducts that bring northern water south, Sierra Nevada run-off west, and Colorado River water across the southern deserts to the coast. It has its own water project, an epic undertaking begun in the early part of the twentieth century

8-002. Shasta Dam, Central Valley Reclamation Project, Sacramento River, Redding Vic., Shasta County, California, 1938–1945. Bureau of Reclamation, 1942. P&P,LC-USZ62-83271.

8-002

that has involved local, state, and federal government investment. One component of this master plan is the Bureau of Reclamation's massive Central Valley Project, largely undertaken to bring surplus water from the northern part of the Central Valley to its southern part through the 450-mile-long California Aqueduct. Water politics have caused significant rifts between northern and southern Californians. Most recently, hydroelectricity generation became a battleground due to the deregulation of electric utilities and the concomitant speculation.

SOUTHERN CALIFORNIA

Los Angeles, Santa Ana, Mojave, San Diego, and Tijuana Rivers

Southern California is largely desert, the eastern stretches of which are known as the Mojave, and in which lie smaller desert preserves such as Death Valley, Anza Borrego, and Joshua Tree. In 1905, a Colorado River flood breached an irrigation canal in the

8-003

8-003. Madera Canal, Central Valley Reclamation Project, San Joaquin River, Madera County, California, 1939–1945. Bureau of Reclamation, 1942. P&P,LC-USZ62-36428.

Imperial Valley at the southern tip of the state, directing the river's flow into the desert basin for two years and creating the Salton Sea in the process. While there are numerous mountain ranges in Southern California, their small rivers are intermittent or run in washes that empty rapidly into internal basins or the ocean. Spanish missionaries built the earliest known dams in this area for their missions, as they moved northward from Mexico City through the region and up the coast to the San Francisco Bay.

Southern California draws its water from three sources. Through the Los Angeles Aqueduct, it entirely captures the Owens River and Mono Lake on the eastern side of the Sierra Nevada. It collects its allocated portion of Colorado River water through a number of dams and canals (see Section 6), including the Colorado River and San Diego aqueducts (1941) and the All-American and Coachella canals. Finally, it uses water all the way from the Sacramento River through the massive California Aqueduct of the State Water Project (1959–1973).

HYDRAULIC MINING TECHNOLOGY

With the discovery of gold in California in 1848, hundreds of thousands of treasure-seekers moved to the state. Early solo efforts at panning for gold were rapidly supplanted

8-004

8-004. Hydraulic gold mining near French Corral, Nevada County, California. Lawrence & Houseworth, 1866. P&P,LC-USZ62-9889.

by cooperative techniques such as the two-man "rocker" box or the "long tom" sluice operated by a crew of miners, both of which required rivers or streams nearby. By 1855, placer, or alluvial, deposits of gold had been depleted across the state, and companies that could mobilize greater organization and capital investment moved in. The most widely employed form of capital-intensive mining in California was hydraulic mining, developed in 1853, which involved shooting water at high pressures onto ore-bearing soils. Entire hillsides were washed away, with the residue directed through sluices to extract the gold before it was dumped it into the nearest river or stream. Some operations went on day and night, lit by torches and later, electric lights.

Technological advances, such as iron-reinforced rubber hoses, dramatically increased the power of the water cannons, known as monitors, allowing miners to attack denser soil and gravel beds. With the construction of high mountain reservoir dams, as well as ditches and flumes that dropped in elevation, it was possible to obtain higher water pressures and mining could continue into the dry season and at greater distances from riverbeds. Water supply became a lucrative enterprise, as companies acquired mining claims when water bills went unpaid. By 1880, the hydraulic industry employed tens of thousands of people and provided enormous returns to investors. Hydraulic mining technology also found a use in large earth-moving projects unrelated to mining—in 1866,

8-005. The working landscape of placer gold mining, Columbia Gulch, Tuolumne County, California. Lawrence & Houseworth, 1866. P&P,LC-USZ62-28150.

8-005

8-006

8-006. Flume and erosion of placer mining, Columbia Gulch, Tuolumne County, California, Lawrence & Houseworth, 1866. P&P,LC-USZ62-17740.

for example, the Central Pacific Railroad used high pressure monitors to open railway cuts near Dutch Flat in Placer County. Seattle used it in the 1890s to level hills blocking the city's northward expansion, and at the turn of the century, engineers used hydraulic monitors in the Panama Canal to open the way to the Pacific.

The effects on mountain landscapes were severe, exemplified in the well-known wastelands of Columbia Gulch in Tuolumne County and Malakoff Diggings in Nevada County. The damage extended downstream as well. As streambeds filled up with silt and tailings from hydraulic operations, floodwaters ran over riverbanks, depositing mining debris over agricultural lands and flooding Central Valley towns. By 1874, public criticism of hydraulic mining was widespread. In 1880, the state engineer reported that California rivers were becoming dangerously shallow due to mining runoff. State and federal governments, sensitive to mining interests, responded only by exploring flood control strategies. But by 1879 the issue had made its way to the courts, as farmers engaged in a series of legal battles to limit hydraulic operations. In 1884, the Federal Ninth Circuit Court ordered an immediate halt to the dumping of tailings into river and streambeds across the state. This brought large-scale hydraulic mining in California to an end.

While gold mining had been the principal industry of the state since 1849, the success of farming interests over the mining consortia was evidence that its future would be in agriculture. Yet the legacy of hydraulic mining would continue in a number of related engineering accomplishments, including dam building, long-distance water transport, and hydroelectric power generation and distribution. A number of dams were built in California using hydraulic-fill techniques, including La Mesa Dam in San Diego County (1895); Snake Ravine Dam in Stanislaus County (1898); Lower San Fernando (1915) and Lake Arrowhead (1922) dams in the greater Los Angeles area; San Pablo (1921) and Calaveras (1925) dams in Alameda County; and Almanor (1914) and Butt Valley (1924; 8-017-8-019) dams in Plumas County. Other significant early hydraulic fill dams across the country include Croton Dam in Michigan (1908; see 2-060–2-062), Conconully Dam in Washington State (1910); and the five flood-control dry dams of the Miami Conservancy District in Ohio (1917–1922; see 2-072–2-078).

8-007. Tulloch Mill Dam, Stanislaus River, Knight's Ferry, Stanislaus County, California, ca. 1854. HABS, 1937. P&P,HABS,CAL,50-KNITF,2B, no. 1.

This timber-crib dam with a stone sluiceway served a grist mill that was built in 1854. The mill ran on mechanical hydropower until 1895, when it was converted to house a small hydroelectric plant. The company, the Stanislaus Water and Power Company, continued in operation until the construction of the New Melones powerhouse in 1920.

EARLY HYDROELECTRIC SYSTEMS IN CALIFORNIA

In California in the 1880s, a number of small companies close to cities with growing populations experimented with power generation and transmission—among these were Visalia Electric Light & Gas Company, Santa Barbara Electric Light Company, San Bernardino Electric Company, and Ventura Land & Power. In 1887, California's first hydroelectric

8-007

plant went into service at Highgrove, near Riverside in the San Bernardino Mountains. Over the following years, many innovative power generation and transmission technologies were developed and demonstrated in California. Stepping up and down the voltage of alternating current, for example, an essential feature of long-distance power transmission, was accomplished in 1892 by San Antonio Light & Power Company to send power 14 miles to Pomona. The same year, Redlands Electric Light & Power Company generated three-phase power, necessary for reliable power supply to industrial operations, in its Mill Creek No. 1 plant in Redlands. In 1899, Southern California Power Company's Santa Ana River No. 1 plant set a record for the highest-voltage, longest-distance transmission of power, 83 miles to Los Angeles.

In the northern part of the state, the moratorium on hydraulic mining forced water supply companies to look for new markets for their extensive dams, ditches, and high-head pressure systems. They quickly moved into irrigation and hydroelectric power generation. During the 1890s, Californian engineers and entrepreneurs adapted these water delivery

8-008. Ruins of dam, headgates, and canal, Folsom Dam and Hydroelectric Plant, American River, Folsom Vic., Sacramento County, California, 1867–1893, powerhouse completed 1895, demolished 1952. Brian Grogan, photographer, 1993. P&P,HAER, CAL, 34-FOLSO. V,2, no. 58.

Built by the Natoma Water & Mining Company, this complex was one of the first plants in the United States to send high voltage alternating current over a long distance, 22 miles to Sacramento. It has been recognized as a national landmark for both civil and mechanical engineering. Originally built to provide mechanical hydropower to run a sawmill and other industries, it was modified in 1893 to feed a hydroelectric plant. The masonry gravity dam, built with labor from nearby Folsom Prison, also supplied a power plant at the prison.

8-008

systems for hydroelectric power, leading the rest of the country in innovation in this field for three decades. In 1895, the Natoma Water & Mining Company added a hydroelectric plant to its Folsom Dam to send high voltage alternating current 22 miles to the state capital, Sacramento. In 1896, the engineer John Eastwood established his San Joaquin Electric Company, which set a number of records for hydroelectric power production, including the world's longest commercial electric power transmission line (at 37 miles to Fresno) and the highest head drop from reservoir to powerhouse (at 1,410 feet). A few years later in 1899, the South Yuba Water & Mining Company added the Colgate Hydroelectric Plant to its Nevada Dam complex, to send electricity 66 miles from the Sierra foothills to Sacramento, powering the city's electric railway. Around the turn of the century, these numerous small companies across the state were consolidated into Pacific Gas & Electric Company, which would operate the largest hydroelectric system in the country, with 174 dams and 68 powerhouses.

8-009. Kern River Hydroelectric Plant No. 1, Kern River, Edison, Kern County, California, 1902–1907. Keystone View Company. P&P,LC-dig-stereo-1s01721.

Located in a placer mining district that housed numerous stamp mills in the latter half of the nineteenth century, this power plant was built by the Edison Electric Company of Los Angeles as one of several hydroelectric facilities along the Kern River. Five miles further up the canyon, a diversion dam directs the Kern River into a culvert that runs along the canyon top. Above the plant, the water drops in a conduit to the powerhouse below. The first plant to use steel towers for a long-distance transmission line, Kern River No. 1 is most significant as an early hydroelectric facility that used three-phase power generation and stepped up transmission in a large-scale application, more than doubling the generation capacity of Southern California Edison when it came on line in 1907.

8-009

8-010

8-012

8-011

8-010. Kern River Hydroelectric Plant No. 1, Kern River, Edison, Kern County, California, 1902–1907. Keystone View Company. P&P,LC-dig-stereo-1s01720.

8-011. Conduit, forebay, and powerhouse, Kern River Hydroelectric Plant No. 1, Kern River, Edison, Kern County, California, 1902–1907, 1930. Gregory L. O'Loughlin, photographer, 1993. P&P,HAER,CAL,15-BAKF.V,1A, no. 1.

8-012. Interior of powerhouse, Kern River Hydroelectric Plant No. 1, Kern River, Edison, Kern County, California, 1902–1907, 1930. Gregory L. O'Loughlin, photographer, 1993. P&P,HAER,CAL,15-BAKF.V,1A, no. 3.

8-013. Van Arsdale Dam, South Fork of the Eel River, Ukiah Vic., Mendocino County, California, 1905–1907, altered 1963. Jack William Schafer, photographer, 1987. P&P,HAER,CAL,23-UKI.V,1-5.

This unusual design is partly a concrete-core earthfill dam and partly a concrete gravity dam. Built as part of a hydroelectric system to serve Ukiah, the dam diverts water into a mile-long tunnel to Potter Valley, where the hydroelectric plant is located. A fish ladder is visible in the foreground.

8-014. Kerckhoff Hydroelectric Dam, San Joaquin River, Madera and Fresno Counties, California, 1920–1924. Philip Brigandi, photographer, 1923. P&P,LC-dig-stereo-1s01725.

The first dam on the San Joaquin River, this structure is a part of Pacific Gas & Electric Company's large hydroelectric network in the central Sierra Nevada that grew out of John Eastwood's Big Creek project of 1902–1913 (8-020–8-022).

8-013

8-014

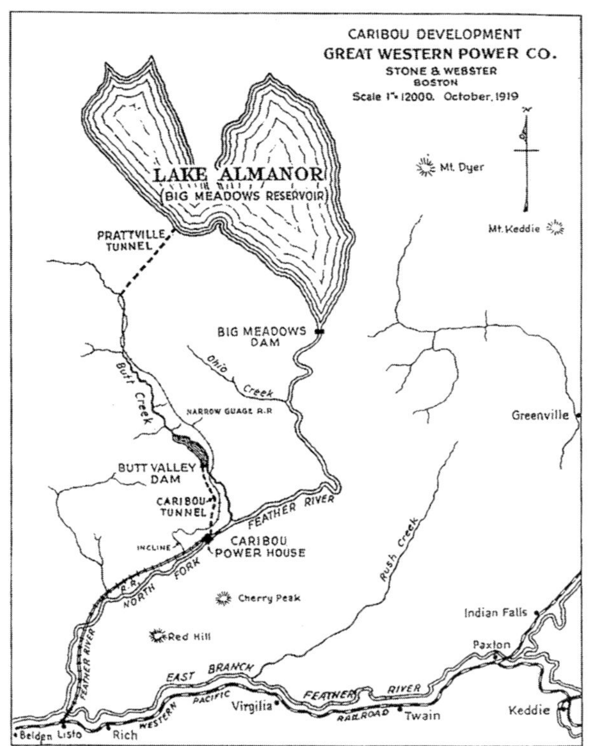

8-015

8-015. Map, Butt Valley Dam and Caribou Hydroelectric Plant, North Fork of the Feather River, Caribou Vic., Plumas County, California, 1919–1921. Stone and Webster, delineator, 1919. P&P,HAER,CAL,32-CARBU.V,1-,p. 12.

The Feather River in northern California drops a precipitous 4,350 feet in 74 miles, making it an attractive site for hydroelectric development. In 1902, plans were drawn up to store water at Big Meadows Valley in order to operate a series of hydroelectric plants downstream. The first powerhouse on this site was Great Western Power Company's Big Bend project (1908), which transmitted power 154 miles to Oakland. The venture was followed by Big Meadows (Almanor) Dam (1910–1914) and Butt Valley Dam (1919–1924), and their associated conduits and powerhouses. The company retained John Eastwood to develop a multiple arch structure for the principal water storage structure, Big Meadows Dam—but, under the persuasion of the nationally respected civil engineer John R. Freeman, the company aborted the project after construction had begun, proceeding instead with a massive hydraulic earthfill structure. After the public dispute, Eastwood turned his energies to promoting his structurally efficient multiple arch dam designs. Butt Valley Dam is a rock-fill dam built in 1921 and added to three years later using a hydraulic fill technique. It serves as a forebay for Big Meadows Dam upstream.

8-016. Plan, Butt Valley Dam and Caribou Hydroelectric Plant, North Fork of the Feather River, Caribou Vic., Plumas County, California, 1919–1921, modified 1924–1927. 1933. P&P,HAER,CAL,32-CARBU.V,-1,p. 13.

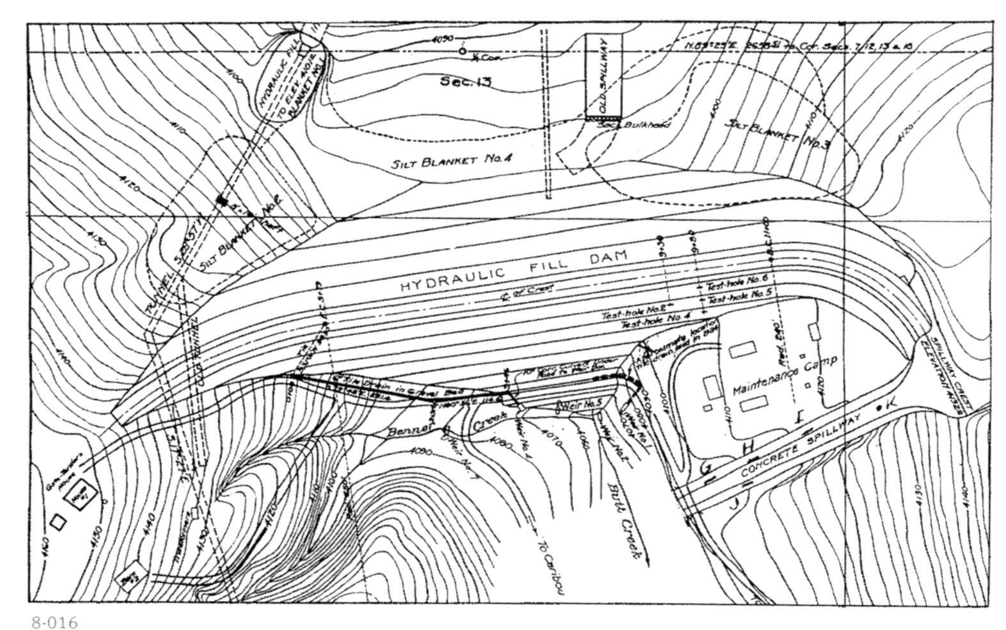

8-016

8-017. Section, Butt Valley Dam, North Fork of the Feather River, Caribou Vic., Plumas County, California, 1919–1921, modified 1924–1927. Great Western Power Company, delineator, 1923. P&P,HAER,CAL,32-CARBU.V,1-p. 15.

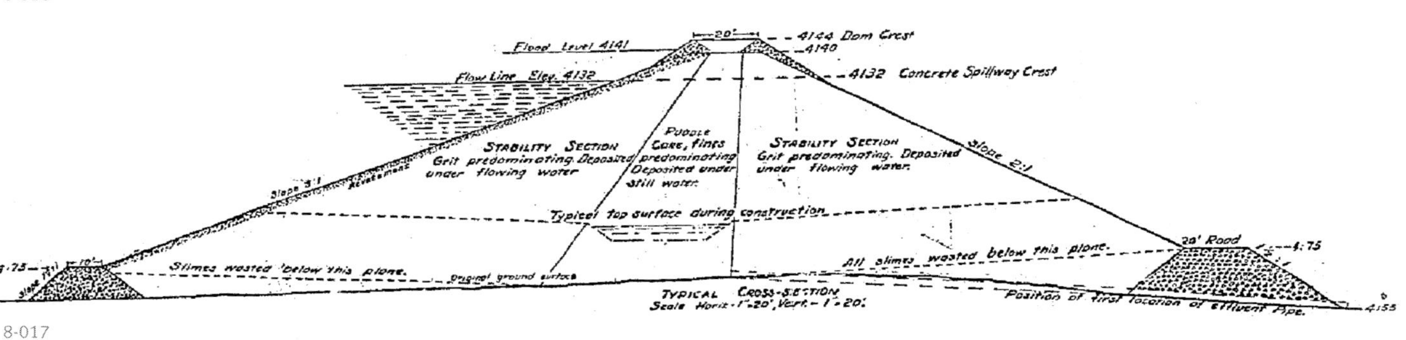

8-017

JOHN EASTWOOD'S MULTIPLE ARCH DAMS

The lightweight dam designs of John Eastwood (1857–1924) were an extraordinary invention that developed in the context of California's experimentation in hydroelectric power. Educated at the University of Minnesota, Eastwood went west to work as a railroad engineer. After settling in Fresno, he worked as a surveyor for logging companies, designing rail lines, flumes, and log ponds in the Sierra Nevada. As that city's first engineer, he began to explore ways of generating power for the town, surveying potential hydroelectric sites in the nearby Sierra Nevada. As a result, he set up the San Joaquin Electric Company. Although it was an engineering success that set a number of records, the company was insufficiently funded and went into bankruptcy. In 1902, Eastwood began work on an

8-018

8-018. John Eastwood surveying site of future Big Creek Hydroelectric Plant No. 1, Big Creek, Fresno County, California, 1902. A.W. Peters, photographer, 1902. P&P,HAER,CAL,19-LITRO.V,1, no. 14.

Eastwood's interest in the power potential of this site in the Sierra Nevada led to his first design for a multiple arch dam. To realize his large-scale vision, in 1902 he approached a group of investors, one of whom was Henry Huntington. Huntington established the Edison Company of California to finance the project, dismissed Eastwood, and instead hired a Boston-based firm to build the dam as a massive curved gravity structure. Eastwood had to look for other clients to realize his innovative designs. By 1913, the project included three dams, two tunnels, two powerhouses, and a 60 kW transmission line that ran 243 miles to Los Angeles, making it the world's largest hydroelectric project.

even more ambitious hydroelectric power system at Big Creek, a mountain stream that dropped 4,500 feet to the San Joaquin Valley. With financial backing from Los Angeles railway magnate Henry Huntington, Eastwood focused on the problem of water storage to develop his first designs for a multiple arch dam. When Huntington refused to proceed with this innovative design, Eastwood looked elsewhere to realize his invention, building dams at Hume Lake (1909) and Big Bear (1911). In 1913, his experimental designs again came under criticism, this time for the Big Meadows Dam on the Feather River, which was never completed. After that, he devoted his energies to arguing for his structurally efficient and economical multiple arch designs. As he found patrons willing to build them, he ultimately completed seventeen of these dams throughout California, Idaho, Arizona, Utah, and British Columbia.

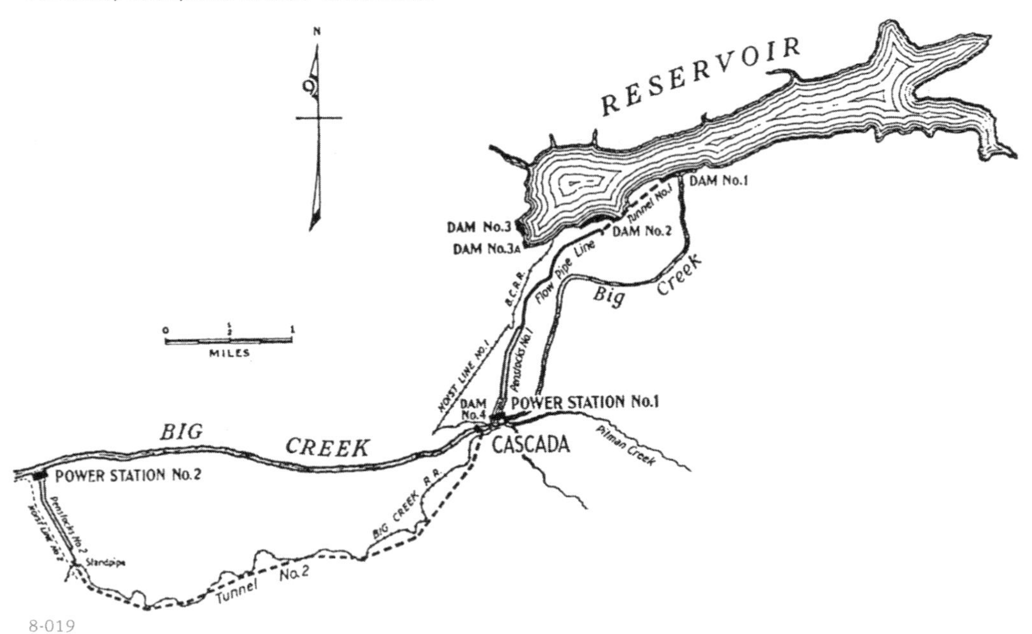

8-019

8-019. Map of initial development at Big Creek, Fresno County, California, 1913. Stone and Webster, delineator, 1920. TC547 J33 1995,p. 80.

8-020

8-020. John's Eastwood's proposed Big Meadows Dam, Big Creek, Fresno County, California, 1912. (Institute Archives and Special Collections, Massachusetts Institute of Technology) TC547 J33 1995,p. 115.

8-021. Hume Lake Dam, Ten Mile Creek, Hume, Fresno County, California, 1908–1909. Richard K. Frear, photographer, 1982. P&P,HAER,CAL,16-HUME,1, no. 1.

This dam was built by a Michigan-based company that had purchased large timber holdings in the Sierra Nevada near Sequoia National Park. Wanting a new sawmill higher in the mountains, the company hired John Eastwood to design a dam for the remote location. The result was Eastwood's first realized multiple arch dam in concrete, an extremely efficient buttress-type design. Hume Lake reservoir stored logs and supplied water for a 60-mile-long flume that delivered cut wood to a railhead in the San Joaquin valley.

8-022. Section through Hume Lake Dam, Ten Mile Creek, Hume, Fresno County, California, 1908–1909. 1909. TC547 J33 1995,p. 86.

8-023. Upstream face showing crest, Hume Lake Dam, Ten Mile Creek, Hume, Fresno County, California, 1909. Richard K. Frear, photographer, 1982. P&P,HAER,CAL,16-HUME,1, no. 19.

8-021

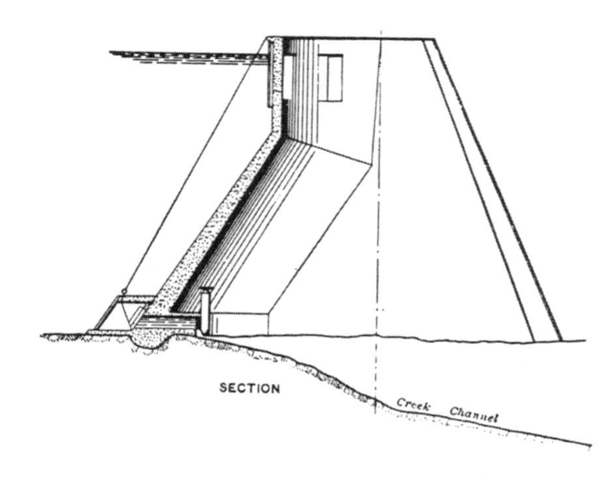

8-022

8-023

8-024

8-025

8-024. Formwork, Hume Lake Dam, Ten Mile Creek, Hume, Fresno County, California, 1908–1909. Unidentified photographer, 1909. (Michigan State University Archives and Historical Collections) TC547 J33 1995,p. 92.

8-025. Buttress and arch, Hume Lake Dam, Ten Mile Creek, Hume, Fresno County, California, 1909. Richard K. Frear, photographer, 1982. P&P,HAER,CAL,16-HUME,1, no. 11.

8-026. Construction, Big Bear Dam, Bear Creek, San Bernardino Mountains, California, 1910–1913. WC, ca. 1912. P&P,LC-dig-ppmsca-17329.

The original Bear Valley Dam of 1884 was the first arch dam in the country, a daringly thin 3 feet at its crest and barely three times that at its base, although it was 65 feet high. After twenty-five years of use, it was submerged in the reservoir created by this 72-foot-high, multiple arch concrete dam designed by John Eastwood. Both served as irrigation reservoirs for orange groves in the San Bernardino Mountains.

8-027. Upstream face, and cameo of John Eastwood, Big Bear Dam, Bear Creek, San Bernardino Mountains, California, 1910–1913. Unidentified photographer, 1912. (Water Resources Center Archives) TC547 J33 1995,p. 10.

8-026

8-027

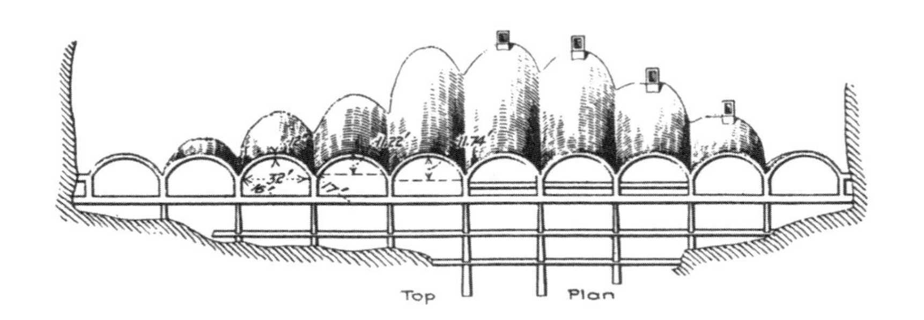

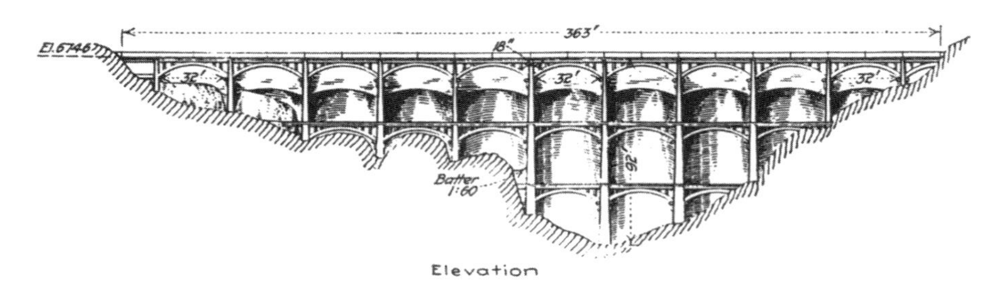

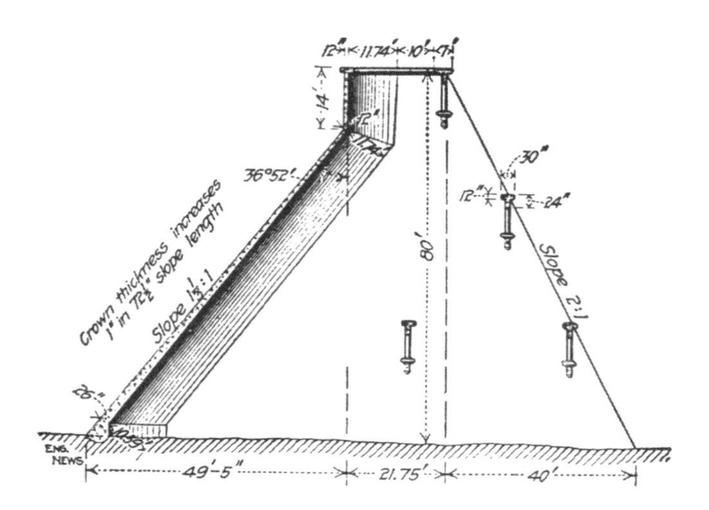

8-028

8-029

8-028. Plan, elevation, and section of Big Bear Dam, Bear Creek, San Bernardino Mountains, California, 1910–1913. 1913. TC547 J33 1995,p. 101.

8-029. Big Bear Lake, Bear Creek, San Bernardino Mountains, California, 1910–1913. WC, ca. 1913. P&P,LC-dig-ppmsca-17328.

8-030

8-031

8-032

8-030. Murray Dam under construction, San Diego County, California, 1915–1919. Unidentified photographer, 1917. (Water Resources Center Archives) TC547 J33 1995,p. 159,detail.

8-031. Murray Dam, San Diego County, California, 1915–1919. Unidentified photographer, 1918. (Water Resources Center Archives) TC547 J33 1995,p. 159,detail.

8-032. John Eastwood and others inspecting Murray Dam, San Diego County, California, 1915–1919. Unidentified photographer, ca. 1917. P&P,HAER,CAL,19-LITRO.V, 1, no. 16.

8-033

8-034

8-033. Lake Hodges Dam, San Dieguito Creek, Rancho Santa Fe Vic., San Diego County, California, 1915–1919. Unidentified photographer, 1919. P&P,Unprocessed items, HAER CA-307.

With the success of the Panama Pacific Exposition of 1915 (held to commemorate the opening of the Panama Canal), San Diego entered a growth phase. This dam was built by a subsidiary of the Santa Fe Railroad company to raise real estate values on company-owned lands north of the city. It is one of four dams built by John Eastwood in the San Diego region. In 1926, the city purchased the 136-foot-high multiple arch dam and its associated flumes and pumping plants.

8-034. Upstream face of Lake Hodges Dam at low water, San Dieguito Creek, Rancho Santa Fe Vic., San Diego County, California, 1915–1919. Unidentified photographer, ca. 1950. P&P,HAER,CAL,19-LITRO.V,1, no. 18.

8-035

8-035. Concrete channel and steel flume, Lake Hodges Dam, San Dieguito Creek, Rancho Santa Fe Vic., San Diego County, California 1915–1919. Unidentified photographer, 1919. P&P,Unprocessed items,HAER Survey No. CA-307-23.

8-036. Concrete trestle and flume, Lake Hodges Dam, San Dieguito Creek, Rancho Santa Fe Vic., San Diego County, California, 1915–1919. Unidentified photographer, 1919. P&P,Unprocessed items,HAER Survey No. CA-307-22.

8-036

8-037

8-038

8-037. Littlerock Dam, Little Rock Creek, Littlerock, Los Angeles County, California, 1922–1924. Jet Lowe, photographer, 1981. P&P,HAER,CAL,19-LITRO.V,1, no. 1.

Built as an irrigation reservoir for fruit farmers in the South Antelope Valley north of Los Angeles, Littlerock Dam was the tallest multiple arch dam in the country at the time of its construction and the largest ever designed by Eastwood. To obtain state approval, Eastwood prepared three variations on the design over a four-year period. The first scheme was a straight-crested dam with parallel buttresses, the second a curved-crested dam with radiating buttresses, and the final design a bent-crested dam. The steeply angled barrel vaults on the upstream face distribute the compressive force of the retained water onto the buttresses, which carry the loads to the foundation.

8-038. Upstream face, Littlerock Dam, Little Rock Creek, Littlerock, Los Angeles County, California, 1922–1924. Jet Lowe, photographer, 1981. P&P,HAER,CAL,19-LITRO.V,1, no. 3.

8-039. Downstream face detail, Littlerock Dam, Little Rock Creek, Littlerock, Los Angeles County, California, 1922–1924. Jet Lowe, photographer, 1981. P&P,HAER,CAL,19-LITRO.V,1, no. 8.

8-040. Plan of first design, Littlerock Dam, Little Rock Creek, Littlerock, Los Angeles County, California, 1922–1924. John Eastwood, delineator, 1918. P&P,HAER,CAL,19-LITRO.V,1-64.

8-039

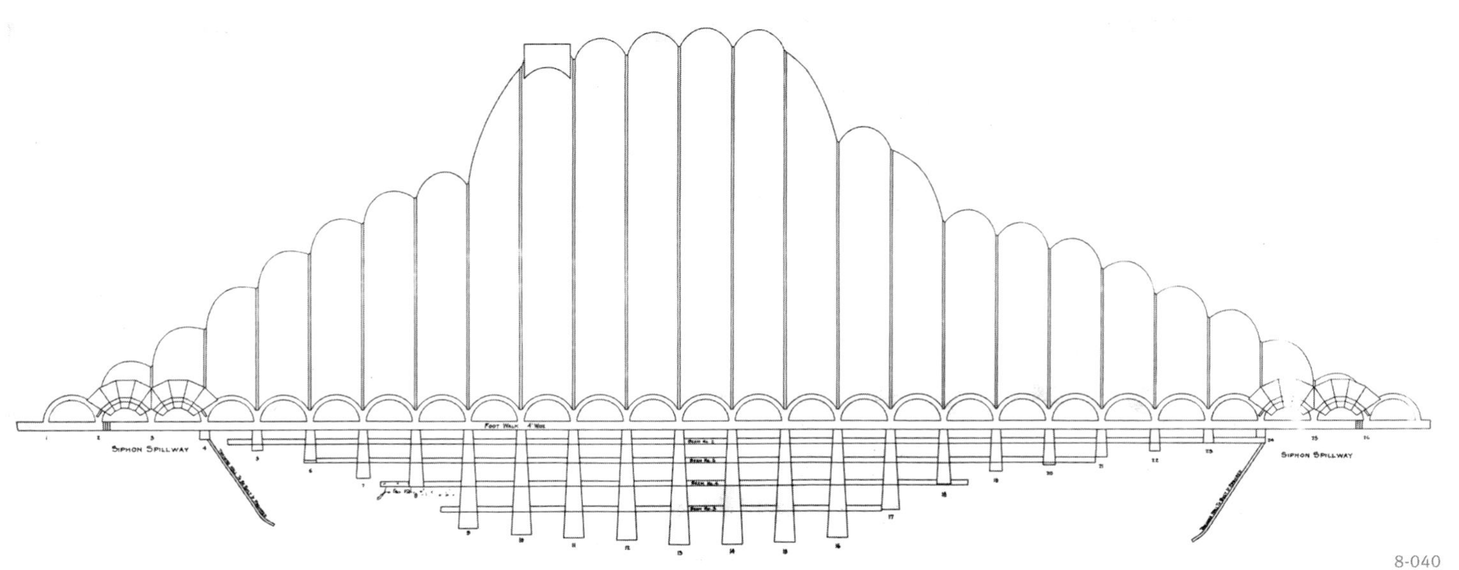

8-040

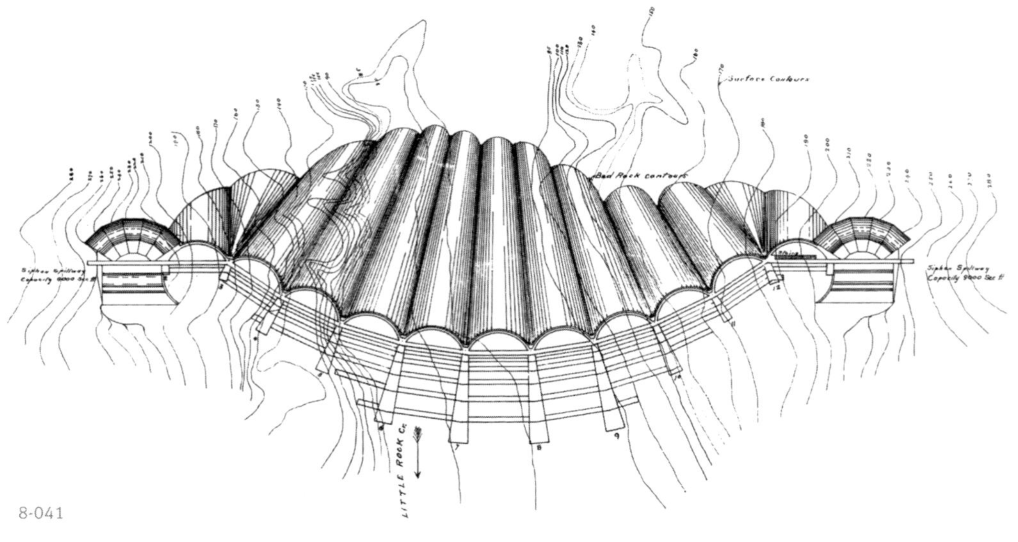

8-041

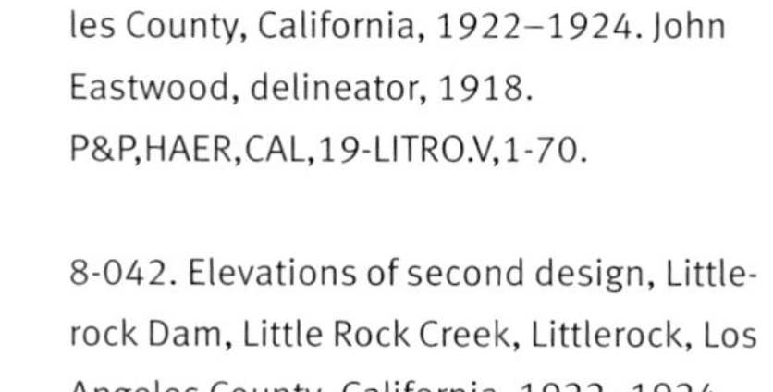

8-041. Plan of second design, Littlerock Dam, Little Rock Creek, Littlerock, Los Angeles County, California, 1922–1924. John Eastwood, delineator, 1918. P&P,HAER,CAL,19-LITRO.V,1-70.

8-042. Elevations of second design, Littlerock Dam, Little Rock Creek, Littlerock, Los Angeles County, California, 1922–1924. John Eastwood, delineator, 1918. P&P,HAER,CAL,19-LITRO.V,1-72.

8-043. Plan of third and final version, Littlerock Dam, Little Rock Creek, Littlerock, Los Angeles County, California, 1922–1924. John Eastwood, delineator, 1919. P&P,HAER,CAL,19-LITRO.V,1-74.

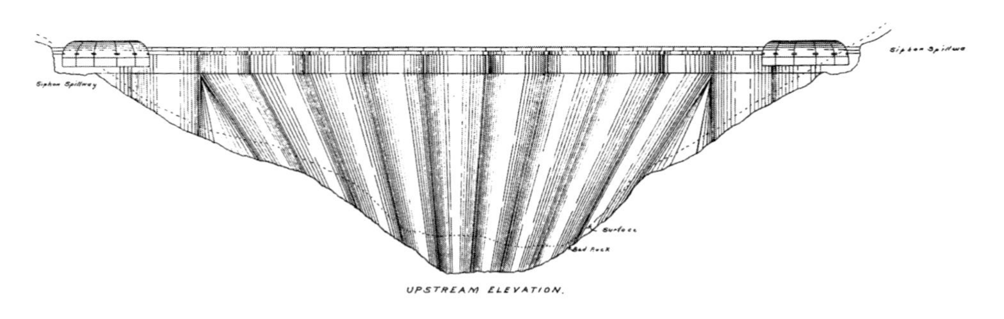

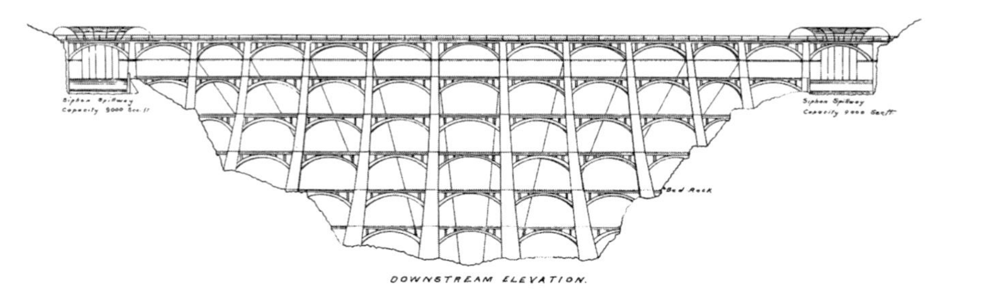

8-042

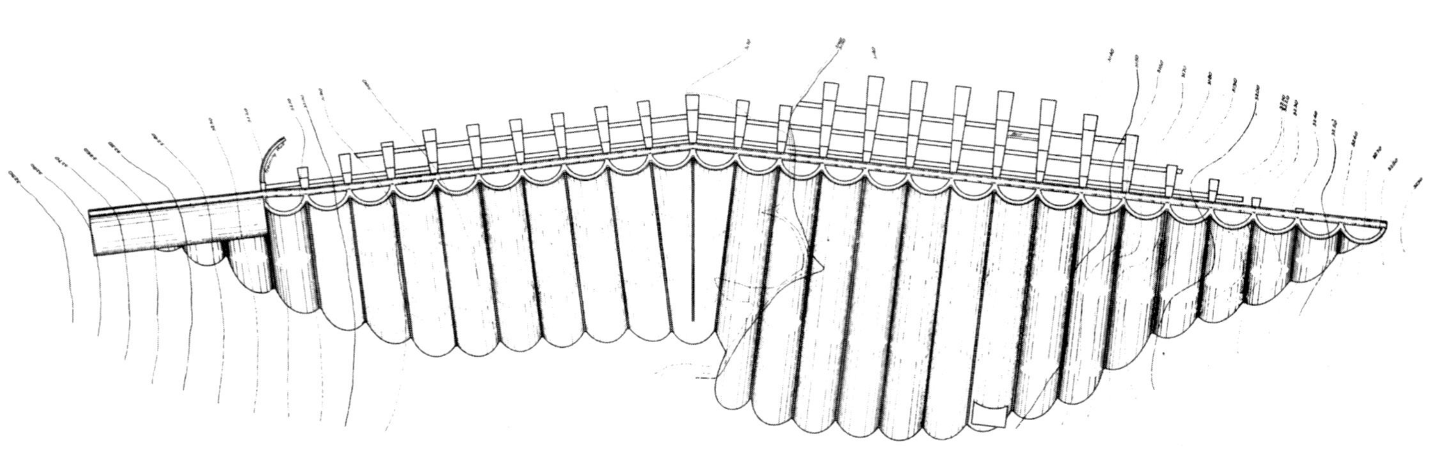

8-043

Early municipal efforts at water supply involved capturing the intermittent streams and small rivers of the coastal mountain ranges that run along the Pacific Ocean. Examples include San Francisco's San Mateo Dam (1889) and San Diego's Sweetwater (1888) and Morena (1912) dams. This approach worked for the growing cities up to a point, but continued growth demanded more water. San Diego citizens even entertained the dubious talents of rainmaker Charles Hatfield.

As populations outgrew limited water supplies, municipal governments began to take more ambitious steps to secure water. Vying with mutual irrigation associations and organized farmers in the agriculturally fertile state, California's cities looked to the snow-capped Sierra Nevada for relief. Vast engineering works, such as San Francisco's Hetch Hetchy project and the many enterprises of Los Angeles's Metropolitan Water District, were required to get the water from the mountains and across deserts to the growing cities on the coast.

8-044. Sweetwater Dam, Sweetwater River, San Diego Vic., San Diego County, California, 1886–1888. WC. P&P,LC-dig-ppmsca-17327.

Water in the San Diego region was at such a premium in the city's early years that it was sold by the bucket. Concerned that the lack of reliable water was discouraging settlement, real estate developers built this dam on the seasonal Sweetwater Creek. Designed as a 60-foot-high arch dam, it was raised an additional 30 feet just before completion. The effectiveness of the arch design proved itself in 1895, when heavy rain overtopped the dam for three days without affecting its integrity. After the flood, a 200-foot spillway was added in the center of the crest, and another on the side.

8-044

8-045

8-045. San Mateo (Crystal Springs) Dam, San Mateo Vic., San Mateo County, California, 1888–1889. Unidentified photographer, 1895. (National Museum of American History) TC547 J33 1995,p. 38.

This concrete curved gravity dam is so massive that its width exceeds its height. The use of concrete was economical only because it could be brought to the site with barges on the nearby San Francisco Bay.

8-046. Morena Dam, Cottonwood Creek, San Diego Vic., San Diego County, California, 1897–1912. WC. P&P,LC-dig-ppmsca-17326.

8-046

8-047. Upper Otay Dam, Otay River, San Diego Vic., San Diego County, 1902. (Water Resources Center Archives) TC547 J33 1995,p. 39.

The elegant arch design of Upper Otay Dam rises 84 feet with a maximum thickness of only 14 feet.

8-048. Lower Otay Dam, Otay River, San Diego Vic., San Diego County, failed 1916, rebuilt 1921. (Water Resources Center Archives) TC547 J33 1995,p. 229.

The arched gravity Lower Otay Dam replaced the original earthfill dam, which washed away in a flood. Both Otay dams—indeed, all the San Diego dams—capture flash flooding on the short rivers that lead directly to the sea.

8-047

8-048

8-049

8-050

8-049. Hetch Hetchy Valley, Yosemite National Park, California. Matt Ashby Wolfskill, photographer, 1911. P&P,PAN US GEOG-California no. 51 (F size).

8-050. Aerial view of O'Shaughnessy (Hetch Hetchy) Dam at low water, Tuolumne River, Yosemite National Park, Tuolumne County, California, 1913–1923. Acme, 1934. P&P,LC-dig-ppmsca-17257.

By the early twentieth century, San Francisco's local reservoirs no longer met the growing city's requirements. Plans were drawn up for an aqueduct that would secure a reliable water supply from the Tuolumne River of the Sierra Nevada. Because the proposed dam would flood the Hetch Hetchy Valley of Yosemite National Park, it was highly controversial and launched one of the major environmental battles of the century. Ultimately, the federal government found in the city's favor, and the curved gravity dam was completed in 1923.

8-051. Construction, O'Shaughnessy Dam, Tuolumne River, Yosemite National Park, Tuolumne County, California, 1913–1923. AP, 1935. P&P,LC-dig-ppmsca-17259.

8-052. Stairway, O'Shaughnessy Dam, Tuolumne River, Yosemite National Park, Tuolumne County, California, 1913–1923, modified 1938. Acme, 1934. P&P,LC-dig-ppmsca-17256.

8-051

8-052

8-053

8-054

8-053. O'Shaughnessy Dam, Tuolumne River, Yosemite National Park, Tuolumne County, California, 1913–1923, modified 1938. Keystone View Company, ca. 1923. P&P,LC-dig-stereo-1s01719.

8-054. El Capitan Dam, San Diego River, San Diego, San Diego County, California, 1932–1934. Unidentified photographer, 1935. P&P,LC-dig-ppmsca-17307.

Los Angeles Metropolitan Water District

Los Angeles Aqueduct

Located in arid southern California, the city of Los Angeles required reliable water supplies to ensure its growth. It had let out a franchise for its water supply in the 1860s, and when this expired near the turn of the century, it began to look for ways to expand the system. One proposal, which was ultimately adopted, involved transporting water from the Owens River, located 240 miles to the north in a deep valley that lay between the Sierra Nevada and the Inyo ranges. The aqueduct would have to cross the Mojave Desert and the San Fernando Mountains to reach the city of Los Angeles. The system, designed by William Mulholland (1855–1935), who was superintendent of the Los Angeles District of Water and Power, diverts water from the Owens River into a canal that runs 60 miles to Haiwee Reservoir, then across the rugged eastern face of the Sierra range and the Mojave Desert through 128 miles of tunnel, conduit, and siphon to Fairmont Reservoir. From there, a number of hydroelectric plants exploit its drop before the water collects in three reservoirs just north of the city.

Owens River was almost completely diverted, drying up inflows into the lake and sinking groundwater levels in the valley. A drought in 1921 further inflamed valley residents against the city. For its part, Los Angeles used hardball tactics, buying up critical water rights and depreciating the values of the remaining farms. By 1924, the farmers and

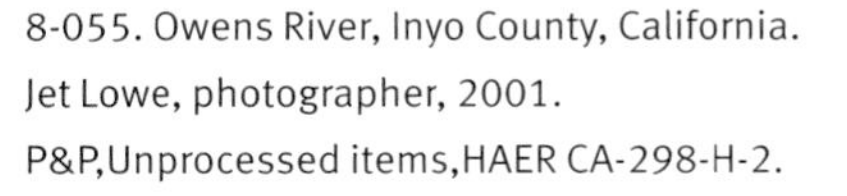

8-055. Owens River, Inyo County, California. Jet Lowe, photographer, 2001. P&P,Unprocessed items,HAER CA-298-H-2.

8-055

ranchers of the valley turned violent, dynamiting sections of the aqueduct. A court decision in 1929 required the city to buy out all landowners who wished to sell, which it did over the following years, emptying Owens Valley of its inhabitants and leaving it a dry alkali flat. In 1938, the aqueduct was extended to Mono Lake, where the process was

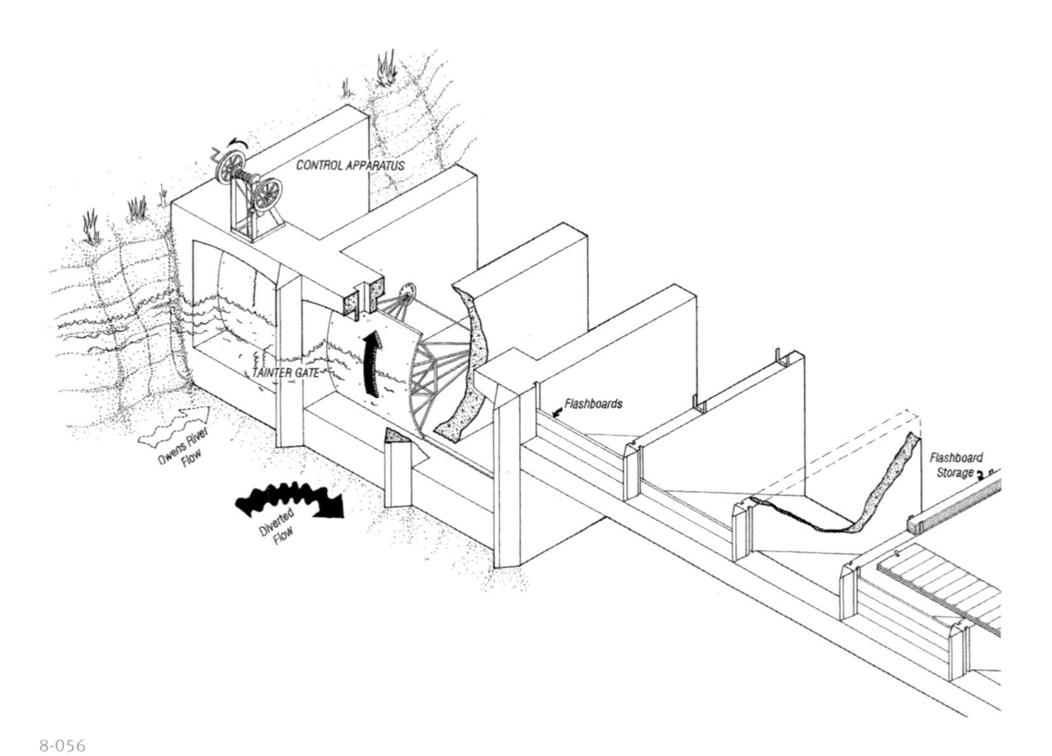

8-056

8-056. Owens River Diversion Dam, Los Angeles Aqueduct, Owens River, Inyo County, California, 1907–1913. Erin E. Ammer, delineator, 2001. P&P,Unprocessed items,HAER CA-298.

The original diversion dam was a 240-foot concrete weir with four tainter gates that allowed the entire flow of the Owens River to be diverted into the aqueduct.

8-057

8-057. Lined channel of the Los Angeles Aqueduct, Lone Pine Vic., Inyo County, California. Jet Lowe, photographer, 2001. P&P,Unprocessed items,HAER Survey No. CA-298-probably 46.

repeated. During World War II, the Japanese-American internment camp Manzanar operated in Owens Valley.

A number of dams served the 233-mile-long aqueduct, including a concrete diversion dam at Owens River and hydraulic earthfill dams at Haiwee Reservoir on the Owens Valley end of the system. At the other end, the Fairmount reservoir in the San Gabriel Mountains overlooking the Los Angeles basin served as a forebay for the San Francisquito power plants. Reservoir dams constructed in the Los Angeles metropolitan region include the earthfill Van Norman (Los Angeles) and Franklin reservoirs; and two curved concrete dams, Mulholland Dam (1924) in the Hollywood Hills, and Saint Francis Dam (1926) in Francisquito Canyon, 40 miles north of Los Angeles. After the Saint Francis Dam collapsed because of faulty geological conditions at its foundation, public pressure required that the downstream face of Mulholland Dam be covered with an earthen embankment. Its name was changed to Hollywood Dam.

8-058. South Haiwee Dam, Los Angeles Aqueduct, Inyo County, California, 1907–1913. Jet Lowe, photographer, 2001. P&P,Unprocessed items,HAER Survey No. CA-298-probably 66.

8-058

8-059

8-060

8-059. Crowd gathered at the Owensmouth Cascades, Sylmar, Vic., Los Angeles County, California, 1913. Los Angeles Department of Water and Power, 1913. HD1694 C2 H83 1992,p. 155.

8-060. San Francisquito Hydroelectric Plant No. 1, Los Angeles Aqueduct, Los Angeles County, California, 1917–1920. Jet Lowe, photographer, 2001. P&P,Unprocessed items,HAER CA-298-AE-1.

8-061. Pelton wheel display at San Francisquito Power Plant No. 1, Los Angeles Aqueduct, Los Angeles County, California, 1917–1920. Jet Lowe, photographer, 2001. P&P, Unprocessed items, HAER CA-298-AE, no. 14.

8-062. Main generator floor showing Pelton wheels, San Francisquito Power Plant No. 1, Los Angeles Aqueduct, Los Angeles County, California, 1917–1920. Jet Lowe, photographer, 2001. P&P,Unprocessed items,HAER CA-298-AE, no. 9.

8-061

8-062

8-063

8-063. Construction, Saint Francis Dam, Los Angeles Aqueduct, San Francisquito Creek, Saugus, Los Angeles County, California, 1926. Unidentified photographer, 1926. P&P,LC-dig-ppmsca-17285.

This concrete arched gravity dam was originally built to supply water to the San Francisquito Hydroelectric Plants Nos. 1 and 2.

8-064

8-064. Saint Francis Dam, Los Angeles Aqueduct, San Francisquito Creek, Saugus, Los Angeles County, California, 1926. Unidentified photographer, 1926. P&P,LC-dig-ppmsca-17286.

8-065. Collapsed Saint Francis Dam, Los Angeles Aqueduct, San Francisquito Creek, Saugus, Los Angeles County, California, 1926, failed 1928. Keystone View Company, 1928. P&P,LC-dig-stereo-1s01718.

The dam's sudden collapse in 1928 was a massive disaster that tore through the San Francisquito Canyon and continued 70 miles down the Santa Clara Valley to the Pacific Ocean, killing over five hundred people. The collapse led to a judicial inquiry and ruined the reputation of William Mulholland.

8-066. Axonmetric views of Mulholland and St. Francis dams, Los Angeles Aqueduct, Los Angeles, Los Angeles County, California. W. Mulholland, delineator, 2001. P&P,Unprocessed items,HAER Survey No. CA-298,sheet no. 14,detail.

8-065

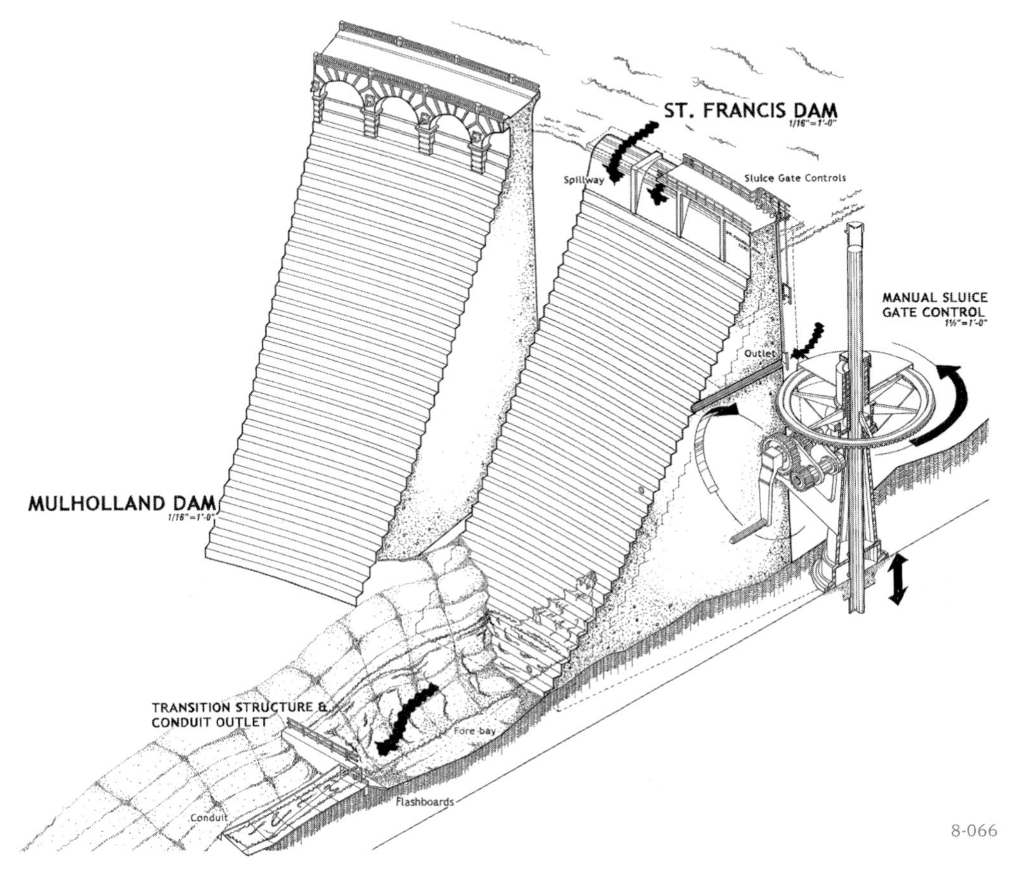

8-066

8-067

8-067. Hollywood (Mulholland) Dam, Los Angeles Aqueduct, Los Angeles, Los Angeles County, California, 1923–1924, earth embankment added 1927, modified 1932. Jet Lowe, photographer, 2001. P&P,Unprocessed items,HAER CA-298-AL, no. 3.

One of the features of the project most visible to Angelenos, this dam celebrated the civic importance of the Los Angeles Aqueduct, with its prominent location in Hollywood, its crest of roman arches, and its decorative California grizzly bear medallions. Because the design of Mulholland Dam so closely paralleled that of the collapsed St. Francis Dam, an earthen berm was added to the downstream side nearly to the crest of the dam.

8-068. Morris Dam, San Gabriel River, Pasadena Vic., Los Angeles County, California, 1934. Acme, 1934. P&P,LC-dig-ppmsca-17270.

8-068

FLOOD CONTROL DAMS

Although the Los Angeles basin is an arid region, its surrounding mountains receive more than 40 inches of rain each year, making floods in the rainy season a regular occurrence. As the city grew, the economic impact of flooding did as well, until public demand led to the formation of the Los Angeles County Flood Control District in 1915. Successful bond issues in 1917 and 1924 financed construction of fourteen storage dams to hold storm waters for later release, including Devil's Gate (1920), San Dimas (1922), Puddingstone (1928), Pacoima (1929), Big Dalton (1929), Big Tujunga (1931), Cogswell (1934), and San Gabriel (1939) dams.

After continuing flood damage in the 1930s, federal assistance was requested and the U.S. Army Corps of Engineers became involved in flood control in the Los Angeles region. Between 1939 and 1960, the Corps built a number of additional dams and channelized the Los Angeles River and its tributaries. Army Corps dams include Hansen (1940), Sepulveda (1941), Fullerton (1941), Brea (1942), Santa Fe (1949), Lopez (1954), and Whittier Narrows (1957).

8-069. Devil's Gate Dam, Arroyo Seco River, Pasadena, Los Angeles County, California, 1920. The Flag Studio, 1920. P&P,LC-dig-ppmsca-17308.

8-069

8-070

8-070. Big Tujunga Dam, Big Tujunga Creek, Sunland Vic., Los Angeles County, California, 1931. Acme, 1938. P&P,LC-dig-ppmsca-17234.

This 200-foot-high concrete variable arch dam served for flood control in the mountains north of the San Fernando Valley. The image shows the spillway in use, releasing flood waters.

8-071. San Gabriel Dam, San Gabriel River, Azuza Vic., Los Angeles County, California, 1932–1939. Acme, 1937. P&P,LC-dig-ppmsca-17281.

At 381 feet high, this dam set a record for earthfill dams. The cross-section is 40 feet wide at the crest and half a mile wide at the base.

8-071

CENTRAL VALLEY RECLAMATION PROJECT/STATE WATER PROJECT

From 1923 to 1932, California developed a comprehensive plan for water use in the entire state. Unable to finance the ambitious scheme, it turned to the Bureau of Reclamation, which authorized the plan in 1935. Stretching the entire length of the state's central valley, it was the largest reclamation project in the country. Essentially, the project transfers water from the northern part of the state to the south, to support agriculture in the arid southern reaches of the San Joaquin Valley. The Central Valley Reclamation Project began with three divisions: (1) Friant Division, which diverts the entire Joaquin River into the Friant-Kern and Madera canals to irrigate the arid southern Central Valley near Fresno; (2) Shasta Division, which impounds Sacramento River water and makes it available for transport; and (3) Delta Division, which pumps the Sacramento River water over the flat

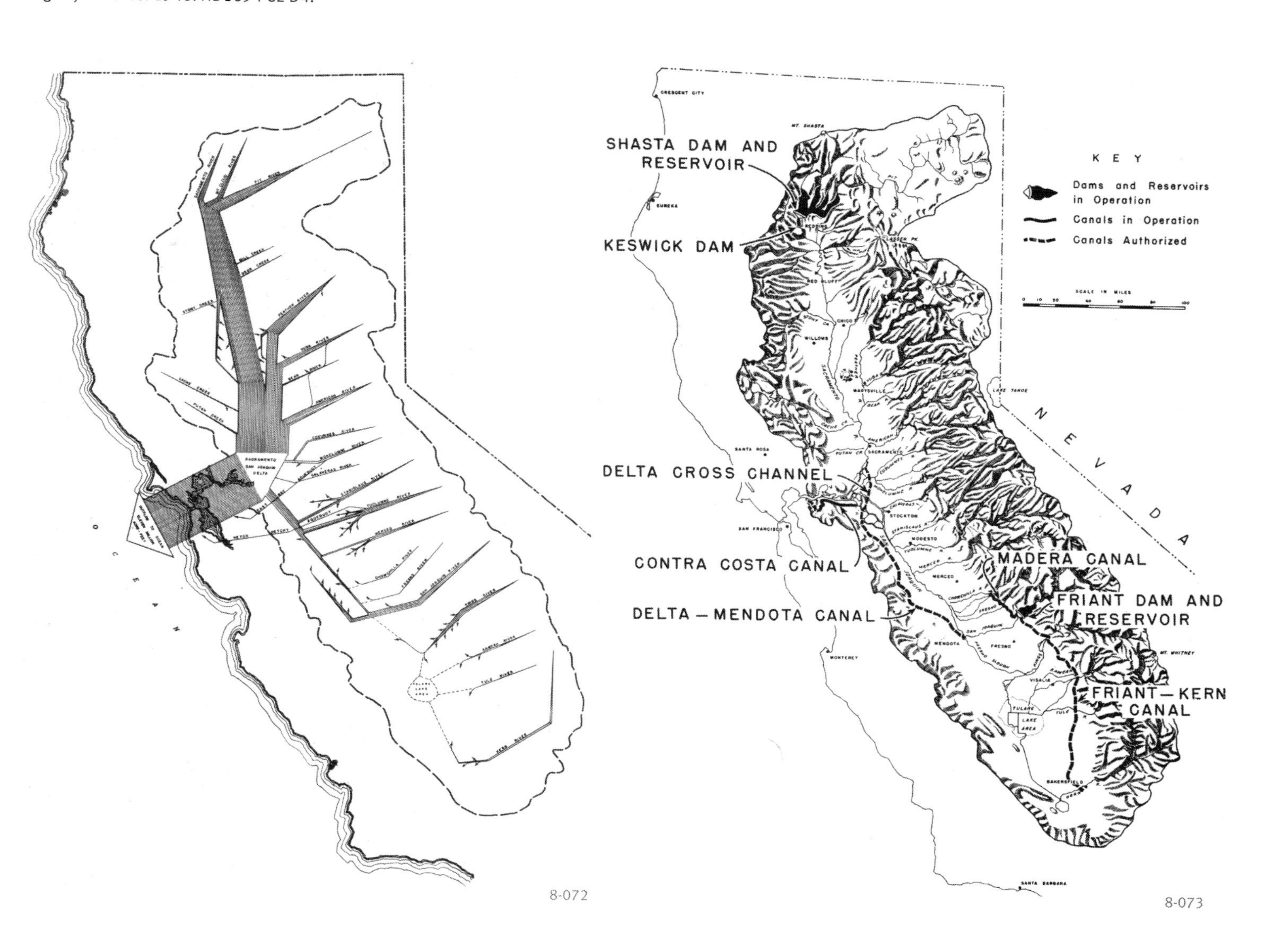

8-072. Map of natural water flows in California. 1948. HD1694 C2 D4.

8-073. Map of Central Valley Project, showing key features. 1948. HD1694 C2 D4.

8-072

8-073

delta and into lands no longer supplied by the San Joaquin River. The California Aqueduct (1959) was a later addition to the system under the State Water Project. It transports water from the Sacramento River delta down the length of the Central Valley and over the Tehachapi Mountains to the Los Angeles basin.

The project has generated its share of controversy, particularly because of the high stakes and powerful players involved—from agricultural industries and large utility companies to state and municipal politicians. Although it gained the support of the California Grange, a movement of small farmers who advocated acreage limitations and public power, the Central Valley Project has ultimately proved a huge benefit to the agricultural corporations that own 80 percent of the state's farmland in holdings of over 1,000 acres. The Central Valley Project and its umbrella, the State Water Project, remain an impressive engineering project, providing water to 23 million people and three-quarters of the irrigated land in the state, which is one-sixth of the irrigated land in the country.

8-074. Friant Dam, Central Valley Reclamation Project, San Joaquin River, Fresno County, California, 1938–1942. Bureau of Reclamation, 1942. P&P,LC-dig-ppmsca-17338.

8-074

8-075. Night view of construction, Friant Dam, Central Valley Reclamation Project, San Joaquin River, Fresno County, California, 1938–1942. Bureau of Reclamation. P&P,FSA/OWI,J 5934.

8-076. Construction, Friant Dam, Central Valley Reclamation Project, San Joaquin River, Fresno County, California, 1938–1942. Bureau of Reclamation. P&P, FSA/OWI,J 5934.

8-077. Concrete finishing, Friant Dam, Central Valley Reclamation Project, San Joaquin River, Fresno County, California, 1938–1942. Bureau of Reclamation. P&P,LC-dig-ppmsca-17396.

8-075

8-076

8-077

8-078

8-079

8-078. Downstream face of spillway section, Friant Dam, Central Valley Reclamation Project, San Joaquin River, Fresno County, California, 1938–1942. Bureau of Reclamation, 1942. P&P,LC-dig-ppmsca-17339.

8-079. Opening day, Friant-Kern Canal, Central Valley Reclamation Project, San Joaquin River, Fresno County, California, 1945–1949. Unidentified photographer, 1949. P&P,LC-dig-ppmsca-17313.

8-080. Construction of a siphon for the Madera Canal, Central Valley Reclamation Project, San Joaquin River, Madera County, California, 1939–1945. Acme, 1941. P&P,LC-USZ62-85997.

8-081. Group of canal structures, Madera Canal, San Joaquin River, Madera County. California, 1939–1945. Bureau of Reclamation, 1942. P&P,LC-dig-ppmsca-17348.

8-080

8-081

8-082

8-083

8-082. Sailors watching construction of Shasta Dam, Central Valley Reclamation Project, Sacramento River, Redding Vic., Shasta County, California, 1938–1945. Russell Lee, photographer, 1942. P&P,LC-USF34-072935-E.

Shasta Dam was the centerpiece of the Central Valley Project. An arched gravity dam 602 feet high and 1,077 feet long, it was one of the largest concrete dams in the country. Labor difficulties dogged its construction from the outset. Before construction even began, numerous labor councils and organizations argued that wages on the Central Valley Project were below the prevailing rates in the state. The principal contractor, Pacific Constructors Incorporated, drew up a contract with the American Federation of Labor, agreeing to employ only union members at union rates for the dam and power plant. Yet construction work in other areas of the project was held up by labor difficulties including disputes with the AFL's rival, the Congress of Industrial Organizations, wage disputes with the Southern Pacific Railroad, and manpower shortages in general. To get the work done, subcontractors had to exceed the minimum wages specified in their contracts. Other impediments to the construction included flooding and material shortages due to the war. In 1941, construction of Shasta Dam was declared a national priority when it became clear that California's defense industry would need the dam's hydroelectricity by the beginning of 1944, more than a year ahead of Shasta's original completion date of 1945. The national press and media closely covered construction of the dam.

8-083. Head tower, Shasta Dam, Central Valley Reclamation Project, Sacramento River, Redding Vic., Shasta County, California, 1938–1945. Bureau of Reclamation, 1940. P&P,LC-dig-ppmsca-17398.

8-084

8-084. Cableway, Shasta Dam, Central Valley Reclamation Project, Sacramento River, Redding Vic., Shasta County, California, 1938–1945. Russell Lee, photographer, 1941. P&P,LC-USF33-013222-M1.

8-085. Gravel stockpiles, Shasta Dam, Central Valley Reclamation Project, Sacramento River, Redding Vic., Shasta County, California, 1938–1945. Russell Lee, photographer, 1941. P&P,LC-USF34-071144-D.

8-085

8-086

8-087

8-088

8-086. Vibrating concrete, Shasta Dam, Central Valley Reclamation Project, Sacramento River, Redding Vic., Shasta County, California, 1938–1945. Russell Lee, photographer, 1942. P&P,LC-USF33-013229-M2.

8-087. Construction, Shasta Dam, Central Valley Reclamation Project, Sacramento River, Redding Vic., Shasta County, California, 1938–1945. Russell Lee, photographer, 1942. P&P,LC-USF34-73055-D.

8-088. First water over the crest, Shasta Dam, Central Valley Reclamation Project, Sacramento River, Redding Vic., Shasta County, California, 1938–1945. Wide World, 1950. P&P,LC-USZ62-83274.

8-089

8-089. Stepped concrete sections, Shasta Dam, Central Valley Reclamation Project, Sacramento River, Redding Vic., Shasta County, California, 1938–1945. Bureau of Reclamation, ca. 1942. P&P,LC-dig-ppmsca-17336.

8-090. Dam near completion, Shasta Dam, Central Valley Reclamation Project, Sacramento River, Redding Vic., Shasta County, California, 1938–1945. Bureau of Reclamation, 1944. P&P,LC-USZ62-11287.

8-090

8-091

8-092

8-091. Spillway, Keswick Dam, Central Valley Reclamation Project, Sacramento River, Redding Vic., Shasta County, California, 1941–1943. Bureau of Reclamation, 1942. P&P,LC-dig-ppmsca-17344.

This run-of-the-river plant nine miles downstream from Shasta Dam functions as an afterbay for that plant. Its construction faced the same difficulties as Shasta's.

8-092. Early scheme for Oroville Dam, Feather River, Oroville Vic., Butte County, California, 1961–1968. UPI,1958. P&P,LC-dig-ppmsca-17275.

Oroville Dam served for flood control and irrigation, with hydroelectric power sold to offset costs of the State Water Project. Most of the water from Oroville Reservoir is sold to agricultural customers in the lower San Joaquin Valley in Southern California, totaling nearly 30 percent of the valley's water purchases. Water is also released to maintain the Feather and Sacramento rivers and to reduce salinity in the San Francisco–San Joaquin delta. The image shows an early scheme for the dam, which was ultimately built as an earthfill dam. At 770 feet high, Oroville Dam is the tallest dam in the United States. A powerhouse with six generators is located inside the massive earthfill structure.

8-093. San Luis Dam, Los Banos Vic., Santa Clara County, California, 1967. UPI, 1962. P&P,LC-dig-ppmsca-17282.

8-093

ANOTHER RECLAMATION SERVICE DAM

8-094. Stony Gorge Dam, Orland Project, Stony Creek, Elk Creek Vic., Glenn County, California, 1928. P&P,LC-dig-ppmsca-17370.

The only early United States Reclamation Service project in California, the Orland Project was begun in 1907 to irrigate land near the town of Orland about 100 miles north of Sacramento. Stony Gorge Dam is one of two reservoir dams built for the project on the western side of the Sacramento River valley. It is the only flat-slab buttress dam built by the Reclamation Service.

8-094

BIBLIOGRAPHY

Banham, Reyner. "Tennessee Valley Authority: The Engineering of Utopia." *Casabella 52*, nos. 542–43 (January–February 1988): 74–81.

Berlow, Lawrence H. *The Reference Guide to Famous Engineering Landmarks of the World: Bridges, Tunnels, Dams, Roads, and Other Structures.* Phoenix, AZ: Oryx Press, 1998.

Billington, David P., Donald C. Jackson, and Martin V. Melosi. *The History of Large Federal Dams: Planning, Design, and Construction in the Era of Big Dams.* Denver: U.S. Department of the Interior, Bureau of Reclamation, 2005.

Bourgin, André. *The Design of Dams.* London: Pitman, 1953.

Briggs, Peter. *Rampage; The Story of Disastrous Floods, Broken Dams, and Human Fallibility.* New York: David McKay, 1973.

Brookings Institution, Institute for Government Research. *The U.S. Reclamation Service: Its History, Activities, and Organization.* New York: AMS Press, 1974. First published 1919 by D. Appleton.

Carson, Rachel. *Silent Spring.* Cambridge, MA: Riverside Press, 1962.

Clark, Ira G. *Water in New Mexico: A History of Its Management and Use.* Albuquerque: University of New Mexico Press, 1987.

Condit, Carl W. *American Building Art: The Twentieth Century.* New York: Oxford University Press, 1961.

Craig, Lois. *The Federal Presence: Architecture, Politics and Symbols in United States Government Building.* Cambridge, MA: MIT Press, 1978.

Creager, William P., Joel D. Justin, and Julian Hinds. *Engineering for Dams*, vols. 1–3. New York: John Wiley & Sons, 1945.

Creese, Walter. *TVA's Public Planning: The Vision, The Reality.* Knoxville: University of Tennessee Press, 1990.

Cronon, William, ed. *Uncommon Ground: Rethinking the Human Place in Nature.* New York: Norton, 1996.

Cross, Whitney R. "Ideas in Politics: The Conservation Policies of the Two Roosevelts." *Journal of the History of Ideas* 14 (June 1953): 421–38.

Cutler, Phoebe. *The Public Landscape of the New Deal.* New Haven: Yale University Press, 1985.

Cutler, William G. *History of the State of Kansas.* Chicago: A.T. Andreas Publisher, 1883.

Davis, Charles, ed. *Western Public Lands and Environmental Politics.* Boulder, Co: Westview Press, 2001.

Opposite: see 2-115.

Davison, Stanley. *The Leadership of the Reclamation Movement,* 1875–1902. New York: Arno Press, 1979.

De Roos, Robert William. *The Thirsty Land: The Story of the Central Valley Project.* Stanford: Stanford University Press, 1948.

Dowell, Cleo Lafoy, and R. G. Petty. *Engineering Data on Dams and Reservoirs in Texas,* vol. 3. Austin: Texas Water Development Board, 1971.

Duffus, R. L., and Charles Krutch. *The Valley and Its People: A Portrait of TVA.* New York: Alfred Knopf, 1944.

Ferriss, Hugh. *Power in Buildings.* New York: Columbia University Press, 1953.

Gutheim, Frederick A. *TVA Architecture.* New York: Museum of Modern Art, 1940.

Hargrove, Erwin C., and Paul K. Conklin. *TVA: Fifty Years of Grassroots Democracy.* Urbana: University of Illinois Press, 1983.

Hay, Duncan. *Hydroelectric Development in the United States, 1880–1940,* 2 vols. Washington, DC: Edison Electric Institute/New York State Museum, 1991.

Hays, Samuel P. *Conservation and the Gospel of Efficiency: The Progressive Conservation Movement, 1890–1920.* Cambridge, MA: Harvard University Press, 1959.

Hundley, Norris, Jr. *The Great Thirst: Californians and Their Water, 1770s–1990s.* Berkeley: University of California Press, 1992.

Huxley, Julian S. *TVA in Planning.* Cheam, Surrey, UK: Architectural Press, 1943.

Hyde, Charles K. *The Lower Peninsula of Michigan: An Inventory of Historic Engineering and Industrial Sites.* Washington, DC: Historical American Engineering Record, 1976.

______. *The Upper Peninsula of Michigan: An Inventory of Historic Engineering and Industrial Sites.* Washington, DC: Historical American Engineering Record, 1978.

Jackson, Donald C. *Building the Ultimate Dam: John S. Eastwood and the Control of Water in the West. Lawrence: University Press of Kansas, 1995.*

Jackson, Donald Conrad. *Great American Bridges and Dams.* New York: John Wiley & Sons, 1988.

Johnson, Leland R. *The Ohio River Division, U.S. Army Corps of Engineers: The History of a Central Command.* Cincinnati: U.S. Army Corps of Engineers, Ohio River Division, 1992.

King, Judson. *The Conservation Fight: From Theodore Roosevelt to the TVA.* Washington, DC: Public Affairs Press, 1959.

Kyle, John H. *The Building of the TVA: An Illustrated History.* Baton Rouge: Louisiana State University Press, 1958.

Lejeune, Jean-Francois. "Democratic Pyramids: The Works of the Tennessee Valley Authority," in "Themes in Architecture: Electricity, United States & USSR, France and Italy," special issue, *Rassegna* 17, no. 63 (1995); 46–57.

Leuchtenberg, William E. *Franklin D. Roosevelt and the New Deal, 1932–1940.* New York: Harper and Row, 1963.

Linenberger, Toni Rae. *Dams, Dynamos, and Development: The Bureau of Reclamation's Power Program and Electrification of the West.* Denver: U.S. Department of the Interior, Bureau of Reclamation, 2002.

Lowitt, Richard. *The New Deal and the West.* Bloomington: Indiana University Press, 1984.

Lowry, William R. *Dam Politics: Restoring America's Rivers.* Washington, DC: Georgetown University Press, 2003.

Macy, Christine, and Sarah Bonnemaison. *Architecture and Nature: Constructing the American Landscape.* London: Routledge, 2003.

Mathur, Anuradha, and Dilip da Cunha. *Mississippi Floods: Designing a Shifting Landscape.* New Haven: Yale University Press, 2001.

McCully, Patrick. *Silenced Rivers: The Ecology and Politics of Large Dams.* London, England: Zed Books, 1996.

McPhee, John. *The Control of Nature.* New York: Farrar, Straus and Giroux, 1989.

Mermel, T.W., ed. *Register of Dams in the United States: Completed, Under Construction and Proposed.* United States Committee on Large Dams. New York: McGraw-Hill, 1958.

Merrill, Lieut. Col. William E. *Davis Island Dam, Ohio River.* Department of War, Corps of Engineers, 1889.

Merritt, Raymond H. *Creativity, Conflict, & Controversy: A History of the St. Paul District US Corps of Engineers.* Washington, DC: U.S. Government Printing Office, 1979.

Moffett, Marian, and Lawrence Wodehouse. *Built for the People of the United States: Fifty Years of TVA Architecture.* Knoxville: University of Tennessee, 1983.

_____. "Noble Structures Set in Handsome Parks: Public Architecture of the TVA." *Modulus: University of Virginia Architectural Review,* no. 17 (1984): 74–83.

Morgan, Arthur E. "Tennessee Valley Authority Becomes Laboratory for the Nation." *New York Times*, March 25, 1934, p. 1.

Nash, Gerald D. *The Federal Landscape: An Economic History of the Twentieth-Century West. Tucson:* University of Arizona Press, 1999.

Nash, Gerald D., and Richard W. Etulain, eds. *The Twentieth-Century West: Historical Interpretations.* Albuquerque: University of New Mexico Press, 1989.

Nye, David E. *American Technological Sublime.* Cambridge, MA: MIT Press, 1994.

_____. *Narratives and Spaces: Technology and the Construction of American Culture.* New York: Columbia University Press, 1997.

Rawson, Michael. "The Nature of Water: Reform and the Antebellum Crusade for Municipal Water in Boston." *Environmental History* 9, no. 3 (July 2004): 411–35.

Richardson, Elmo. *Dams, Parks & Politics: Resource Development & Preservation in the Truman-Eisenhower Era.* Lexington: University Press of Kentucky, 1973.

Robinson, Michael C. *Water for the West: The Bureau of Reclamation, 1902–1977.* Chicago: Public Works Historical Society, 1979.

Ross, Malcolm. *Machine Age in the Hills.* New York: Macmillan, 1933.

Rothman, Hal, ed. *Reopening the American West.* Tucson: University of Arizona Press, 1998.

Rowley, William D. *The Bureau of Reclamation: Origins and Growth to 1945*, vol. 1. Denver: U.S. Department of the Interior, Bureau of Reclamation, 2006.

Schodek, Daniel L. *Landmarks in American Civil Engineering.* Cambridge, MA: MIT Press, 1987.

Sherow, James E. "The fellow who can talk the loudest and has the best shotgun gets the water." *Montana: The Magazine of Western History* (Spring 2004).

Smith, Karen. *The Magnificant Experiment: Building the Salt River Project, 1870–1917*. Tucson: University of Arizona Press, 1986.
Smith, Norman. *A History of Dams*. London: Peter Davies, 1971.
Spears, Ross. *The Electric Valley*. Johnson City, TX: James Agee Film Project, 1984. Documentary film.
Tennessee Valley Authority. *The Twenty Major Dams Built by the Tennessee Valley Authority and Wilson Dam*. Knoxville, TN: TVA, 1958.
Twee, Roald. *Rock Island District: 1866–1983*. Rock Island, IL: U.S. Army Corps of Engineers District, 1984.
U.S. Army Corps of Engineers. *Hydraulic Gates and Dams in the Ohio River. Letter from the Secretary of War, Relative to a Report upon the Applicability of Movable Hydraulic Gates and Dams to the Improvement of the Ohio River*. Washington, DC: GPO, 1875.
______. *National Dam Inventory. http://crunch.tec.army.mil/nid/webpages/nid.html.*
______. *Report of Examination of Ohio River with a View to Obtaining Channel Depths of 6 and 9 Feet, Respectively, Made by a Board of Engineers, Transmitted by the Chief of Engineers, War Department*. Washington, DC: GPO, 1908.
______. Omaha District. *The Federal Engineer: Damsites to Missile Sites*. Omaha: U.S. Army Corps of Engineers, 1985.
U.S. Department of the Interior, Bureau of Reclamation. *Design of Small Dams*, 2nd ed. Washington, DC: GPO, 1973.
U.S. Public Works Administration. *America Builds: The Record of the PWA*. Washington, DC: GPO, 1939.
Walker, Ward. *The Story of the Metropolitan Sanitary District of Greater Chicago: The Seventh Wonder of America*. Chicago: Metropolitan Sanitary District of Greater Chicago, 1956.
Water Science and Technology Board (Committee on Privatization of Water Services in the United States), National Research Council. *Privatization of Water Services in the United States: An Assessment of Issues and Experience*. Washington, DC: National Academy Press, 2002.
White, Richard. *The Organic Machine: The Remaking of the Columbia River*. New York: Hill and Wang, 1995.
Wilson, Richard Guy, Dianne Pilgrim, and Dickran Tashjian. *The Machine Age in America 1918–41*. New York: Brooklyn Museum/Harry Abrams, 1986.
Wolff, Jane. *Delta Primer: A Field Guide to the California Delta*. San Francisco: William Stout Publishers, 2003.
Work Projects Administration, California, Writers' Program. *The Central Valley Project*, compiled by workers of the Writers' Program of the Work Project[s] Administration in northern California. Sacramento: California State Department of Education, 1942.
World Commission on Dams. *Dams and Development: A New Framework for Decision-Making*. London, England: Earthscan Publications, *2000*.
Zucker, Paul. *American Bridges and Dams*. New York: Greystone Press, 1941.

GLOSSARY

Acequia. Spanish for "irrigation canal."

Afterbay. The body of water immediately downstream from a power plant. Also called a tailrace.

Ambursen dam. See *Flat-slab dam.*

Apron. A short ramp of concrete, timber, or rip-rap at the toe of a dam or spillway, installed to prevent undercutting of the structure.

Arch dam. A concrete or masonry dam that is shaped like an arch in plan. This shape works in compression to transmit the water load to the abutments. An arch dam is most commonly used in a narrow site with steep walls of sound rock.

Arroyo. An intermittent creek in an arid region.

Bear trap gate. A crest gate with two leaves that are hinged at their bases and join to form a triangle in cross-section, with a sliding seal at their juncture. When lowered, the leaves slide over each other to rest in a horizontal position. The gate is raised by admitting water from the upper pool into the space beneath the leaves, raising the buoyant lower leaf, which in turn lifts the upper leaf. This was probably the first gate to operate on the principal of application of headwater pressure.

Buttress dam. A dam supported on the downstream side by buttresses. Buttress dams can take many forms, including flat-slab buttress dam, arch buttress dam, or multiple arch dam.

Chanoine wickets. Moveable wicket dams invented by Jacques Chanoine in 1852. Chanoine wickets are wooden boards hinged to a masonry foundation. Assembled in a row, they can be raised to impound water or lowered to allow water to pass freely. The U.S. Army Corps of Engineers modified this design on the Ohio River.

Clamshell gate. A high-pressure gate of two curved leaves that open and close over the end of a conduit. Used for free discharge into air with minimal cavitation damage.

Cofferdam. A temporary barrier, usually an earthen dike, constructed around a work site so construction of the dam can proceed in dry conditions.

Conduit. A closed channel to convey water through or around a dam.

Crest. The top of a dam.

Crest gates. See *Spillway gate.*

Crevasse. A natural break in a naturally formed levee, or an unintended break in a constructed levee.

Crib dam. A gravity dam made of timber boxes (cribs) that are filled with earth or rock.

Distributary. The watercourse a river flows into.

Ditch. The term commonly used in western states to describe an open canal used for irrigation or hydraulic mining.

Diversion dam. A dam built to divert water from a river into (usually) a canal.

Downstream face. The surface of a dam away from the reservoir.

Driving dam. A dam, usually temporary, built to raise water levels on a riverbed so logs could be floated downstream. Also called a splash dam.

Drum gate. A spillway gate in the form of a long hollow drum hinged at either the upstream or the downstream end. The gate is operated by reservoir pressure. The drum may be held in its raised position by the water pressure in a flotation chamber beneath the drum. This design allows water to flow over the gate.

Dry dam. Built for flood control, dry dams are designed to retain flood water only during periods of intense rainfall. The concept was first used in the Miami Conservancy District of the Ohio River valley.

Earthfill dam. A dam of compacted earth. A homogeneous earthfill dam is of one material, while a zoned earthfill has a watertight core embedded in an outer shell. Seepage through the dam is controlled by upstream blankets or an impermeable internal core. Usually compacted by rollers (rolled earthfill), but sometimes created by using hydraulic earthfill.

Fish ladder. A gradually inclined channel that bypasses a dam so that fish can swim upstream. Some fish ladders have baffles to reduce the flow, while others are a series of cascading boxes.

Flap gate. A gate hinged on one edge.

Flashboards Temporary barriers of timber, concrete, or steel that are anchored to the crest of a spillway to increase the reservoir storage. Flashboards are designed to be removed, lowered, or carried away during a flood.

Flat-slab dam A dam in which buttresses support a flat slab of reinforced concrete that forms the upstream face of the dam. Also called an Ambursen dam, after their inventor, Nils Ambursen (1903).

Floating weir. A weir dam designed to rest on silt, developed for sites with thick alluvial deposits where excavation to bedrock is impractical.

Floodway. A flow path, natural or constructed, that carries significant volumes of floodwaters during a flood.

Flume. A channel built to carry rapidly flowing water, often elevated above the ground.

Forebay. The basin immediately upstream from a dam or intake structure. The term is applicable to all types of hydroelectric developments (storage, run-of-river, and pumped-storage). Also called a headrace.

Francis turbine. A highly efficient inward flow turbine developed by James B. Francis in 1848. Water is directed through a spiral-shaped inlet, called a scroll case, that wraps around the turbine. Guide vanes direct the water tangentially to the curved blades of the runner, which is the rotating part of the turbine. This causes the runner to spin. The guide vanes (also called wicket gates) can be adjusted for varying rates of flow. The rotating motion of the runner is transmitted by a shaft to a generator. As the water moves through the runner its spinning radius decreases, further acting on the runner, and is spent by the time it exits. The turbine's exit tube is called a draft tube.

Generator. A machine that converts mechanical energy into electrical energy.

Gravity dam. A concrete or masonry dam that relies on its weight and internal strength for stability. Gravity dams are generally used on rock foundations.

Head of water. The vertical distance between two water surface elevations, such as that in a reservoir and that below a dam. The difference in pressure between the two levels causes water to flow. The higher the head, the greater the velocity of flowing water.

Headrace. See *Forebay*.

High line canal. The principal canal of an irrigation system, so called because it is at a higher elevation than the secondary canals it supplies.

Hurdy gurdy wheel. A primitive turbine used in the California gold fields in the 1860s.

Hydraulic mining. A large-scale form of placer mining in which water is blasted at high pressures to liquefy ore-bearing soils and separate out the ores. Widespread in western mining regions, its use was restricted after 1884.

Hydraulic earthfill. Dredged materials transported and deposited in water.

Hydroelectric plant. A power plant that produces electricity from water turning turbine-generators.

Hydroelectricity. Electricity produced by flowing water.

Hydropower. Mechanical energy produced by flowing water.

Levee. A natural or constructed barrier along the length of a river that keeps floodwaters within the riverbanks.

Lock. A section of canal that can be closed to control the water level, it is used to lift or lower vessels from one water level to another. Dammed rivers require locks to permit navigation.

Log boom. Floating logs tied together in rafts for transport. In a reservoir, the term also refers to a chain of logs secured upstream from spillways and intakes to prevent the entrance of floating debris.

Log chute. A sluiceway designed for the passage of logs.

Mill race. A canal of forebay leading to a hydropower mill.

Monitor. A high-pressure nozzle, or water cannon, used in hydraulic mining.

Morning glory spillway. A hollow tower or shaft, usually funnel-shaped at the crest, set in a reservoir to serve as a spillway.

Movable dam. A dam that can be raised to impound water flow in a river or lowered to allow free passage. It may use flap, wicket, bear trap, or other gates.

Multiple arch dam. A type of buttress dam made of a series of barrel vaults (arches in plan view) on the upstream side.

Multipurpose dam. A dam constructed for multiple purposes, such as water storage, flood control, navigation, power generation, and/or recreation.

Needle valve. A valve that regulates flow through the movement of a needle in an orifice.

Ogee. The curved shape of a spillway crest or weir that follows the path made by water flowing over a sharp edge.

Outlet works. A series of components located in a dam that permit the controlled release of water. These may include conduits, regulating gates, and stilling basin.

Pass-through. See *Run-of-the-river*.

Peak demand. The maximum electrical demand in a specific period of time.

Pelton wheel. An efficient form of turbine, or water wheel, named after its inventor, Lester Pelton, who developed it in California in 1879. Water directed tangent to the path of the runner strikes a series of spoon-shaped buckets that reverse the flow of the water, transferring the impulse to the wheel. It is most efficient in high head, or high pressure, applications.

Penstock. A conduit that leads from a forebay or reservoir to the turbines of a powerhouse. Because of the height difference, the water is under pressure.

Placer mining. The removal of ore from glacial or alluvial deposits containing valuable minerals. The word is derived from French, for "deposit."

Powerhouse. A structure that houses turbines, generators, and associated control equipment.

Puddling. A method of compacting soil whereby it is deposited into a pool of water and stirred. This method is used in hydraulic earthfill dams.

Pumped storage. A pumped-storage power plant pumps water into an elevated reservoir during off-peak periods when electricity is readily available, and releases it later to generate power when demand is peaking.

Raceway. Any canal for a current of water.

Reclamation. As used by the Bureau of Reclamation, this term refers to the irrigation of wilderness or arid land for agriculture and settlement.

Regulating gate. A gate used to regulated the rate of flow through an outlet works or spillway.

Reservoir A body of water impounded by a dam.

Roller gate A hollow cylindrical gate that can be drawn up or down on inclined tracks set on each side of the gate, using mechanical hoists. Developed in 1900 by the Swedish engineer M. Karstanjen, they operate well under icy conditions, accommodating spans of 80 feet or more between the supporting piers.

Run-of-the-river. In run-of-the-river power plants, the outflow from one power plant is equal to the inflow of the plant downstream. Such hydroelectric plants use stream flows as they occur and have little reservoir capacity for storage. Also called pass-through dams.

Siphon. A pipe used to convey water over an obstruction from a reservoir to a lower point.

Slack water navigation system. A natural watercourse (such as a river) impounded by dams to create a series of pools or small lakes of a depth sufficient for navigation. Locks allow boats to pass through the system.

Sluice. An opening to release water from below the water level.

Sluice gate. A gate that slides in supporting guides.

Sluiceway. An opening in a diversion dam that allows floating debris to bypass the dam.

Spillway. An overflow channel designed for the release of excess water while protecting the integrity of the dam. When fitted with mechanical gates, it is a controlled spillway.

Spillway gate. A gate on the crest of a spillway to control the reservoir water level.

Stilling basin. A pool located below a spillway, gate, or valve that dissipates the energy of the discharged water to avoid erosion.

Tail race. See *Afterbay.*

Tainter gate. A pivoted crest gate, with a convex upstream face, that is hinged on its sides to piers or other supporting structures. The rush of water helps to open and close the gates. Developed by Jeremiah Tainter in 1886, on the Red Cedar River in Michigan, as an easily operating gate to release water from a mill pond during lumber drives.

Toe of dam. The point where the upstream or downstream face of a dam meets the natural ground. In a concrete dam, the upstream toe is called a heel.

Transformer. A device that transforms current flowing from one circuit into another through electromagnetic induction, usually changing its voltage.

Tributary. A smaller river that flows into a larger one.

Turbine. A machine for generating rotary mechanical power from flowing water. Turbines convert kinetic energy into mechanical energy. See *Francis turbine* and *Pelton wheel.*

Watershed. The area of land that drains into a river or basin.

Weir. An overflow structure built across a river to raise the water level or to measure the flow of water.

Wing dam. A partial dam that projects into a body of water to divert or direct the flow.

Zoned earthfill. See *Earthfill dam.*

INDEX

Locators that include a section designator followed by a hyphen (e.g., IN-010, 3-020, 8-159) refer to numbered captions. All other locators are page numbers. Individuals who created the original images used in this collection are identified as follows: ph. = photographer; del. = delineator.

O

P

T

U

V

W

Y

ABOUT THE CD-ROM

The CD-ROM includes direct links to four of the most useful online catalogs and sites, which you may choose to consult in locating and downloading images included on it or related items. Searching directions, help, and search examples (by text or keywords, titles, authors or creators, subject or location, and catalog and reproduction numbers, etc.) are provided online, in addition to information on rights and restrictions, how to order reproductions, and how to consult the materials in person.

1. The Prints & Photographs Online Catalog (PPOC) (http://www.loc.gov/rr/print/catalogabt.html) contains over one million catalog records and digital images representing a rich cross-section of graphic documents held by the Prints & Photographs Division and other units of the Library. It includes a majority of the images on this CD-ROM and many related images, such as those in the HABS and HAER collections cited below. At this writing the catalog provides access through group or item records to about 50 percent of the Division's holdings.

SCOPE OF THE PRINTS AND PHOTOGRAPHS ONLINE CATALOG

Although the catalog is added to on a regular basis, it is not a complete listing of the holdings of the Prints & Photographs Division, and does not include all the items on this CD-ROM. It also overlaps with some other Library of Congress search systems. Some of the records in the PPOC are also found in the LC Online Catalog, mentioned below, but the P&P Online Catalog includes additional records, direct display of digital images, and links to rights, ordering, and background information about the collections represented in the catalog. In many cases, only "thumbnail" images (GIF images) will display to those searching outside the Library of Congress because of potential rights considerations, while onsite searchers have access to larger JPEG and TIFF images as well. There are no digital images for some collections, such as the Look Magazine Photograph Collection. In some collections, only a portion of the images have been digitized so far. For further information about the scope of the Prints & Photographs online catalog and how to use it, consult the Prints & Photographs Online Catalog *HELP* document.

WHAT TO DO WHEN DESIRED IMAGES ARE NOT FOUND IN THE CATALOG

For further information about how to search for Prints & Photographs Division holdings not represented in the online catalog or in the lists of selected images, submit an email using the "Ask a Librarian" link on the Prints & Photographs Reading Room home page or contact: Prints & Photographs Reading Room, Library of Congress, 101 Independence Ave., SE, Washington, D.C. 20540-4730 (telephone: 202-707-6394).

2. The American Memory site (http://memory.loc.gov), a gateway to rich primary source materials relating to the history and culture of the United States. The site offers more than seven million digital items from more than 100 historical collections.

3. The Library of Congress Online Catalog (http://catalog.loc.gov/) contains approximately 13.6 million records representing books, serials, computer files, manuscripts, cartographic materials, music, sound recordings, and visual materials. It is especially useful for finding items identified as being from the Manuscript Division and the Geography and Map Division of the Library of Congress.

4. Built in America: Historic American Buildings Survey/Historic American Engineering Record, 1933–Present (http://memory.loc.gov/ammem/collections/habs_haer) describes and links to the catalog of the Historic American Buildings Survey (HABS) and the Historic American Engineering Record (HAER), among the most heavily represented collections on the CD-ROM.